A Comprehensive Study of Surface Energy

A Comprehensive Study of Surface Energy

Edited by Grace Young

STATES
ACADEMIC PRESS
www.statesacademicpress.com

States Academic Press,
109 South 5th Street,
Brooklyn, NY 11249, USA

ISBN: 978-1-63989-003-3

Cataloging-in-Publication Data

A comprehensive study of surface energy / edited by Grace Young.
p. cm.
Includes bibliographical references and index.
ISBN 978-1-63989-003-3
1. Surfaces (Technology). 2. Surface energy. I. Young, Grace.
TA418.7 .C66 2022
620.44--dc23

For information on all States Academic Press publications
visit our website at www.statesacademicpress.com

STATES
ACADEMIC PRESS

Contents

Preface

Surface energy represents the excess energy that is present at the surface of a material as compared to the whole material. When a solid body is cut into two pieces, its bonds get disrupted and the surface area increases. As a result, the surface energy also increases. Surface energy can be determined by measuring it with contact angle experiments. In this method, the contact angle of liquids, such as water and diiodomethane, is measured with the surface and surface energy is calculated on the basis of surface tension and contact angle. Surface energy can also be determined by calculating the surface energy of a deformed solid and by calculating the surface formation energy of a crystalline solid. This book covers in detail some existent theories and innovative concepts revolving around surface energy. It is an upcoming field of science that has undergone rapid development over the past few decades. For someone with an interest and eye for detail, this book covers the most significant topics in the field of surface energy.

The researches compiled throughout the book are authentic and of high quality, combining several disciplines and from very diverse regions from around the world. Drawing on the contributions of many researchers from diverse countries, the book's objective is to provide the readers with the latest achievements in the area of research. This book will surely be a source of knowledge to all interested and researching the field.

In the end, I would like to express my deep sense of gratitude to all the authors for meeting the set deadlines in completing and submitting their research chapters. I would also like to thank the publisher for the support offered to us throughout the course of the book. Finally, I extend my sincere thanks to my family for being a constant source of inspiration and encouragement.

Editor

1

DFT Investigations on the CVD Growth of Graphene

Meicheng Li, Yingfeng Li and
Joseph Michel Mbengue

Abstract

The chemical vapor deposition technique is the most popular for preparing high-quality graphene. Surface energy will dominate the nucleation process of graphene; thus, the surface energy problems involved in thin film growth are introduced first. The experimental tools to describe the growth process in detail are insufficient. So, a mass of simulation investigations, which can give out a very fine description of the surface atomic process, have been carried out on this topic. We mainly summarized the density functional theory works in unearthing the graphene nuclei process and mechanisms. In addition, some studies using molecular dynamics methods are also listed. Such a summary will be helpful to stimulate future experimental efforts on graphene synthesis.

Keywords: Surface energy, graphene, density functional theory, chemical vapor deposition

1. Introduction

After the graphene has been prepared perfectly on a nickel surface by the chemical vapor deposition (CVD) method, [1] the 2D crystal formation process on a perfect cleavage plane suddenly becomes a very important issue that needs to be deeply understood and penetrated. Similar to the general surface absorption phenomenon, the basic physical laws under graphene's formation on a metal surface should be the minimization of surface energy. What's special is that the classic surface energy theory, to some extent, seems to be slightly "broad-brush." To explain and describe the perfect morphology of the obtained graphene crystals on a metal surface, we need to provide a very fine description of the process of graphene growth on a metal surface using CVD technology. This description should contain two main aspects:

1. The real-time surface energy variations from the absorption of a single carbon atom on the metal surface, to the nucleation of graphene, to the formation of a large-scale single-layer graphene, and finally to the formation of multilayers of graphene

2. The stationary state on-surface atomic structures and morphologies of the carbon phase on the metal surface, as well as the surface energies of the different surface carbon morphologies.

The main results are listed in Table 1. The aim of this chapter is to introduce the specific surface energy problems in the formation of 2D crystal graphene on metal surfaces, which are helpful for materials selection and process optimization in the fabrication of 2D carbons and other materials.

Classification	Method	Brief descriptions
Stationary state researches	DFT	Investigated the decomposition of CH_4 on Cu, and declared the active species should mainly be CH_x for graphene growth [2]
	DFT	Identified a carbon atom approaching induced bridging-metal structure formation on Cu [3]
	DFT	C atoms should be unstable on Cu surface but diffuse directly to the subsurface [4]
	DFT	Chain configuration is superior and will dominate the Ni surface until the number of C reaches 12 [5]
	DFT	Linear C chain owns 6 atoms should be the representative structure on copper surface [6]
	DFT	With increased concentration, C atoms will undergo a sinking and up-floating process on Cu surface [7]
	DFT	Graphene nucleation is very likely to start from the stepped regions, but can easily grow over the step [8]
	DFT	Revealed an energetic preference for the formation of stable 1D carbon nanoarches consisting of 3–13 atoms when compared to 2D compact islands of equal sizes [9]
Real-time kinetic researches	Exp.	Analyzed the activation energy paths for graphene nucleation and growth on Cu [10]
	Theo.	Paved the way for graphene growth on Cu based on Langmuir adsorption and 2D crystallization theory [11]
	Exp.	Using carbon isotope labeling, clearly verified that graphene growth on Ni occurs by a C dissolving–precipitating process whereas it grows by a surface adsorption process on Cu [12]
	MD	On Ni surface, high C concentration leads to the formation of graphene island and at ~1000 K, the graphene quality can be significantly improved [13]

Classification	Method	Brief descriptions
	MC	Showed the morphological evolution process of graphene growth on Pt surface [14]

DFT, density functional theory; MD, molecular dynamics; MC, Monte Carlo; Exp., experimental characterization; Theo., theoretical analysis.

Table 1. Summarization of researches on the surface process of graphene growth

By the present experimental technologies, it is still a great challenge to obtain the abovementioned knowledge especially under high-temperature conditions. So, it naturally becomes the best choice to investigate the graphene on-surface growth process by the accurate, reliable theoretical approach: the density functional theory (DFT). [15] Except for some controllable numerical approximations, the DFT approach can be recognized as a rigorous accurate theory, and since it is based on the maximum advanced of physical theory today, the results obtained by the DFT approach is of great reliability. Using the DFT approach, the graphene on-surface growth process can be described even at the level of electron distribution. In recent years, DFT researches on this aspect have been in full swing and some key progresses have been obtained, e.g., the surface energy evolution curve with the carbon atom number increases during the graphene growth on cupper surface has been revealed in detail by Li et al. [6] Parts of these theoretical results have been verified in experiments, and thus can provide reliable guides to optimize the preparation of graphene.

In this chapter, the surface energy problems involved in thin film growth will be first introduced from the aspect of the classic surface energy theory. Then, we will introduce the approaches that have been exploited in describing the graphene CVD growth process from the energy aspect using the DFT approach. Thirdly, we will review the achievements using these approaches on the growth process of graphene.

2. Surface energy problems in thin film growth

2.1. General fundamentals of surface energy

Consider the atoms in the bulk and surface regions of a crystal, the atoms in the bulk possess lower energy since they are more tightly bound, while the ones on the surface will possess higher energy since they are less tightly bound. The sum of all the excess energies of the surface atoms is the surface energy. Therefore, surface energy is generally defined as the excess energy needed when a new surface is created. [16]

For a liquid, the surface energy density is identical to the surface tension (force per unit length), e.g., water has a surface energy density of 0.072 J/m^2 and a surface tension of 0.072 N/m (the units are equivalent). For a solid, if we cut a body into pieces, it will consume energy. Under the hypothesis that the cutting process is reversible, according to the conservation of energy, the energy consumed by the cutting process will be equal to the energy inherent in the two new surfaces created. Therefore, the unit surface energy of a solid material would be half of its energy of cohesion. In practice, this is only an approximate description. Surfaces often change their form away from the simple "cleaved bond" model implied above, and readily rearrange or react to reduce their energy.

Surface energy is the essence of "energy." According to statistical thermodynamics, the surface energy can be described by the Gibbs isotherm. [17] From the expression of Gibbs free energy in differential form, $dG \equiv -SdT + VdP + \gamma dA$, it can be defined as $\gamma \equiv \left(\frac{\partial G}{\partial A}\right)_{T,P}$. Therefore, surface energy can be considered as a criterion at equilibrium. When a system is reaching the equilibrium state, it tends to reduce its free energy. In some cases, this stable state is achieved by the reduction of the surface energy. This principle of surface energy determines many physical phenomena. For example, the smaller drops will aggregate into larger ones; in the absence of gravity, a droplet will tend to become spherical to reduce its surface energy.

Most importantly, the relative magnitude of surface energy determines the wettability of one material on another material. A material having very low surface energy can easily wet a material having high surface energy and form a uniform adhesive layer on it. Conversely, if the deposition material has a higher surface energy, it readily forms an atomic group on the substrate with lower surface energy. Such a function of the surface energy has a wide range of practical applications. We can coat surface-modified materials (e.g., pigments) on the surface of buildings, automobiles or mechanical components to modify their surface energy and thus adapt to the working environment. A good example is waterproofing. Since organic material has low surface energy, coating wax on the surface of a vehicle can prevent the formation of water film when the automobile is wet with rain. Another important application, which is mainly discussed in this chapter (i.e., wettability), is that it determines the process of thin film growth and the quality of the samples obtained.

2.2. Determination of the surface energy

To measure the surface energy of a liquid, commonly, there are two methods. The first one is the so-called stretching film method, as illustrated in Figure 1. By this method, the surface energy is measured by stretching a liquid membrane (which increases the surface area and hence the surface energy density). In that case, in order to increase the surface area of a mass of liquid by an amount, $2\delta L$ (the liquid membrane has two surfaces), a quantity of work, $F\delta$, is needed. The increase of the surface energy is $2\gamma\delta L$ (where γ is the surface energy density of the liquid). So, $2\gamma\delta L = F\delta$, the surface energy density can be calculated by $\gamma = F/2L$.

The other commonly used method to measure the surface energy of a liquid is the capillary tube method, as illustrated in Figure 2. When inserting a capillary tube of greater surface energy into the liquid, the adhesive force of the liquid to the tube wall will point upward. Thus, to reduce the amount of energy, the liquid will rise along the tube till a new equilibrium is reached: $f\cos\theta = mg$. f is the adhesive force, which is equal to $2\pi R\gamma$ (R is the radius of the capillary, γ is the surface energy density of the liquid). θ is the contact angle between the liquid and the capillary. mg is the gravity of the liquid column, which equals to $\pi R^2 h\rho g$ (h is the height of rise in the capillary and ρ is the density of the liquid). Therefore, the equality can be rewritten

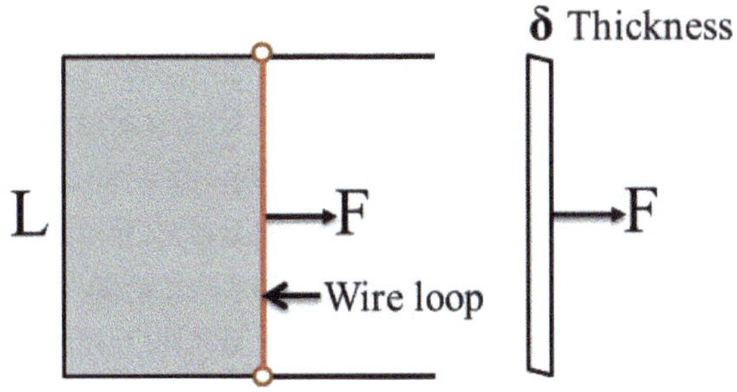

Figure 1. A liquid membrane stretched by an external force.

as $2\pi R\gamma\cos\theta = \pi R^2 h\rho g$, then the surface energy of the liquid can be calculated by $\gamma = R\rho gh / 2\cos\theta$.

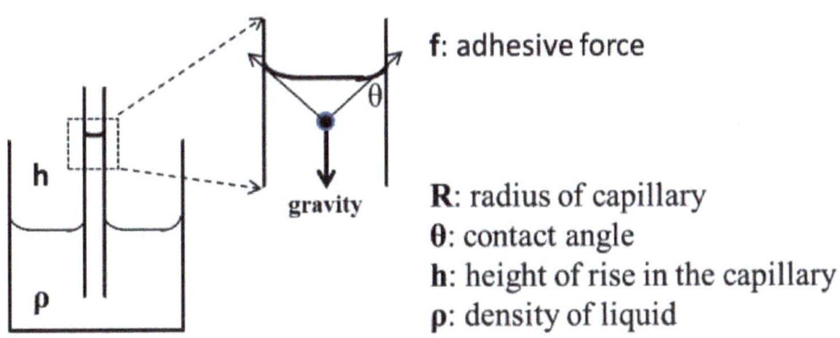

Figure 2. Illustrations for the capillary tube method in measuring the surface energy of a liquid.

For the measurement of the surface energy of a solid, the above two methods cannot be used. But there exists one way which is approximate to the capillary tube method in the case of liquids. This is called the zero creep method. At some high temperatures, the solid creeps and even though the surface area changes, the volume remains approximately constant. If overhanging a cylindrical line of radius r and length l, at equilibrium, the variation of the total energy vanishes and we have

$$dE = 1/2\rho g\pi\left(2l^2 rdr + 2r^2 ldl\right) - 2\pi\gamma\left(rdl + ldr\right) = 0$$

where ρ is the density, and γ is the surface energy density of the solid line. Since the volume of the solid line remains constant, the variation of the volume is zero, i.e.

$$\pi r^2 l = \pi\left(r + dr\right)^2\left(l - dl\right)$$

From these two equations, the expression of the surface energy for a solid can be deduced as $\gamma = r\rho gl/2$, which is similar as that of the liquid deduced from the capillary tube method. So, the surface energy density of the solid can be obtained by measuring the radius and length of a cylindrical line, at equilibrium.

This method is only valid for isotropic solids, whose surface energy is the same in all orientations. Such a condition can be strictly satisfied only for amorphous solids. If the sample is a metal or made by powder sintering (like ceramics), isotropy is also a good approximation. In the case of single-crystal materials, which are obviously anisotropic, this method will become invalid. In addition, since the measurement conditions for the surface energy of a solid requires high temperature, the obtained results sometimes will be inaccurate and thus unreliable in practical applications. So, in determining the surface energy of a solid, theoretical methods are of special importance. The traditional theoretical methods in estimating the surface energy of a solid include the thermodynamics calculation and mechanical calculation methods.

In the thermodynamics calculation method, the surface energy of a solid at a given temperature is derived by the thermodynamic relations from the measured value under a melting state. So, the calculated surface energy is also only valid for isotropic solids. In the mechanical calculation method, the surface energy of a solid is related to its Young's modulus. But since the Young's modulus for many materials is nearly the same, the obtained values are quite coarse with about six to seven multiples of errors. Additionally, both the thermodynamics and mechanical calculation methods can't reflect the relationships between the surface energy and the crystallographic orientations. So, they are mostly used for some particular systems or in cases that require less precision.

The third theoretical method is called the atomic calculation method. In this method, the surface energy is determined by multiplying the energy of one bond and the number of bonds being broken. According to the different approaches in obtaining the energy of one bond, this method can be split into two branches. In the first one, the energy of one bond is estimated by the sublimation energy. For the crystal owning one mole (N_A) atoms, at least 0.5 N_A bonds will be formed. In consideration of the coordination number Z, the number of bonds in one-mole crystal will be 0.5 N_A*Z. So, the average energy per bond can be estimated as $\varepsilon = \frac{\Delta H_s}{0.5 N_A * Z}$. ΔH_s is the molar enthalpy of sublimation. In the other one, the energy of one bond is calculated by the interatomic potential, which can be calculated using the modern molecular simulation method or the first principle method. Take the FCC crystal as an example, the coordination number of every atom is 12. While the atoms at the (111) plane only possess nine coordination atoms. This means that 3 bonds per atoms are broken during the formation of a (111) plane in the FCC crystal. So, the energy required to form one surface atom can be given as: $E(111) = \varepsilon * 3 \equiv \frac{\Delta H_s}{2N_A}$. The surface energy γ can be defined as $\gamma \equiv E(111) \times$ number of surface atom / surface area. For the (111) plane in FCC crystal, the number of surface atom / surface area equals $\frac{4}{\sqrt{3}a_0^2}$. So, $\gamma = \frac{\Delta H_s}{2N_A}\left(\frac{4}{\sqrt{3}a_0^2}\right) = \frac{\Delta H_s}{\sqrt{3}N_A a_0^2}$.

2.3. Thermokinetics process in thin film growth

During the thin film growth process, an old surface will be covered and a new surface will be created. Thus, this process should be mainly controlled by the surface energy theory. Therefore, the growth thermokinetics as well as the quality of the obtained samples will be controlled by the surface energy. In the first, the critical size of the surface nucleation in thin film growth and the nucleation rate are determined by the surface energy. In this section, the thermodynamic process of thin film growth on the flat surface is illustrated to introduce this determination relationship.

In order to study step generation on a flat surface, we consider the case of forming a disc with radius r and height a (an atomic layer), as illustrated in Figure 3. We assume that the material of the disc is the same as the flat surface and the disc is grown epitaxially, the additional surface energy caused by the formation of the disc is only the surface energy on the side region of the disc,

$$E_{\mathrm{d}} = 2\pi r a \gamma$$

This increased surface energy will exert a compressive force on the disc,

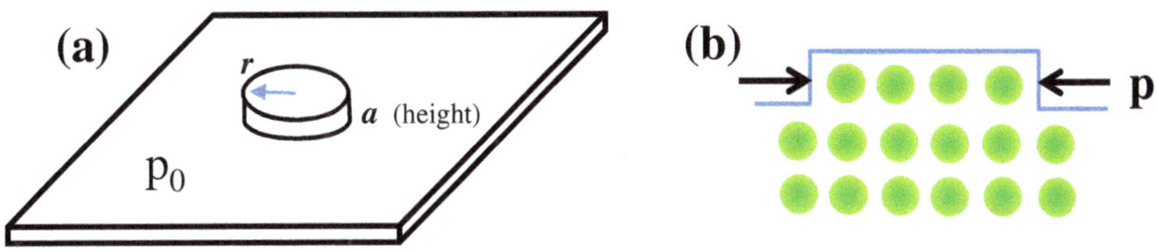

Figure 3. (a) The disc nucleation on flat surface with height a and radius r; (b) schematic cross-sectional view of the disc (the side of the disc exerts a compressive force on it).

$$p = \frac{1}{A}\frac{dE_{\mathrm{d}}}{dr} = \frac{2\pi a \gamma}{2\pi r a} = \frac{\gamma}{r}$$

where A is the surrounding area of the disc. Under such a compressive force, the energy of each atom in the disc will increase

$$p\Omega = \frac{\gamma\Omega}{r}$$

where Ω is the atomic volume. Due to the energy increase, the atoms in the disc become much easier to sublimate than those in the flat substrate. The sublimation rate can be written as

$$J_{\mathrm{c}}^{'} = N_0 \upsilon_{\mathrm{s}} \exp\left(-\frac{\Delta G_{\mathrm{des}}}{kT} + \frac{\gamma\Omega}{rkT}\right) = J_0 \times \exp\left(\frac{\gamma\Omega}{rkT}\right)$$

where N_0 is the number of atoms being absorbed on the flat surface per unit area, υ_{s} is the RMS velocity of atoms on the flat surface, ΔG_{des} is the activation energy during the desorption process of the surface atom. At a given temperature, $J_c^{'}$ and J_0 are both proportional to the pressure, so the pressure ratio on the disc and the flat surface is

$$\frac{p}{p_0} = \exp\left(\frac{\gamma\Omega}{rkT}\right)$$

From this equation, the energy changes per atom from the gas phase to the solid phase can be calculated as $\Delta\mu = kT\ln(\frac{p}{p_0})$, thus the sublimation heat per unit volume ΔE_s can be calculated by $\Delta E_s = \frac{\Delta\mu}{\Omega}$.

Consider the energy change from the gas phase to the solid phase during the disc growth, including the surface energy being increased and the sublimation heat being released,

$$\Delta E = E_s - V\Delta E_s = 2\pi ra\gamma - \pi r^2 a\Delta E_s = 2\pi ra\gamma - \pi r^2 a\frac{\Delta\mu}{\Omega}$$

From this equation, the critical size (here means the critical radius r_c) of the surface nucleation can be defined. Let $\frac{d\Delta E}{dr} = 0$, we can obtain

$$r_c = \frac{\gamma\Omega}{\Delta\mu}$$

If the radius of the disc reaches r_c, the energy change ΔE_c will equal

$$\Delta E_c = \pi r_c a\gamma$$

This energy is defined as the nucleation activation energy, which determines the nucleation rate

$$\text{nucleation rate} \propto \exp\left(-\frac{\Delta E_c}{kT}\right) = \exp\left(-\frac{\pi a\gamma^2\Omega}{kT\Delta\mu}\right)$$

In the second, the growth mode of thin film means that the layered growth or island growth is also controlled by the surface energy. Such control is illustrated in the heteroepitaxial growth of a thin film on a flat substrate, taking the (001) facet growth of simple cubic crystal as an example, and only the nearest-neighbor interactions are considered. Let thin film of material A grow on surface B, the interaction potential between two A atoms and two B atoms is indicated by U_{AA} and U_{BB}, respectively, and the interaction potential between atoms A and B is indicated by U_{AB}. According to the definition of the surface energy, the surface energies of material A and B, and the interface energy between material A and B can be calculated as

$$\gamma_A = u_{AA} / 2a^2$$
$$\gamma_B = u_{BB} / 2a^2$$
$$\gamma_{AB} = \left[\frac{u_{AA}}{2a^2} + \frac{u_{BB}}{2a^2} - \frac{u_{AB}}{a^2} \right]$$

where a is the lattice constant (the same for A and B). If $u_{AB} \geq u_{AA}$, which means $\gamma_B \geq \gamma_A + \gamma_{AB}$ (the film growth will reduce the energy of the system), the material A will wet the surface of material B completely, which results in a layered thin film growth process. If $u_{AB} < u_{AA}$, which means $\gamma_B < \gamma_A + \gamma_{AB}$ (the film growth will increase the energy of the system), material A can't completely wet material B, which results in an island thin film growth process.

In the initial stage of the thin film growth, there are only very few atoms. The concept of surface energy, which is a collection quantity for a given material, seems too big to be used in analyzing the growth process. So, in practice, for fine studies on the thin film nucleation process, energy evolution on the atomic and even on the electric level based on the theoretical method (specially the molecular simulation technology) is always used.

3. Approaches from energy aspect in investigating graphene growth

Graphene is the first truly 2D crystal ever observed in nature,[18] as shown in Figure 4. This is remarkable because 2D crystals were predicated to be unstable and thus inexistent in the past, due to the Mermin–Wagner theorem. This theorem states that a 2D crystal will melt at any temperature but zero due to thermal fluctuations. From its discovery, graphene has grabbed appreciable attention due to its exceptional electronic and optoelectronic properties: it is reported as one of the best electronic materials. The reported properties and applications of graphene have opened up new opportunities for future devices and systems.

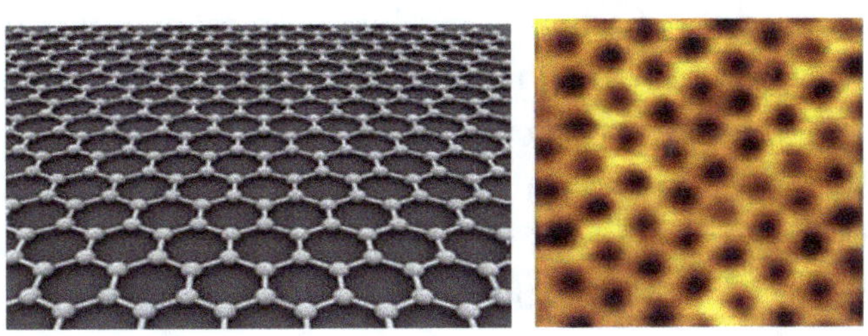

Figure 4. Illustration and the scanning probe microscopy image of graphene. (Haider I. Rasool, Emil B. Song, Matthew Mecklenburg, et al., JACS, 2011)

Because of the great application potential of graphene, investigations on the synthesis of single sheets of graphene have also attracted a large number of researchers and companies. Till now, the synthesis routes of graphene can be broadly categorized into five main different sections[19]: the exfoliation and cleavage route, the chemical vapor deposition techniques, thermal decomposition of SiC or other substrates, unzipping CNTs, and chemical methods. Among them, the chemical vapor deposition technique is the most popular for preparing large-scale and high-quality graphene samples. Most of the chemical vapor deposition growth of graphene uses nickel, iron, and copper foils as the substrate. While nickel and iron have great carbon solubility, the graphene growth will go through a carbon dissolving and precipitating process. [12] Therefore, on nickel and iron foils, the obtained graphene films are often inhomogeneous with many defects and multilayer flakes. Such disadvantage can be overcome by using copper foil as the substrate due to its low carbon solubility. The growth process of graphene on copper foil is self-limited, thus large-scale high-quality single-layer samples can be obtained. So, here, we focus on the formation of graphene on copper substrate. To investigate the nucleation and growth mechanism of graphene on copper surface systematically is of great importance for exploiting and optimizing the fabrication of graphene by the chemical vapor deposition technique.

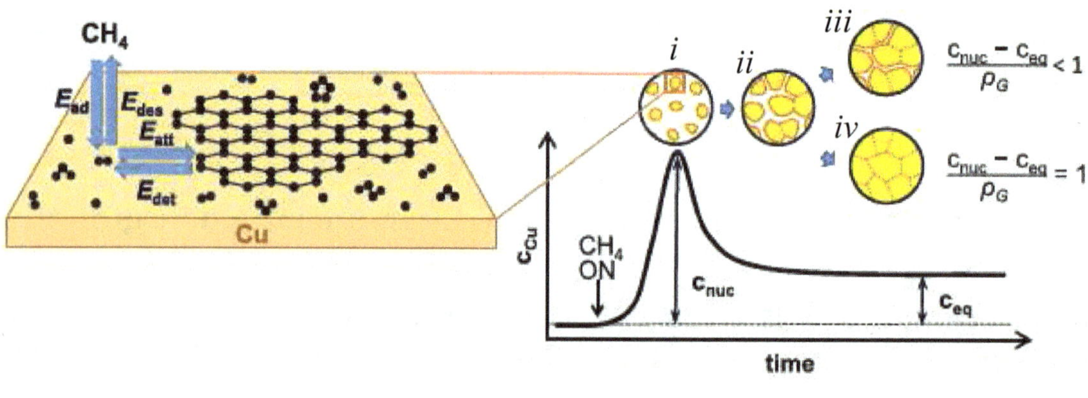

Figure 5. Growth mechanism of graphene on Cu surface. Methane is adsorbed on the surface and then decomposed to carbon adatoms. When the supersaturation of the carbon adatoms reaches a critical value (C_{nuc}), graphene begins to nucleate and grow step by step.

Through experimental method, by characterizing the graphene nuclei grown on copper for different growth temperatures and times by high-resolution scanning electron microscope, Kim et al. [10] have analyzed the nucleation and growth mechanism of graphene on copper substrates (as shown in Figure 5). Methane molecules are first chemisorbed on the copper surface. Such adsorbed methane can be decomposed to carbon adatoms. The concentration of these carbon adatoms, C_{cu}, increases with time increases, until a critical point, i.e., the supersaturation of the carbon adatoms reaches a critical value (C_{nuc}). Now, graphene begins to nucleate. Such process and the growth of the nuclei will deplete the carbon surrounding them. So, C_{cu} decreases quickly. In this decreasing process, the nucleation rate becomes negligible, while the growth of formed nuclei continues. Till the C_{cu} is reduced to a stable level, C_{eq}, the equilibrium between graphene, surface carbon, and CH_4/H_2 will be reached.

Analyses based on the experimental characterizations can give the framework of graphene growth similar to the existing theories for two-dimensional nucleation and growth of thin films. However, encumbered by the growth conditions of graphene and the lack of effective characterization methods, some fine messages, like the exact nature of the active carbon species adsorbed on the Cu surface, the adsorption and desorption energy of the carbon species on the Cu surface, the surface diffusion energy barrier, as well as the bonding process of the carbon species, cannot be well obtained yet. These messages are of great importance. The first-principle method, especially the density functional theory (DFT), is much suited for the investigation of such messages. By this method, a lot of simulation studies have been carried out on the atomic process of graphene growth on copper substrates.

DFT is a computational quantum mechanical modelling method, which is among the most popular and versatile methods available in physics, chemistry, and materials science. With this method, the properties of the system are determined by using functionals, i.e., functions of another function, which in this case is the spatially dependent electron density. DFT has broad applications in the chemical and material sciences for the interpretation and prediction of complex system behaviors. Specifically, DFT computational methods are applied for the study of systems exhibiting high sensitivity to synthesis and processing parameters. In this section, we will briefly introduce the primary approaches that have been exploited in investigating the growth mechanisms of graphene on copper surface.

3.1. The approaches that have been exploited in investigating graphene growth

As mentioned above, the system tends to reduce its free energy as it reaches the equilibrium state. So, the most direct approach in investigating the growth process of graphene is the so-called geometry optimization approach, to study the geometry configuration evolution of the system with carbon atoms on copper surfaces by minimizing free energy. By DFT technology, during such evolution processes, the bond breaking and formation can be presented virtually, and the optimized configuration close to the real nucleation form can be obtained. For example, through carrying out configuration evolution calculations on the system containing 4–6 carbon atoms on the copper(111) surface, Li et al. [6] found that, at the very first stage, linear chains will be formed and dominate the copper surface, as illustrated in Figure 6.

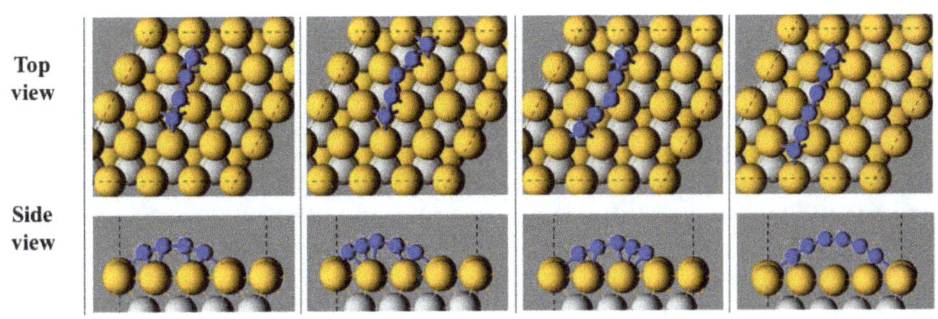

Figure 6. Top and side views for the stable configuration of system containing 4–6 carbon atoms on copper(111) surface.

It is easy to imagine that, if several carbon atoms are absorbed on the copper surface, there will be some different nucleation forms, e.g., four carbon atoms will have four different forms as illustrated in Figure 7. These four configurations are obtained by geometry optimizations starting from different initial configurations with carbon atoms arranged on different locations. Depending on the laws of statistical thermodynamics, the occurrence probabilities of these different configurations will be dramatically determined by their configuration energy. So, calculating and comparing the stable energies of different possible configuration is also the commonly used approach in investigating the graphene growth process.

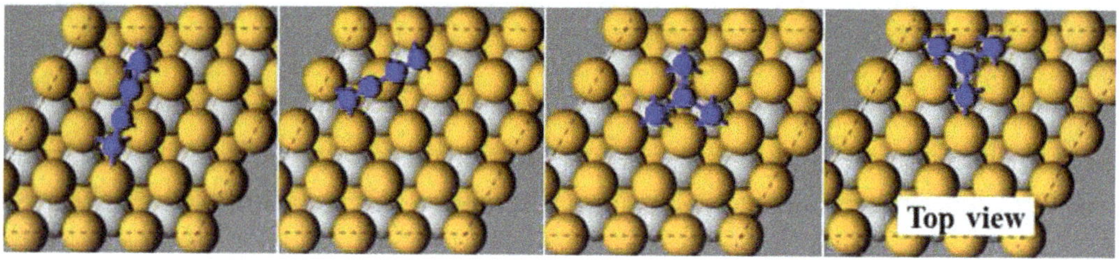

Figure 7. Different nucleation forms of four carbon atoms on Cu(111) surface.

The growth process of graphene is a dynamic chemical reaction process, thus it is not only determined by the energies of the reactants (initial configurations) and products (stable configurations), but also deeply influenced by the activation energy. The transition state search technology in DFT method can offer these messages quite well. For example, Wu et al. [3] have used the climbing image nudged elastic band method to search the transition state of a 1 + 1 (the bonding of two carbon atoms on Cu(111) surface) reaction, and identified the minimum energy path (MEP) of this reaction, as shown in Figure 8. First, the carbon atom remaining on the surface (carbon B) rotates around the bridging Cu to its neighboring site (Figure 8c), with a 0.51 eV barrier. Then, it rotates further toward carbon A with an activation energy of 0.64 eV. Finally, by conquering a 0.37 eV barrier, carbon B drags carbon A to the surface and forms a dimer (Figure 8e). This MEP with several barriers thus gives a very rugged part of the two-adatom potential energy surface.

To reveal the realistic nucleation process of graphene on copper surface, the quantum mechanics/molecular mechanics (QM/MM) method should be more suitable than the above approaches since it can consider the real reaction temperature and model the growth kinetics and nonequilibrium processes. However, since it contains quantum mechanics calculations, the computational efficiency is very low and thus this method can't handle large systems and can't model long enough times to reproduce the real reaction process. So, till now, few reports [2] using such approaches have been found.

The next best approach in revealing the dynamics nucleation process of graphene is the molecular dynamics (MD) method, based on empirical atomic force field. The MD method can simulate the real physical movements of every atom in a system of interacting atoms. The movement trajectories of every atom are determined by solving Newton's equations of motion numerically, where the interatom forces are defined by interatomic potentials (atomic force field). The precision and reliability of the MD method is much lower than the DFT method

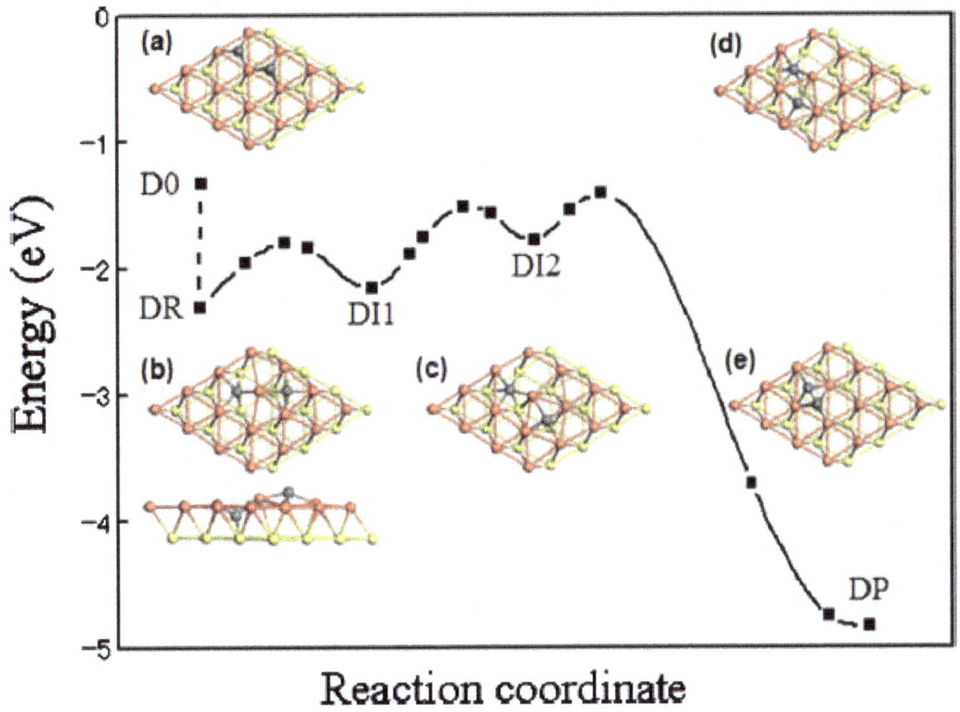

Figure 8. Minimum energy path of the 1 + 1 carbon atom reaction on copper(111) surface.

since they dramatically depend on the force field, which is fixed and thus cannot take the chemical conditions of atoms into account. While, just because of the coarse graining of the MD method compared with the DFT method, it can deal with systems consisting of a vast number (tens of thousands) of atoms. Using MD simulation based on the ReaxFF force-field, Ding et al. [13] have found that after 100 ps MD simulation at 1000 K, high C concentration leads to the formation of graphene islands, as shown in Figure 9.

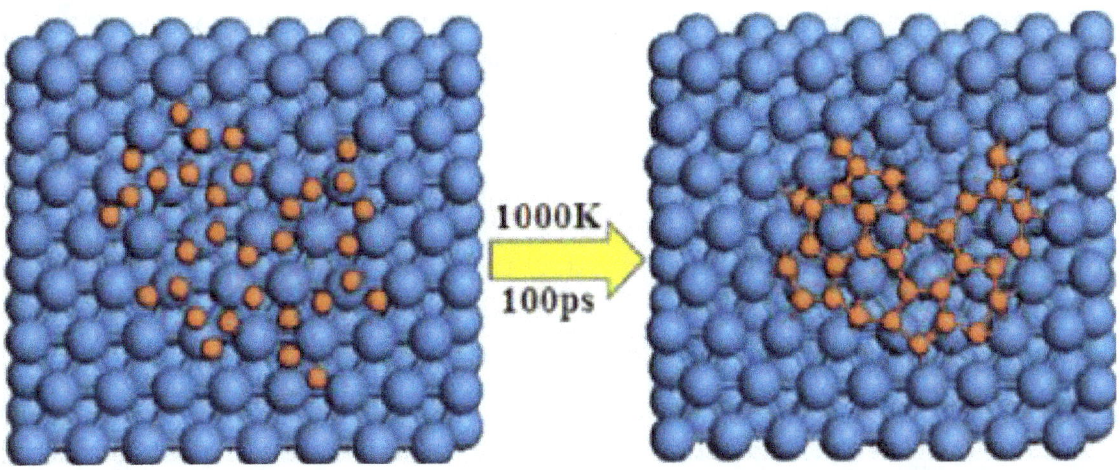

Figure 9. Initial and final structures obtained in 100 ps MD simulations at 1000 K for 32 carbon atoms.

In summary, upon investigating the nucleation behaviors of carbon atoms to form graphene, four main approaches have been exploited. These approaches can provide intuitive images in the nucleation process of graphene, and some very useful messages in analyzing the surface reaction path. Besides these four approaches, the Monte Carlo simulation method[14, 20] has been also used. Since the Monte Carlo method provides similar messages as the DFT method with less accuracy, we did not discuss it here.

4. Achievements on investigations of graphene growth

4.1. The exact nature of the active carbon species in graphene growth

Hydrocarbon decomposition is the first step in the growth of graphene on copper surface, which determines the exact nature of the active carbon species. By the DFT calculations, Zhang et al. [2] investigated the decomposition process of CH_4 on a five-layer $p(3 \times 3)$ copper slab by the transition state search technology. The initial state is an adsorbed CH_4 molecule, and the final product is a C atom plus four H atoms on the surface. There are three intermediates, namely, methyl (CH_3), methylene (CH_2), and methylidyne (CH). As shown in Figure 10, all four dehydrogenation steps are endothermic, and the corresponding activation energy barriers are about 1.0–2.0 eV. The final product C + 4H is already 3.60 eV higher in energy than the adsorbed CH_4, which suggests that atomic carbon is energetically very unfavorable on Cu surface. According to their results, the active species for graphene growth on the copper surface should not be carbon atoms but mainly CHx (especially CH). This is quite different from the case of other active metal surfaces, such as Pd, Ru (where the decomposition of CH_4 is exothermic), and Ni (decomposition of CH_4 is slightly endothermic).

Zhang's work is quite helpful to understand the growth mechanisms of graphene at the very initial stage. While, in spite of these insights, in most DFT investigations on the nucleation of graphene, atomic carbons are used. Surely, one of the main reasons for using atomic carbons is for simplicity (the situations will become very complicated if CHx is used as the active species). Besides, the more reasonable reason for using atomic carbons as the active species in graphene growth is that it is still hard to determine certain active carbon species for graphene growth on copper surface: since there exist complex surface morphologies on real copper substrates in experiments, the active site for the dehydrogenation of CHx might not be located at the plane copper surface but perhaps near the step regions, which should introduce significant influences on the dehydrogenation energy. In addition, to grow graphene on copper surfaces, dehydrogenation should finally be completed. To determine the exact nature of the active carbon species in graphene growth, further investigations are quite necessary.

4.2. The stable configuration of 1-2 carbon atoms on the copper surface

By experimental methods, the growth mechanism of graphene on copper substrate has been demonstrated to be a surface adsorption process. So, naturally, the carbon atoms should be more stable on the surface than in the bulk of the copper lattice. However, by comparing the absorption energy of carbon atoms on different locations in the 4 × 6 copper(111) slab, as shown

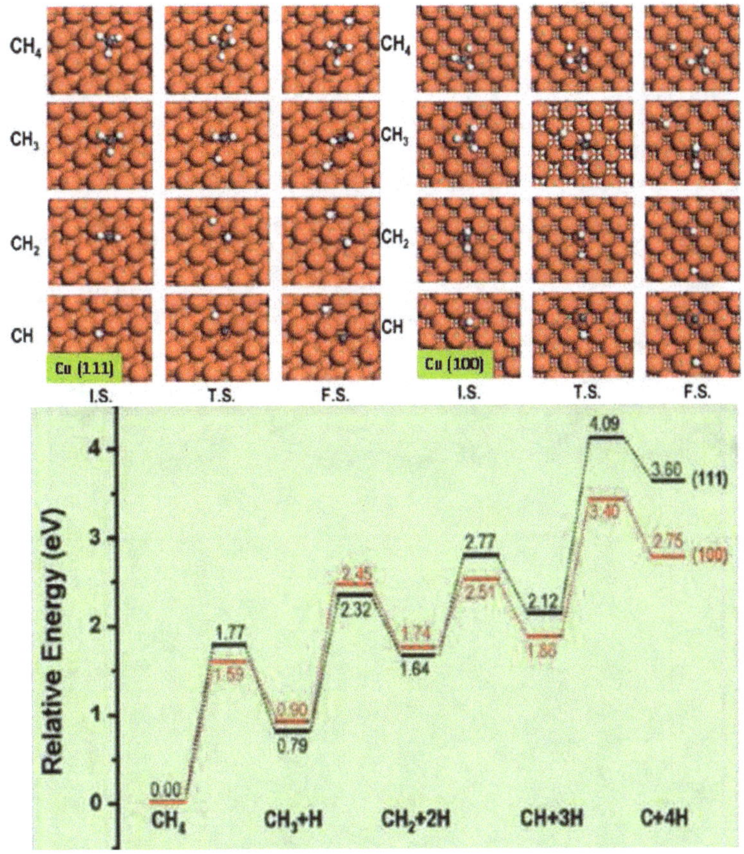

Figure 10. Geometric structures of the initial state (I.S.), transition state (T.S.), and final state (F.S.) of the four steps of CH_4 dehydrogenation on Cu(111) and Cu(100) surface (top); their energy profile is also shown (bottom).

in Figure 11, Riikonen et al. [4] found that the HCP adsorption site is unstable. The FCC and BRI are also metastable sites; at finite temperatures, carbon diffuses directly to the subsurface A site. According to their analysis, the stabilization of carbon interstitials in the copper subsurface area can be understood with a few simple arguments. They state that, the copper atoms at the topmost layer are easily pushed toward the vacuum due to their low coordination numbers. Such greater flexibility of the topmost copper atoms compared with the bulk ones opens a gate to the carbon atoms, which can thus sink into the subsurface and form octahedrally symmetric copper surroundings.

The occurrence of sinking carbon atoms on the copper subsurface has been also found in the work of Wu et al. [3] through transition state searches for two next nearest neighboring carbon atoms to form a dimer on the surface (Figure 8). In their report, an almost linear C-Cu-C configuration (Figure 8b) was formed. They named it the bridging-metal (BM) structure. Through geometry optimization for the system having two carbon atoms on the Cu(111) surface, Li et al. [7] also found this so-called BM structure, as shown in Figure 12.

In conclusion, despite the growth mechanism of graphene on copper substrate being a surface adsorption process, the carbon atoms still have a chance to penetrate into the subsurface (solve in copper), especially under very low carbon atom concentrations. This means that under special process conditions, carbon atoms can also be implanted into the copper foil. Thus,

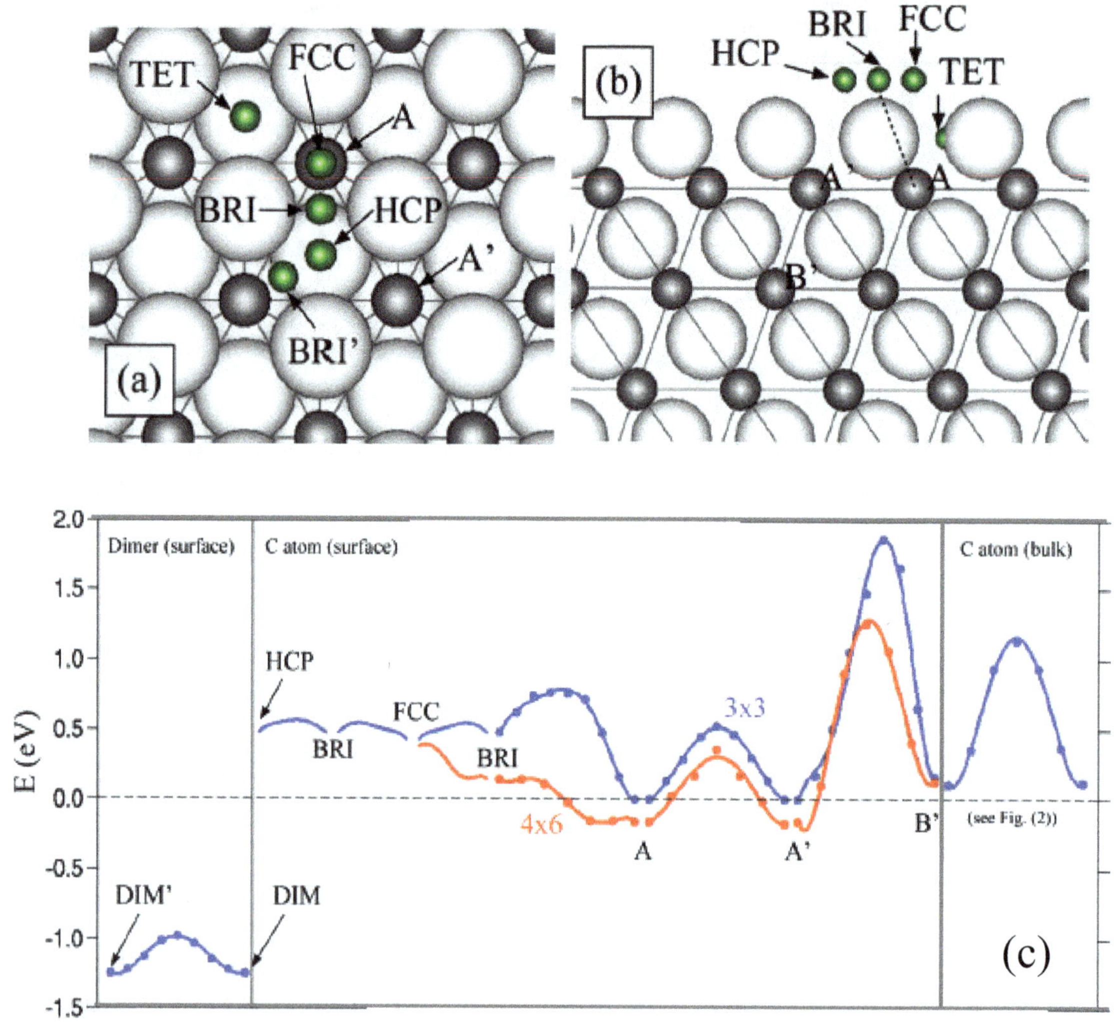

Figure 11. The Cu(111) surface, top (a) and side (b) views; carbon energetics and minimum energy paths on the Cu(111) surface, subsurface and in the bulk (c). The small green and dark spheres denote the surface and octahedral adsorption sites, respectively. Surface adsorption sites contain four inequivalent styles: FCC, HCP, TET, and BRI. BRI' is a neighboring equivalent site to BRI.

graphene can be formed by following high-temperature annealing, which results in the migration of implanted atoms to the surface and eventually bonding to each other. Such a technique has successfully been used.

In the usual fabrication process of graphene by CVD technology, such a "sinking" of carbon atoms into the copper subsurface will have nearly no influence on the surface adsorption growth of graphene, firstly because dimer formation and the subsequent graphene growth is by far the most favorable reaction in both energetic and kinetic terms. [12]. Secondly, Li et al. [7] have investigated the following behavior of the sunken carbon when more carbon atoms are absorbed around it. They found that, the sunken carbon atom will spontaneously form a dimer with one of the newly adsorbed carbon atoms, and the formed dimer will up-float on the top of the surface again, as shown in Figure 13.

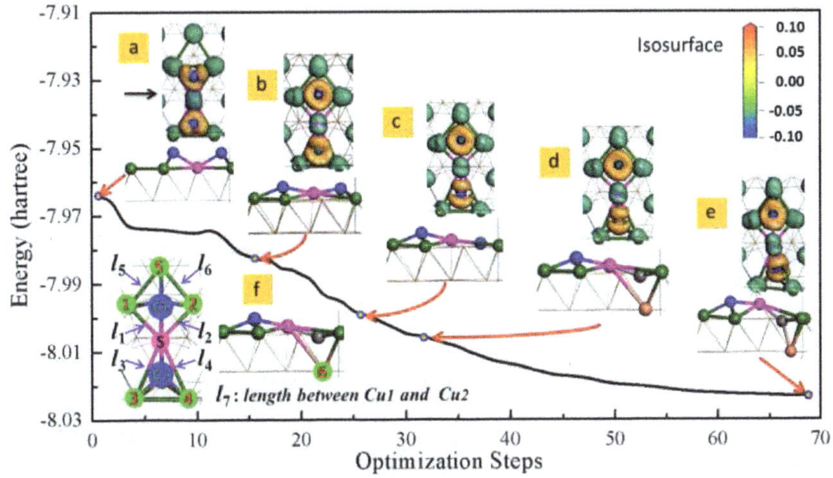

Figure 12. Geometry optimization paths from two on-surface carbon atoms to the BM structure. The deformation electrodensities are mapped to characterize the bonding situations.

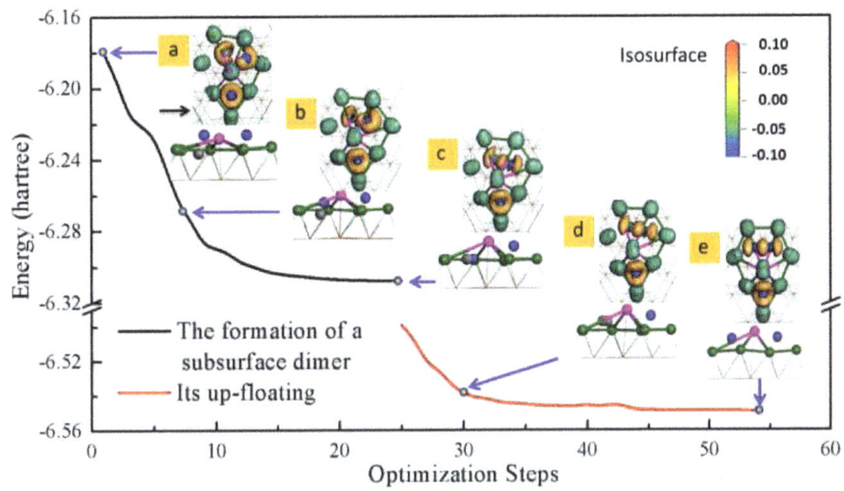

Figure 13. Geometry optimization paths for the system with one carbon atom absorbed near the BM structure.

4.3. Configuration selectivity of the initial carbon clusters

Through DFT calculations, Gao et al. [5] have investigated the stable configurations of carbon clusters containing 1 to 24 atoms on the Ni(111) surface. For different configurations of the carbon clusters with the same size, they analyzed their stability by comparing the absorption energies (so-called formation energy), as given in Figure 14. They found that, within the entire size range of their calculations, carbon chains on the Ni(111) surface are always more stable than ring configurations of the same size. The crossover between the carbon chains and the C-sp2 network occurs at N = 12, beyond which the energy difference between chain and sp2 configurations becomes larger and larger. This means that, during the graphene growth on Ni(111) surface, until N = 12, the C chain configuration is superior and will dominate the metal surface. A ground state structural transition from a C chain to a C-sp2 network (graphene island) occurs at N = 12.

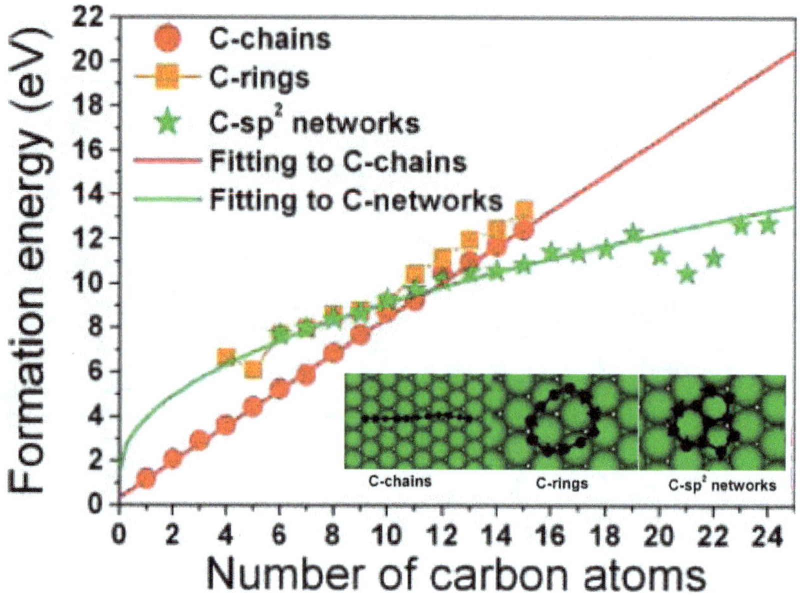

Figure 14. Formation energies of chains, rings, and most stable sp2 networks on the Ni(111) surface versus the number of carbon atoms.

For the case of copper surfaces, Li et al. [6] have investigated the graphene nucleation path by importing carbon atoms step by -step. At every step, they exhausted all possible configurations and discussed their stability. Based on careful configuration and energy analyses, an overall path of graphene nucleation has been proposed in Figure 15. At the very first stage, the linear chains containing 4 to 10 carbon atoms will be formed and dominate the copper surface, while both the Y-type and circular carbon species are energetically repelled. Then, the growth of the carbon cluster encounters an energy barrier at about 0.25 eV. By conquering such a barrier, the carbon clusters will present Y-type (furcate) structures. Then, by adsorbing new carbon atoms step by step, ring-containing carbon structures and graphene nuclei will be formed, with energetic preference. Their results suggest that, it will be difficult to form furcate and ring-containing carbon structures at the very initial stage of graphene nucleation, but it should be formed when the linear chains have grown to some length.

Based on analysis of the deformation electrodensity maps of the linear carbon chains containing 4 to 10 atoms (Figure 16), Li et al. [6] have also discussed the bonding situation between the linear chain and the copper surface. The green color in Figure 16 indicates that net electrons remain and stable chemical bonds are formed. Through bonding situation analyses, they point out L6 should be a representative structure. In the linear carbon chains, when the number of carbon atoms is less than 6, not only the end- but also the mid-carbon atoms bond stably with the copper surface; if the number of carbon atoms reaches 6, the mid-carbon atoms are completely detached from the copper surface, and thus an arc is formed.

4.4. Continuous growth of graphene over steps

In the chemical vapor deposition fabrication of graphene on copper surface, it has been found that the growth of macroscopic pristine graphene is not limited by the underlying copper

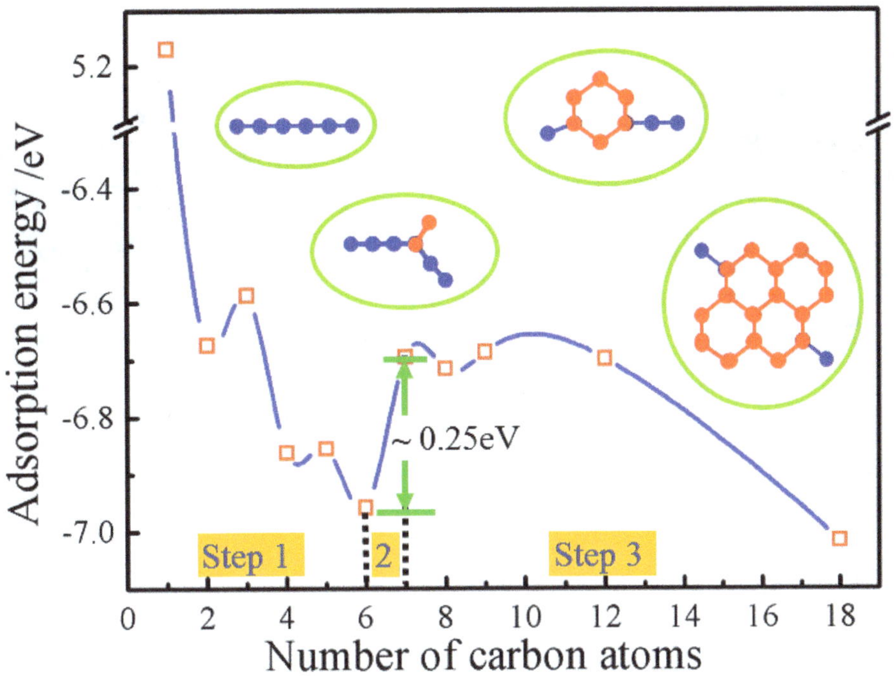

Figure 15. Energy evolution curve of the overall nucleation path of graphene growth.

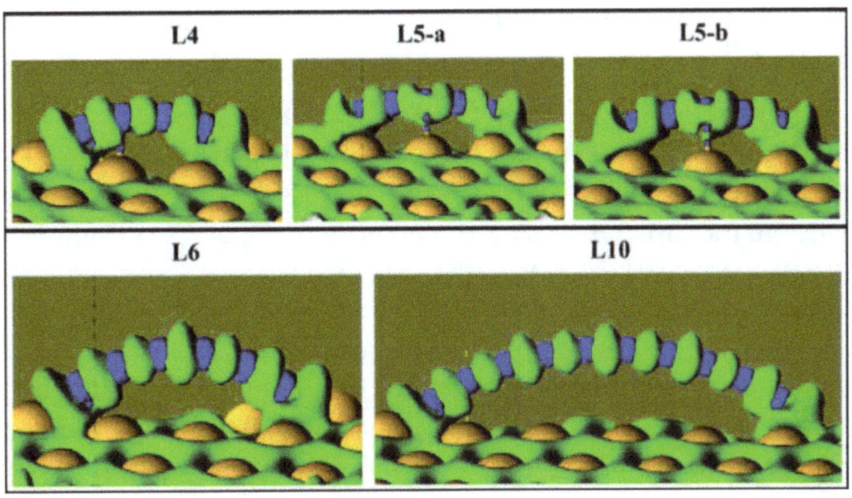

Figure 16. Deformation electrodensity maps for the linear carbon chains.

structure. Haider et al. [21, 22] have characterized the surface morphology of the copper substrate and the graphene grown by scanning tunneling microscopy (STM), as shown in Figure 17. They revealed that the atomic arrangement of graphene was not affected by the morphology and atomic arrangement of the copper substrate. This feature implies practical value for the mass production of high-quality graphene on rough copper substrates.

Inspired by this experimental phenomenon, Li et al. [8] have investigated the coalescence of carbon atoms over a copper monatomic step by the DFT calculations. They constructed a monatomic step as shown in Figure 18a, and carefully explored how the carbon atoms bond together over the step. Firstly, they put some carbon atoms on and under the steps separately.

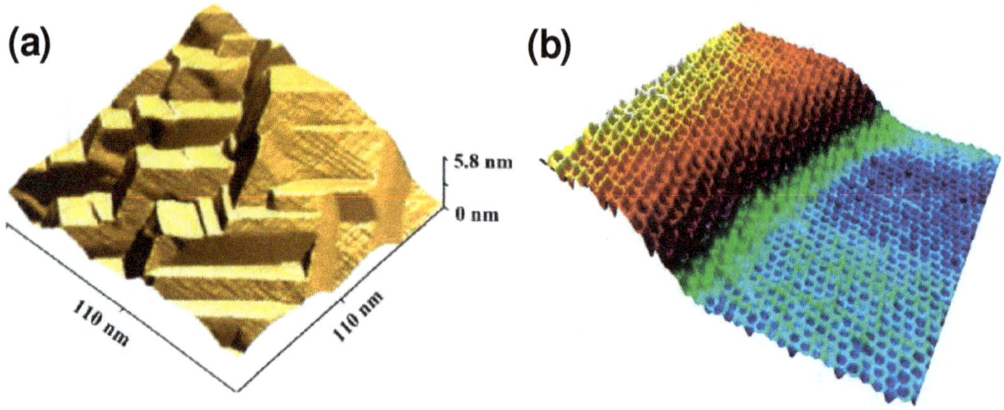

Figure 17. (a) STM topograph of a highly corrugated region of the sample. (b) Atomic resolution STM topograph over a copper monatomic step.

In some special cases, as shown in Figure 18b, after geometric optimizations, the separated carbon atoms can successfully bond together. Additionally, this over-step coalescence of the carbon atoms is spontaneous as the energy evolution curve descends monotonically throughout all the geometry optimization steps. They attributed this success to the energy barrier preventing the bonding of the two carbon atoms, which is reduced significantly, since the two copper atoms between them are both shared by three carbon atoms and thus have weak interactions with the two bonded carbon atoms. However, the dimer formed finally moves up to the upper terrace but is not located on the over-step position. In their following DFT calculations, by importing another carbon atom to the "left hole" of the atom being drawn up to the upper terrace, they found that this over-step coalescence process is unrepeatable (Figure 18c). As a conclusion, the direct over-step coalescence of the carbon atoms separated by the steps is very difficult, and thus should not be the main pattern of graphene's continuous growth over the steps.

By importing additional carbon atoms between the existing ones separated by the steps, they found that the main way in which graphene grows over the steps continuously is that the carbon atoms, adsorbed additionally on the locations between the already existing ones which are separated by the steps, link them (these carbon atoms separated by the steps) together. They first imported one additional carbon atom, as illustrated in Figure 19a. The obtained configurations after geometry optimization show a positive trend to a successful over-step coalescence of carbon atoms. Then, one more carbon atom is imported again in the optimized configurations as shown in Figure 19b. Finally, the new carbon atom links the separated ones together, and a cambered over-step carbon chain was formed.

During the calculations of the absorption energies, they found that the adsorption energy of the single carbon atom near the steps (5.91–6.48 eV) is about 1.0 eV higher than that adsorbed on a flat copper surface (5.13–5.17 eV). This indicates that carbon nucleation should be very likely to start from the stepped regions. The thermodynamic reason is that around the step regions on the substrate, the surface energy is higher than that on the flat regions.

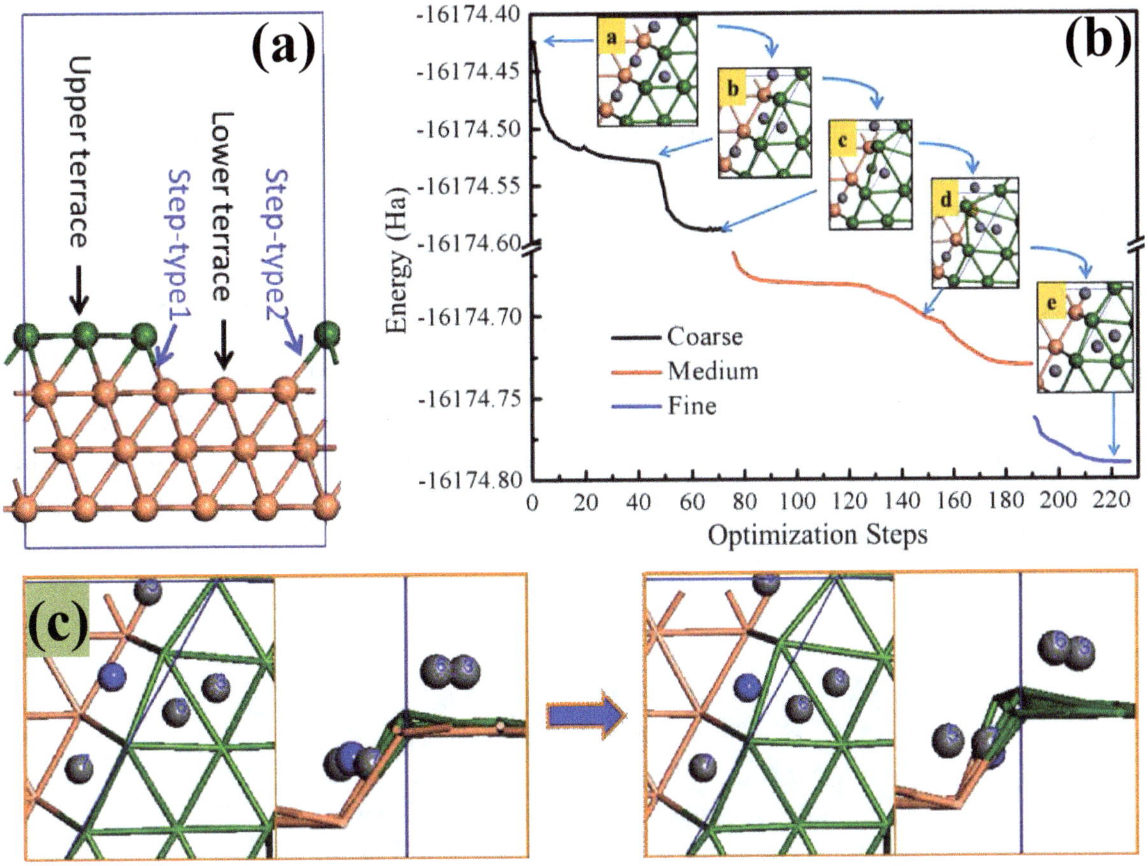

Figure 18. (a) Side view of Cu(111) surface separated on terraces by monatomic steps. (b) Energy evolution curves and corresponding key configurations during the direct over-step coalescence of the carbon atoms. (c) The direct over-step coalescence of the carbon atoms is unrepeatable.

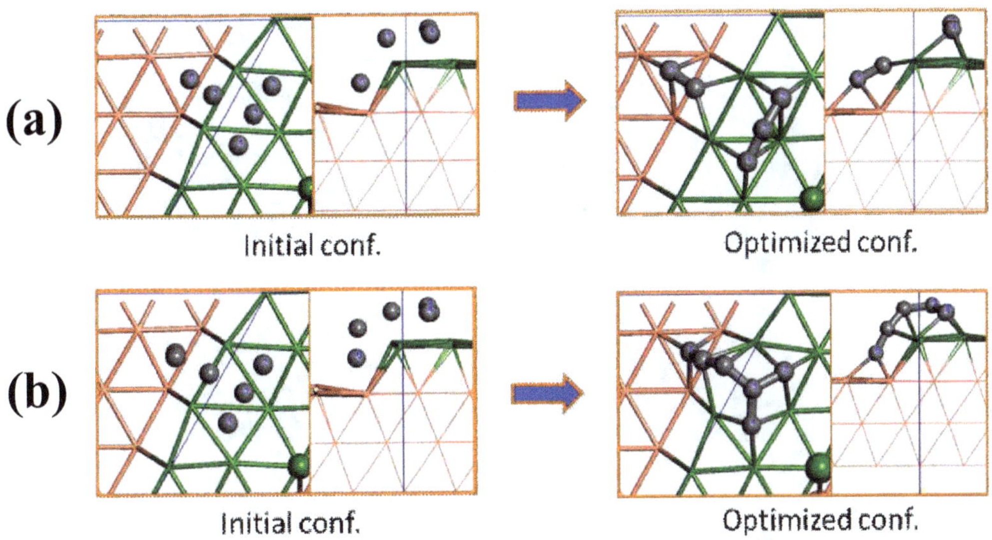

Figure 19. (a) Initial and optimized configurations after importing one additional carbon atoms on the step. (b) Those after importing the second additional carbon atoms on the step.

4.5. Molecular dynamics simulations on graphene growth

Since graphene's growth is typical of a kinetic and nonequilibrium process. Molecular dynamics is a powerful tool to explore such processes at the atomic level. Based on the reactive force-field (ReaxFF), Ding et al. [13] have investigated the evolution of carbon structures and the growth kinetics of graphene on Ni(111) surface. Taking into account that the carbon concentration is raised gradually during the chemical vapor deposition experiments on graphene growth, they firstly investigate the effect of carbon concentration on the nucleation of graphene, as illustrated in Figure 20. By arranging 16 and 32 carbon atoms on the Ni surface, namely, 1/8 and 1/4 monolayers, they found that at the low concentration (16 carbon atoms), the carbon monomers readily enter the subsurface, which is identical to the DFT results above. As the carbon concentration increases, after 100 ps annealing, nearly all of the carbon atoms, which are arranged coincidentally to contain long chains or large polygonal rings, eventually form a sp2 network of pentagons, hexagons, and heptagons. As a summary, low concentrations do not allow the formation of large sp2 network, and high concentrations are required to induce the formation of graphene islands.

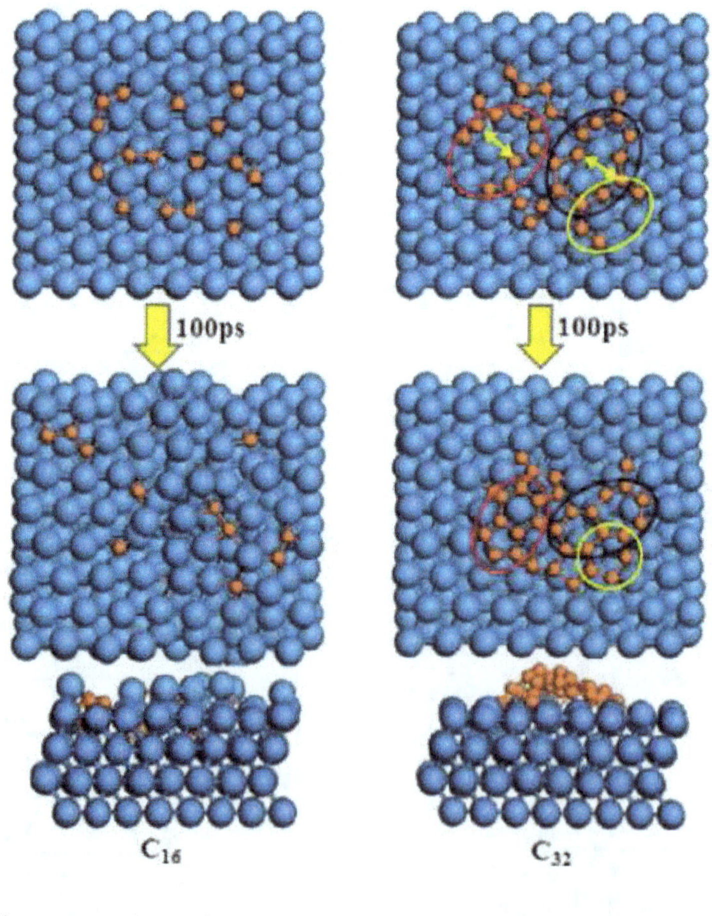

Figure 20. Initial and final structures obtained in 100 ps MD simulations at 1000 K for C16 and C32. Cyan and orange spheres represent Ni and C atoms, respectively.

Then, they turned to the influence of temperature on the formation of carbon structures. Under high carbon concentrations with 64 carbon atoms on the Ni surface, they analyzed the final configuration (Figure 21) after 100 ps molecular dynamics simulation at four temperatures 800, 1000, 1200, and 1400 K. They summarized the number of i-membered rings (MRs) ($i = 3, 4, \ldots, 9$) in the C structures formed at different temperatures in Table 2. With the increases in temperature, the fact that the number of 3- and 4-MRs are greatly reduced indicates that they are very unstable; the increased numbers of 5-MRs and 7-MRs show that they are very stable. The number of 6-MRs denotes the quality of graphene fabricated. According to their simulation, the number of 6-MR reaches the maximum at 1000 K and decreases as the temperature further increases. Their results indicate that the optimal growth temperature of graphene should be around 1000 K, as most CVD experiments of graphene growth were applied.

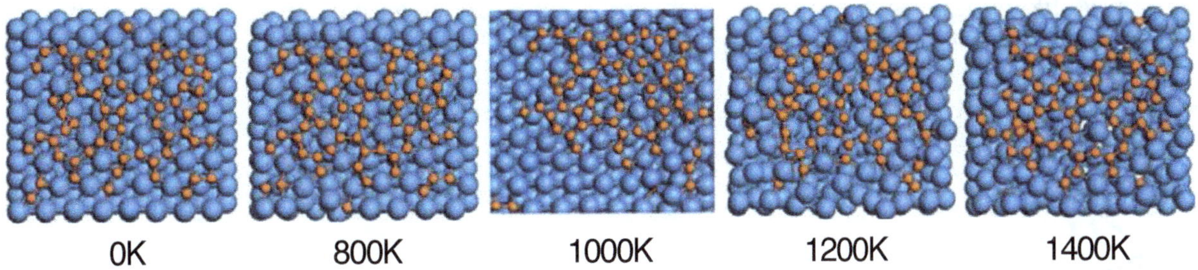

Figure 21. Equilibrium configurations at 0, 800, 1000, 1200, and 1400 K.

T (K)	3-MR	4- MR	5- MR	6- MR	7- MR	8- MR	9- MR
0	6	3	1	2	1	0	1
800	1	0	2	3	2	1	1
1000	1	1	2	8	1	1	0
1200	1	0	2	5	1	0	0
1400	0	0	4	3	1	1	1

Table 2. Number of various polygons in the equilibrium configurations

Through simulating the growth of graphene islands by adding carbon atoms around it, they found that graphene islands can grow larger by capturing deposited C atoms and forming more hexagons on the edge with its self-healing capabilities during growth.

5. Summary

The fundamental theory of surface energy and the corresponding problems involved in thin film growth have been briefly introduced in this chapter. For the special issue of graphene growth on metal surface under the frame of chemical vapor deposition technology, the growth process and the quality of samples obtained is also determined by the surface energy theory. But due to the high-temperature growth conditions of graphene and the lack of effective real-time characterization methods, the fine messages of key importance in analyzing the thermo-kinetics process of graphene growth are difficult to obtain by experimental measurements. Therefore, DFT investigations on the nucleation process of graphene under chemical vapor deposition growth have been carried out very prosperously. The key approaches being exploited in describing the graphene CVD growth process have been introduced. And some main achievements, on investigating the growth process of graphene, using the DFT method and the molecular dynamics method, have been reviewed. Till now, in spite of high-quality graphene samples being fabricated successfully using the chemical vapor deposition technology, to develop a wider range of applications, the preparation of graphene with special structures, like single crystal graphene sheets, graphene nanoflakes, graphene nanoribbons, graphene nanomesh, and graphene quantum dots, still needs further investigations on their growth mechanisms. We hope the above contents are helpful for the improvement in the fabrication and application of graphene.

Author details

Meicheng Li*, Yingfeng Li and Joseph Michel Mbengue

*Address all correspondence to: mcli@ncepu.edu.cn

State Key Laboratory of Alternate Electrical Power System with Renewable Energy Sources, School of Renewable Energy, North China Electric Power University, Beijing, China

References

[1] K. S. Kim, Y. Zhao, H. Jang, S. Y. Lee, J. M. Kim, K. S. Kim, J. H. Ahn, P. Kim, J. Y. Choi, and B. H. Hong, Nature 457, 706 (2009).

[2] W. H. Zhang, P. Wu, Z. Y. Li, J. L. Yang, J Phys Chem C 115, 17782-17787 (2011).

[3] P. Wu, W. H. Zhang, Z. Y. Li, J. L. Yang, J. G. Hou, J Chem Phys 133 071101-1:4 (2010).

[4] S. Riikonen, A. Krasheninnikov, L. Halonen, R. Nieminen, J Phys Chem C 116, 5802–5809 (2012).

[5] J. F. Gao, Q. H. Yuan, H. Hu, J. J. Zhao, F. Ding, J Phys Chem C 115, 17695-17703 (2011).

[6] Y. Li, M. Li, T. Wang, F. Bai, Y.-X. Yu, Phys Chem Chem Phys 16, 5213-5220 (2014).

[7] Y. Li, M. Li, T. Gu, F. Bai, Y. Yu, M. Trevor, Y. Yu, Appl Surf Sci 284, 207-213 (2013).

[8] Y. Li, M. Li, T. Gu, F. Bai, Y. Yu, T. Mwenya, Y. Yu, AIP Adv 3 052130 (2013).

[9] R. G. Van Wesep, H. Chen, W. Zhu, and Z. Zhang, J Chem Phys 134, 171105 (2011).

[10] H. Kim, C. Mattevi, M. R. Calvo, J. C. Oberg, L. Artiglia, S. Agnoli, C. F. Hirjibehedin, M. Chhowalla, E. Saiz, ACS Nano 6, 3614-3623 (2012).

[11] HoKwon Kim, Eduardo Saiz, Manish Chhowalla, and Cecilia Mattevi, arXiv preprint arXiv:1302.0179 (2013).

[12] X. S. Li, W. W. Cai, L. Colombo, and R. S. Ruoff, Nano Lett 9, 4268 (2009).

[13] L. J. Meng, Q. Sun, J. L. Wang, F. Ding, J Phys Chem C 116 6097-6102 (2012).

[14] J. Y. Guo, C. X. Xu, F. Y. Sheng, Z. L. Shi, J. Dai, Z. H. Li, and X. Hu, J Appl Phys 111, 044318 (2012).

[15] Koch W, Holthausen M C, Holthausen M C. *A Chemist's Guide to Density Functional Theory*. Weinheim: Wiley-Vch, 2001.

[16] J. W. Mayer, L. C. Feldman *Electronic Thin Film Science: for Electrical Engineers and Materials Scientists*. Prentice Hall, 1992.

[17] D. P. Woodruff, D. A. King *The Chemical Physics of Solid Surfaces. 10. Surface Alloys and Alloy Surfaces*. Elsevier, 2002.

[18] K. S. Novoselov, A. K. Geim, S. V. Morozov, D. Jiang, Y. Zhang, S. V. Dubonos, I. V. Grigorieva, A. A. Firsov, Science 306, 666-669 (2004).

[19] W. Choi, I. Lahiri, R. Seelaboyina, Y. S. Kang, Crit Rev Solid State Mater Sci 35, 52–71 (2010).

[20] F. Ming, A. Zangwill, J Phys D Appl Phys 45 154007:1-6 (2012).

[21] H. I. Rasool, E. B. Song, M. Mecklenburg, B. C. Regan, K. L. Wang, B. H. Weiller, J. K. Gimzewski, J Am Chem Soc 133, 12536-12543 (2011).

[22] H. I. Rasool, E. B. Song, M. J. Allen, J. K. Wassei, R. B. Kaner, K. L. Wang, B. H. Weiller, J. K. Gimzewski, Nano Lett 11, 251-256 (2011).

2

A General-Purpose Multiphase/ Multispecies Model to Predict the Spread, Percutaneous Hazard, and Contact Dynamics for Nonporous and Porous Substrates and Membranes

Navaz Homayun, Zand Ali, Gat Amir and Atkinson Theresa

Abstract

A computational model to solve the coupled transport equations with chemical reaction and phase change for a liquid sessile droplet or the contact and spread of a sessile droplet between two approaching porous or non-porous surfaces, is developed. The model is general therefore it can be applied to toxic chemicals (contact hazard), drug delivery through porous organs and membranes, combustion processes within porous material, and liquid movements in the ground. The equation of motion and the spread of the incompressible liquid available on the primary surface for transfer into the contacting surface while reacting with other chemicals (or water) and/or the solid substrate are solved in a finite difference domain with adaptive meshing. The comparison with experimental data demonstrated the model is robust and accurate. The impact of the initial velocity on the spread topology and mass transfer into the pores is also addressed.

Keywords: Liquid spread in porous materials, gaseous spread in porous material, evaporation in porous material, phase change in porous materials, multiphase process in porous materials, multi-species process in porous materials

1. Introduction

Many hazardous chemicals are dispersed in their liquid phase as droplets and reside on the surface of the liquid shortly after dispersion. In this state they are referred to as sessile droplets. Many coupled processes commence after this stage, resulting in different forms of threats, depending on the surface and physical properties of the liquid droplet. For nonporous surfaces, if the vapor pressure of the liquid is high, the threat of exposure is through the respiratory system, and if the vapor pressure is low, a percutaneous or direct contact hazard can be expected. Furthermore, should liquid droplets enter a chemical reaction with the surface, the products or the remaining reactants can reduce or augment the threat. In general, the following processes can be expected on a nonporous surface:

a. Surface evaporation

b. Chemical reaction or adsorption at the contact surface only

c. Formation of liquid bridge when a secondary surface come into contact with the sessile droplet on the nonporous or primary surface

d. Spread of the liquid bridge as the gap between the two surfaces closes.

For porous surfaces the following processes can be simultaneously present:

a. Surface evaporation

b. Capillary transport

c. Evaporation through the pores referred to as secondary evaporation

d. Chemical reaction and adsorption inside the pores

e. Possible formation of liquid bridge when a secondary surface comes into contact with any possible sessile droplet left due to capillary transport on the primary substrate in the postcontact phase

f. Spread of the liquid bridge if the conditions for its formation are met

g. Transfer of chemicals from the primary surface to the contacting surface occurs when the droplet on the primary substrate is observable or nonobservable. Therefore, contact hazard can exist even though the liquid droplet cannot be seen.

Although nonporous and porous surfaces share some processes, there are some differences in their outcome. Furthermore, in case of temperature variation, phase change can occur, thereby necessitating the inclusion of the energy equation in exposure risk models.

It is evident that the processes such as capillary transport (for porous surfaces), surface evaporation, in-pore or secondary evaporation, adsorption, and chemical reaction are initiated simultaneously as soon as a liquid droplet contacts a surface. In the case of a contact, a liquid droplet can be transported to the secondary (or contacting surface). This can occur when the droplet is sandwiched between the two surfaces, with the droplet forming a liquid bridge, or long after the droplet has disappeared from the surface. Therefore, an additional coupled process, which is mainly physical (liquid bridge spread), can be added to the above processes. The end result of all these processes determines the available chemical for atmospheric transport in the form of vapor or percutaneous hazard on surfaces available for transfer to a secondary surface upon contact.

When a wetting fluid is introduced into a porous medium, in the absence of external forces, the distribution of the liquid in pores is influenced by pressure differential, capillary, gravitational, and viscous forces. The main driver for the spread of this wetting fluid is capillary pressure. As the porous medium imbibes the liquid, the liquid flow rate decreases due to the domination of viscous forces, causing a reduction in the flow capillary number. The spread dynamics are influenced by the porous medium thickness, where two- and three-dimensional flows occur. For thin porous media, the liquid flows in a radial direction and can be modeled by the method of "common-lines" [1, 2]. In these solutions, two parameters need to be experimentally determined. In thick porous media, the flow becomes three-dimensional, although a cylindrical symmetry has been assumed in the past [3, 4] The experimental results of Denesuk et al. [5] verified this three-dimensional nature of the flow. Holman et al. [6] studied the surface spread in a porous medium as a function of permeability and have demonstrated that the surface spread is the dominant mechanism as compared to droplet penetration dynamics. In the above studies, a well-defined interface between the wetted and nonwetted volumes existed. It is known that spread dynamics becomes more complex when the saturation is less than unity, as demonstrated via MRI experiments by Mantle et al. [7]

When a droplet spreads into a porous substrate, saturation gets distributed between 0 and 1, necessitating a multiphase flow approach. Therefore, the governing equations need to account for the presence of all the phases [8, 9]. In these equations, two additional transport parameters are present: a relative permeability and a capillary pressure function that includes the interaction parameters between phases as shown by Chen et al. [10]. The relative permeability can be modeled using a power function as indicated by Brooks and Corey, [11] or van Genuchten equations, [12] in which each equation has a large number of parameters expressing the relative permeability. The capillary pressure is often modeled using the Leverett [13]and Udell [14] *J*-function.

In solving the droplet spread on or into a porous substrate, the shape of the droplet on the porous substrate surface (droplet free volume) and the shape of the wetted porous substrate imprint must be determined. Starov et al. [2, 15] have solved the governing equations to obtain the shape of a droplet free volume that spreads into a thin porous medium where he found a large variation of the droplet base radius. Using full numerical calculations for a three-dimensional porous medium, Alleborn et al. [4] have come to a similar conclusion. In respect to the imprint shape, the spread of the droplet is usually modeled by clearly separating the fully saturated and non-saturated regions. However, the spread dynamic is more complex due to the existence of partially saturated regions and varying transport parameters within the pores as indicated by Lenormand et al. [16]

The change of the flow mechanism, and the formation of partially saturated regions, requires solving the droplet spread into porous medium as a multiphase problem for which the multiphase flow parameters have to be determined. The multiphase parameters depend on the type of process and the predominance of gravity, viscous, or capillary forces. [16] Valavanides and Payatakes [17] summarized these influences on the multiphase parameters, where, in addition to phase content or phase saturation and emerging forces, solid–fluid contact angles, ratios of phase viscosities and flow rates, and the flow history (e.g., drainage or imbibition) altered the behavior of multiphase parameters. Discrete pore network models provide an alternative approach to elucidate the transfer phenomena and to evaluate transport parameters. In network models, an actual porous medium is represented as a network of pores that are connected by throats. [18] The phase progression is set using a potential threshold and the flow pattern is determined thereafter, as outlined by Prat. [19] From the phase distribution, the multiphase parameters are calculated (see, e.g., Constantinides and Payatakes [20]).

Further studies using a discrete pore network split the flow into primary and secondary regimes. The primary flow occurs when the volume of the sessile drop is greater than zero and the secondary flow happens after the sessile droplet volume becomes zero [21, 22, 23, 24]. In the secondary flow regime, the flow front at the surface becomes an irregular flow front, forming a heterogeneous porous medium at the surface. [25]

In spite of all previous progress in the field of porous media flow, there was still a need for a production code (or model) that included all the physiochemical processes in order to calculate the amount of a hazardous liquid droplet that poses a threat. The first problem encountered, in a realistic scenario, is the evaporation of a sessile droplet that is deposited onto an impermeable surface and exposed to outside wind conditions. This is called "convective evaporation" and its driving force is the outdoor wind velocity. As mentioned previously, a convective evaporation (laminar or turbulent) model needs to be developed in the absence of all other processes, i.e., for non-permeable non-reacting surfaces. An evaporation model for a sessile droplet on a non-porous surface with stagnant air above it is discussed by Popov [26]. Hu et al. [27] used finite element method to mesh the domain and then solved the mass conservation equations for the vapor concentration to find the evaporation rate. The above models assume no motion of air above the sessile droplet eliminating the convective effects. Baines et al. [28] developed a very simple convective evaporation model that did not compare with experimental results very well and in some cases overpredicted the evaporation rate by 90%.

A comprehensive literature review of the existing models is presented by Winter et al. [29] All of the previous studies focus on capturing the physics of evaporation and do not address the role of turbulence on evaporation. The effect of turbulence on evaporation of droplets that are moving in the core flow has been addressed by Navaz et al. [30] and Ward et al. [31] for spray combustion problems where high density and temperature gradients will reduce the evaporation time to the scale of turbulence fluctuations. However, for smaller or more moderate temperature gradients, the turbulence time scale will be smaller than the evaporation time scale. Therefore, a new approach should be developed to examine the effects of turbulence on evaporation of sessile drops. There are numerous works that examine atmospheric turbulence. The effect of the free stream Reynolds number on shear stress and friction velocity at a wall under outdoor conditions is discussed by Metzger and Klewicki. [32] Their study was performed on the Great Salt Lake Desert and spans over Re=2000 – 5e^6 or Re=2000 - 5× 10^6. There is also a dataset for friction velocity measured over a period of several months compiled from the outdoor wind data of Klinger et al. [33]. It has been mentioned by Weber [34] that the main scaling parameter in similarity theory of atmospheric boundary layer is friction velocity. He concludes that there could be a significant difference in friction velocity depending on the type of air flow. Several studies have focused on the varying turbulence intensity in wind-driven flows under outdoor conditions [35, 36]. Their approach relies on the validity of some wall function that relates the friction velocity to the mean free stream velocity and height. Upon knowing the friction velocity, the shear stress and turbulence intensity are correlated. For the evaporation problem, the previous studies have correlated the free stream velocity (or Reynolds number) to the friction velocity. However, they have not addressed conditions where turbulence intensity can change while the mean free stream velocity stays constant, or the turbulence intensity can be altered without changing the free stream velocity (or mean velocity in a channel flow).

Once a droplet spreads into a porous substrate, secondary evaporation (from the porous substrate) occurs. This mass transport and vapor diffusion is hindered by granules of the porous medium, and the vapor may face liquid parcels of the agent that act as additional obstructions. These obstructions greatly alter the transport properties of the vapor inside the pores and the transport rate toward the surface. Basically, the transport properties for the vapor phase become a function of porosity and saturation. These properties are yet to be determined.

As mentioned earlier, capillary transport of sessile droplet of a chemical agent on a porous substrate can be further complicated by the presence of chemical reaction. The chemical reaction can occur between the chemical agent and the constituents of a porous substrate or the chemical agent and pre-existing chemicals inside the pores, such as water or other chemicals. It is known that some chemical agents undergo hydrolysis reaction when they encounter moisture in the soil or sand, as it will be shown later. A simple set of one-dimensional mass and energy equations for a medium with no motion (conductive only) with chemical reaction was solved by Sadig [37, 38] who also demonstrated how the local properties are affected by chemical reaction, as the composition of constituents vary in time. Xu et al., [39] Lichtner, [40] and Steefel [41] have solved a very simple reactive flow with diffusion in geological settings dealing with ion transport. To have a general-purpose model, it is required that the conservation equations for all the phases (solid, liquid, and gaseous) and existing constituents, in forms of reactants and products with a chemical reaction, be solved simultaneously in time. The variation of species concentration in any of the existing phases will alter the local properties of liquid, gas, or solid mixture. Furthermore, the production or destruction of any solid phase will also change the local porosity of the medium.

A low-volatile chemical in its liquid form poses a threat upon contact with skin or other materials (vehicles, etc.). Mass transfer to the secondary surface occurs during this process. However, the contact may occur before a sessile droplet completely sinks in the primary substrate, if porous, or after a sessile droplet is not observable on the surface. In the first case, a liquid bridge is formed between the two approaching surfaces and may spread depending on the surface properties. Furthermore, if the primary surface or substrate is non-porous, a liquid bridge is definitely formed and spreads as the distance between the approaching surfaces decreases. The forces applied by liquid bridges connected to static supporting surfaces were studied extensively by Butt et al. [42] However, in many applications, it is common that at least one of the bodies is moving, and thus, the motion of the solid body may be influenced by the forces associated with the liquid bridge [43, 44, 45]. Pitois et al. [46] studied the forces applied by a liquid bridge connecting two spheres moving at a constant speed relative to each other. Similarly, Meurisse and Querry [47] studied the effects of liquid bridges connecting two parallel flat plates, moving perpendicularly to the plane of the plates, at a constant speed, or at a constant force applied on the liquid. Both Pitois et al. [46] and Meurisse and Querry [47] observed a rapid change from an attractive force due to capillary effects to a repulsive force due to viscous effects for the case of surfaces approaching each other at a constant speed. De Souza et al. [48] studied the effect of contact angle hysteresis on the capillary forces in the absence of significant viscous forces and obtained good agreement between the experimental data and the existing models. The capillary force measurement of liquid bridges was examined and shown to be a reliable method for estimating advancing and receding wetting angles.

The absorption of chemical into the skin is of primary interest due to threat from contact with hazardous materials. Experimental studies of Ngo, O'Malley, and Maibach [49] have attempted to quantify mass absorption into the skin. This is a complex problem because skin chemistry and perspiration rate may vary between individuals affecting the absorption rate. Although some simplified models based on linear regression exist to estimate the absorption rate, a more comprehensive and scientifically sound approach provides the solution of coupled equations for multi-species systems, as discussed hereinafter.

However, there was still a need for a model that can predict the available threat from a liquid chemical toxicant release. This issue has been addressed by Kilpatrick et al. [50], Savage [51, 52], and Munro et al. [53], and prompted the Department of Defense (DoD), Edgewood Chemical and Biological Center (ECBC), and Defense Threat Reduction Agency (DTRA) to initiate a comprehensive research project called Agent Fate Program that was geared toward chemical agents of warfare. This project led to the development of the first-generation production code COMCAD (COmputational Modeling of Chemical Agent Dispersion) by Navaz et al. [54, 55], and the next generation for contact hazard MOCHA (Modeling Of Contact HAzard) codes. [56] Since the model is general, it can not only be applied to chemical agents of the warfare, but also any other liquid that can be disseminated into the environmental substrates in a sessile droplet form (e.g., pesticides, toxic industrial chemicals (TICs), and toxic industrial materials (TIMs)). This chapter summarizes the result of this research and its application to a variety of problems.

1.1. Mathematical model

The species mass and momentum, and energy equations for a multicomponent system in a porous medium with all phases being active is rather complex and are given in Navaz et al. [57], Navaz et al., [55] Vafai, [9] Navaz et al., [58] and Kuo. [59] The continuity equation for a multicomponent system for all phases includes any loss of mass due to surface and secondary evaporation, and cross-over mass due to chemical reaction, phase change, and adsorption. The momentum equations in porous media are considered for liquid and gaseous phases. Local thermal equilibrium is considered for the energy equation. That is to say that the local temperatures of all phases are equal ($T_s = T_\ell = T_g$). Note that the solid phase is stagnant, but the porosity will be changing in time if any solid constituent is formed due to phase change or chemical reaction. The continuity equation for the porosity is eliminated, but the porosity is updated in each time step (lagged by one time step). Although for an open system under the usual atmospheric conditions the gaseous phase pressure normally stays at the environmental conditions, the pressure term is included in the equations to extend the applicability of the model for the future applications, in which the pressure may change (closed systems or existence of explosions). All the conservation equations are solved explicitly on a finite difference mesh as will be described later. The sub-models describing the source terms (e.g., evaporation), chemical kinetics, and adsorption models are described in separate sections. We start with explaining the evaporation rate term.

Mass Conservation for Multiple Solid System (Substrate – Solid "k")

$$\frac{\partial \rho_{sk}\varphi_s}{\partial t} = -\dot{\omega}_{sk}^{Reaction} - \dot{\omega}_{sk}^{Adsorption} \tag{1}$$

Mass Conservation for Multiple Solid System (Liquid "i")

$$\frac{\partial\left(\varphi \rho_{\ell i} s_{\ell i}\right)}{\partial t} + \nabla\left(\varphi \rho_{\ell i} s_{\ell i} \vec{V}_{\ell i}\right) = \left(-\dot{\rho}_{\ell i}^{Secondary\ \ Evaporation} - \dot{\rho}_{\ell i}^{Surface\ \ Evaporation} - \dot{\omega}_{\ell i}^{Reaction} - \dot{\omega}_{\ell i}^{Adsorption}\right) \tag{2}$$

$$\rho_{\ell-mixture} = \sum_{i=1}^{N(Liquids)} \rho_{\ell i} C_{\ell i} \qquad C_{\ell i} = Mass\ \ Fraction = \frac{\rho_{\ell i} s_{\ell i}}{\sum_{i=1}^{N} \rho_{\ell i} s_{\ell i}},\ s_\ell = \sum_{i=1}^{N(Liquid)} s_{\ell i} \tag{3}$$

$$\vec{V}_{\ell i} = -\frac{Kk_{\ell i}}{\mu_\ell}\left(\nabla P_{\ell i} - \rho_{\ell i} g s_{\ell i}\right) \tag{4}$$

Mass Conservation for Multiple Gaseous System (Gas"j")

$$\frac{\partial\left(\varphi \rho_{gj} s_g\right)}{\partial t} + \nabla\left(\varphi \rho_{gj} s_g \vec{V}_{gj}\right) = \nabla \bullet \left(\rho_g D_{j-mix} \nabla C_j\right) + \dot{\rho}_{\ell,i=j}^{Secondary\ \ Evaporation} - \dot{\omega}_{gj}^{Reaction} - \dot{\omega}_{gj}^{Adsorption} \tag{5}$$

$$\vec{V}_{gj} = -\frac{Kk_{gj}}{\mu_g}\left(\nabla P_{gj} - \rho_{gj} g s_g\right) - D_{j\rightarrow mixture} \nabla C_{gj} \tag{6}$$

k_{gj} = Relative Permeability to jth constituent of the gas phase

$$\mu_g(\text{gas mixture viscosity}) = \sum_{j=1}^{M(Gases)} \mu_{gj} C_{gj}, \tag{7}$$

$$\rho_{gj} = \text{Density of } jth \text{ constituent of the gas phase}$$
$$\vec{V}_{gj} = \text{Velocity component of } jth \text{ gas phase constituent}$$

where

$$s_g = 1 - \sum_{i}^{N(Liquids)} s_{\ell i} \quad \text{and} \quad \rho_g = \sum_{j=1}^{M(Gases)} \rho_{gj} \tag{8}$$

$$D_{j-mix} = \textit{Effective Diffusion Coefficient of gas species "j" into the mixture}$$

$$C_j = \textit{Mass Fraction} = \frac{\rho_{gj}}{\rho_g} \tag{9}$$

$$P_{\ell i} = P - P_{ci} \text{ (Capillary Pressure)} + \rho_{\ell i} g h^* \tag{10}$$

h^* = Local height of the droplet as a function of time (hydrostatic pressure)
$\dot{\omega}$ = Inter-Phase Chemical Species production or destruction
K = Saturation permeability
$k_\ell = k(s_\ell)$ Relative permeability is a function of the local saturation = s_ℓ^2
$\varphi = \varphi(\rho_{sk}, x, y, z)$ Local porosity is a function of local density of the solid phase

$$k_{gj} = 1 + s_\ell^2 (2s_\ell - 3) \tag{11}$$

Energy Equation

$$\begin{aligned}
&\frac{\partial}{\partial t}\left[\rho_s\varphi_s H_s + \rho_\ell\varphi_\ell H_\ell + \varphi_g\left(\rho_g H_g - P\right)\right] + \\
&+\nabla \bullet \dot{\mathbf{q}} + \nabla \bullet \left(\rho_\ell\varphi_\ell H_\ell \vec{V}_\ell + \rho_g\varphi_g H_g \vec{V}_g\right) + \nabla \bullet \left(-\bar{\bar{\boldsymbol{\tau}}}_{mn\ell}\vec{V}_{n\ell} - \bar{\bar{\boldsymbol{\tau}}}_{mng}\vec{V}_{ng}\right) \\
&\rho\left[-\upsilon_\ell g - \upsilon_g g\right] + \sum_{i=1}^{N_{solids}} \dot{\rho}_{s-\ell i}\Delta h_{s-\ell i} + \sum_{i=1}^{N_{liquids}} \dot{\rho}_{\ell-gi}\Delta h_{\ell-gi} + \sum_{i=1}^{N_{solids}} \dot{\rho}_{s-gi}\Delta h_{s-gi} = 0
\end{aligned} \tag{12}$$

where

$$\varphi_s = (1-\varphi), \qquad \varphi_\ell = s_\ell \varphi, \qquad \varphi_g = (1-s_\ell)\varphi, \qquad V^2 = \sqrt{u^2+\upsilon^2+w^2} \tag{13}$$

$\dot{q} = \text{Heat flux} = -k_{eff}\nabla T,$ $\Delta h = \text{Latent heat of phase change}$

$\dot{\rho} = \text{rate of phase Change}$ $H = \text{Total enthalpy} = c_p T + \dfrac{V^2}{2}$

$c_p = \text{Specific heat}$ $\bar{\bar{\tau}} = \text{Shear stress tensor given by:}$

$$\begin{aligned}
&\tau_{xx} = \mu\left[2\frac{\partial u}{\partial x} - \frac{2}{3}\left(\frac{\partial u}{\partial x}+\frac{\partial \upsilon}{\partial y}+\frac{\partial w}{\partial z}\right)\right], && \tau_{xy} = \tau_{yx} = \mu\left[\frac{\partial u}{\partial y}+\frac{\partial \upsilon}{\partial x}\right],\\
&\tau_{xz} = \tau_{zx} = \mu\left[\frac{\partial w}{\partial x}+\frac{\partial u}{\partial z}\right] && \tau_{yy} = \mu\left[2\frac{\partial \upsilon}{\partial y} - \frac{2}{3}\left(\frac{\partial u}{\partial x}+\frac{\partial \upsilon}{\partial y}+\frac{\partial w}{\partial z}\right)\right],\\
&\tau_{yz} = \tau_{zy} = \mu\left[\frac{\partial \upsilon}{\partial z}+\frac{\partial w}{\partial y}\right], && \tau_{zz} = \mu\left[2\frac{\partial w}{\partial z} - \frac{2}{3}\left(\frac{\partial u}{\partial x}+\frac{\partial \upsilon}{\partial y}+\frac{\partial w}{\partial z}\right)\right]
\end{aligned} \tag{14}$$

1.2. Evaporation models

When surface evaporation of a sessile droplet on a non-porous substrate occurs, one or both of the following mechanisms determines the topology of the sessile droplet. Either surface evaporation takes place with the droplet's base area (radius) remaining constant while the height of the droplet decreases, or the surface evaporation causes the shrinking of a droplet while the ratio *h/r* remains constant. It has been observed that for the evaporation of nerve agents (HD, VX, and GD) the process starts with the first mechanism and when the contact angle reaches about 12°, the second mechanism is initiated.

Irrespective of the evaporation mechanism (*r=const.* or $\lambda_r = h/r = const.$), the volume and surface area of a sessile droplet in a form of a spherical cap is given by:

$$\begin{aligned}
&\forall = \frac{1}{3}\pi h^2(3R_s - h) = \frac{1}{6}\pi h(3r^2 + h^2)\\
&A_s = 2\pi R h
\end{aligned} \tag{15}$$

The dynamics and topology of this model is shown in Figure 1. where the evaporation is initiated with the first mechanism and then is switched to the *h/r=const.* case.

Having the mass rate of evaporation ($\dot{m}$) equal to the negative change of the mass (m) left on the surface at time (t) and it can be written as:

$$\dot{m} = -\frac{d}{dt}m = -\frac{d}{dt}\left(\rho_\ell V\right) = -\rho_\ell \frac{\pi}{2}\left(r^2 + h^2\right)\frac{dh}{dt} \tag{16}$$

where (ρ_ℓ) is the liquid density. From Equation (16) the height of the droplet is calculated as:

$$\frac{dh}{dt} = \frac{-\dot{m}}{\rho_\ell \frac{\pi}{2}\left(r^2 + h^2\right)} \tag{17}$$

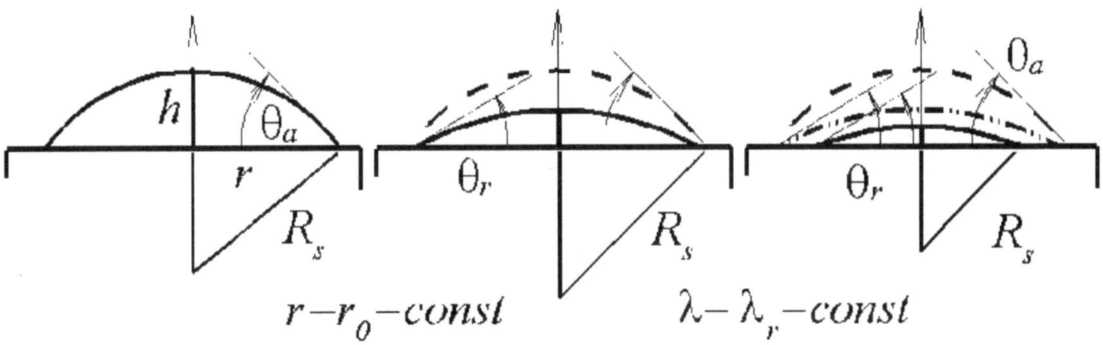

Figure 1. Schematic for the Topology and Dynamics of HD Evaporation on a Non-Permeable Surface.

The instantaneous droplet height can be obtained by integrating Equation (17), provided that the forcing function $\dot{m}$ is known. There are numerous expressions in the literature for the evaporation rate $\dot{m}$ given by researchers in spray combustion field [58, 59, 60, 61] for spherical droplets following a gas trajectory in a rocket engine. The current model follows the same process with a few exceptions. Toxic chemicals are generally tested in small wind tunnels mainly for better control of contamination. The flow in these wind tunnels is basically a channel flow and a boundary layer plate cannot be installed (due to size constraints) to create a scalable model to the open air scenarios. To overcome this issue, the Reynolds number in Equation (18) was based on the friction velocity instead of the free stream velocity. Embedded in the friction velocity is the impact of turbulence on evaporation. This is discussed by Navaz et al. [57] and is the focus of Section 1.3. Further modification was necessary to define a new transfer number, B. The transfer number in combustion is defined as: $\frac{c_p(T_g - T_{\text{boiling point}})}{h_{fg}}$ where h_{fg} is the latent heat of evaporation. Although this definition will work in a combustion chamber, it does not apply to the evaporation of toxic chemicals under environmental conditions that mainly involve the vapor pressure of the sessile droplet and convective mass transfer. It is assumed that the evaporation process takes place isothermally and any minimal change of temperature at the interface and its conduction through the sessile droplet can be ignored. Two different length scales are defined depending on the droplet topology. [57] The following surface evaporation model, Equation (18), provides the source term ($\dot{\rho}_{\ell}^{surface\ evaporation}$) in Equations (2) and (5) for sessile droplets, which is derived, calibrated, and used for this study.

$$\dot{m}_{surface\ evaporation} = 2\pi L_c \frac{\mu}{\Pr}\left(C_f + C_1 \mathrm{Re}^{c_m} \Pr^{c_n}\right) \ell n\left[1+B\right]$$

$$L_c = \text{Length scale} \begin{cases} \text{Mechanism 1: } = 2\pi R_s \lambda_r \\ \text{Mechanism 2: } = 2\pi R_s \left[1\text{-}\sin\left(\frac{\pi}{2} - \theta_a\right)\right] \end{cases}$$

Re = Reynolds number based on the radius of curvature and friction velocity

ρ_g = Gas mixture density

$$B = \text{Transfer Number } = \left(\frac{y}{1-y}\right)^{\varepsilon}, \quad y = \frac{P_{\upsilon j}}{P} \tag{18}$$

$P_{\upsilon j}$ = Vapor pressure of species "j"

where: m is the mass of liquid species "j" at each node

$C_f, C_1, c_m, c_n, \varepsilon$ = Model constants to be determined experimentally

The evaporation model for flow through the pores is also based on the methodology discussed by Navaz et al. [55, 58]. A numerical investigation of this process on hot surfaces has been published by Nikolopoulos et al. [62] However, in the context of finite difference or finite volume framework, a new model needs to be developed. It is assumed that the liquid phase at each node is spherical and the instantaneous length scale is found by knowing the mass and density of each liquid species at this node. The evaporation rates in the species continuity equation (Eq. 2) are obtained as indicated by Equation (19).

$$\dot{m}^{\text{Secondary evaporation}} = 4\pi\rho_g D_{j-Mix} L_c \left(C_1 + C_2 \mathrm{Re}^m Sc^n\right) \ell n \left[1 + \frac{\left(\frac{P_{\upsilon j}}{P}\right)^{\varepsilon}}{1-\left(\frac{P_{\upsilon j}}{P}\right)^{\varepsilon}}\right]$$

$$\dot{\rho}^{\text{Secondary Evaporation}} = \dot{m} / \forall_{cell}$$

$$Sc = \text{Local Schmidt Number} = \frac{\nu_j}{D_{j-Mix}}$$

ρ_g = Gas mixture density

D_{j-Mix} = Effective Diffusion Coefficient of species "j" into the mixture (19)

$P_{\upsilon j}$ = Vapor Pressure of species "j"

P = Gas mixture pressure

$$L_c = \text{Characteristic Length} = \left(\frac{m_{\ell j}}{\rho_{\ell j}}\right)^{1/3}$$

$$\text{Re} = \text{Reynolds Number} = \frac{L_c\sqrt{u_\ell^2 + \upsilon_\ell^2 + w_\ell^2}}{\nu_j}$$

where: $m_{\ell j}$ is the mass of liquid species "j" at each node

$C_1, C_2, C_3, m, n, \varepsilon$ = Model constants to be determined experimentally

1.3. Scalability of surface evaporation model

As mentioned earlier, the Reynolds number in the evaporation model is based on friction velocity. This is due to the fact that a similar solution can be found for all cases involving the evaporation of a sessile droplet. Furthermore, the turbulence also speeds up the evaporation and it was postulated that the turbulence effects will be embedded in the friction velocity.

The model from Equation (18) suggests that the evaporation is influenced by geometry (because of the exposed surface to convective motion), convective transfer (turbulence and free stream velocity), and driving force (temperature and/or mole fraction). Based on analysis performed by similitudes, non-dimensional groups that lump all these effects and can influence the evaporation rate were derived. It was further postulated that the characteristic velocity representing the convective evaporation would be the friction velocity and not the traditional mean flow velocity. Based on these analyses, a geometrical factor of V/r^2 defining the length scale was defined. Basically, a sessile droplet will "see" the friction velocity rather than the mean velocity of the flow. Furthermore, the turbulence effect that has an impact on evaporation rate is embedded in the friction velocity. Therefore, expressing the Reynolds number, which is the driver for convective evaporation in terms of the friction velocity, seems like a plausible proposition. Based on the similitude analysis that was explained earlier, a non-dimensional time scale is defined as:

$$tu^{*}\left(V/r^{2}\right)^{-1}\ell n\left(1+B\right)=tu^{*}\left(r^{2}/V\right)\,\ell n\left(1+B\right) \tag{20}$$

The model was used to calculate the evaporation rate and the normalized amount of mass left as a function of the non-dimensional time shown in Equation (20) for a variety of droplet initial volume, friction velocity, and initial contact angle (or base radius). The HD properties were used for this study. Figure 1.3.1 shows the collective effects of this study and it is observed that regardless of the droplet initial topology, air velocity, and turbulence intensity, a similar solution exists. This verifies the validity of our earlier postulate about the friction velocity and its ability to scale wind tunnel data to open air evaporation results. An identical experimental study was conducted with HD and the same similar solutions were calculated and plotted analogous to Figure 1.3.1 to create Figure 2.. It is observed that the experimental results verify the results obtained by the mathematical model.

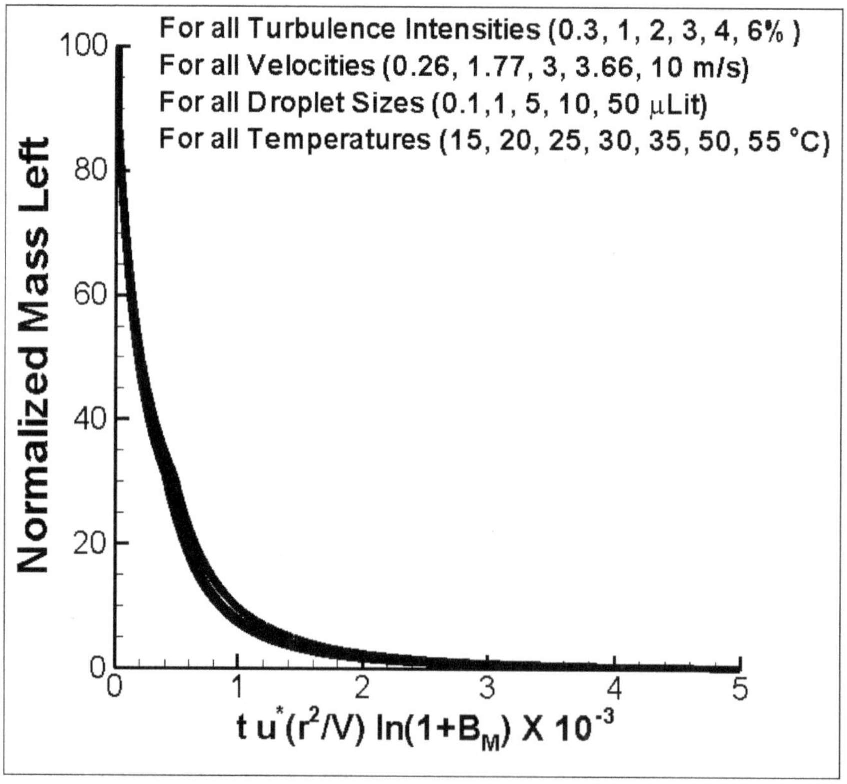

Figure 2. Similar solutions in the form of the percentage of mass left as a function of dimensionless time that is reduced with respect to geometrical, convective, and driving force parameters.

The significance of this result is in identifying the similar solution that lumps the effects of free stream momentum and turbulence (u^*), the geometrical factor (r^2/V), and heat or mass transfer (transfer number) into a single scalable variable. The variation in the initial contact angle makes a very small difference in the curve. Therefore, if experiments are conducted in different wind tunnels with different droplet sizes and different transfer numbers (liquids), the same evaporation rate curve should be obtained for all wind tunnels as long as the results are

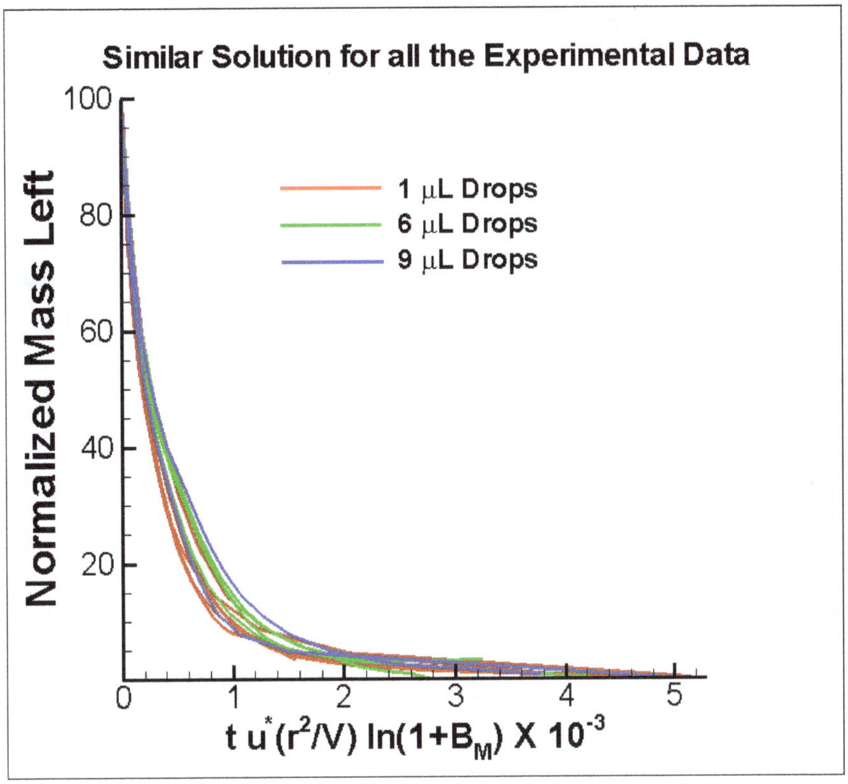

Figure 3. Similar solutions for the available experimental data in the form of the percentage of mass left as a function of dimensionless time.

expressed in the non-dimensional time scales. In doing so, the friction velocity should be measured in each experiment. This is an important conclusion and should be directly tested to verify the results of Figures 2 and 3.

It is evident that the free stream velocity and turbulence intensity should be treated as independent variables, i.e., one can be altered while the other remains constant. A series of boundary layer measurement made at Caltech's 6'× 6 ' wind tunnel altered the turbulence intensity while the mean velocity was maintained constant. The turbulence intensity was altered by the number of bungee cords in the wind tunnel. A schematic of the wind tunnel and the cords for changing the turbulence level is shown in Figure 4. The wall shear stress was measured by the oil film technique in which a drop of olive oil is placed on a surface under a monochromatic light source. The combined reflection from oil and the glass surface create an interference image pattern as shown in Figure 5. The fringe spacing growth rate can be correlated to the wall shear stress as $\tau_w = \frac{2n\mu}{\lambda_{wave}} \frac{dw}{dt}$ where n is the index of refraction for oil, μ is the dynamic viscosity, and λ_{wave} is the light wavelength.

The turbulence intensity was measured by hot wire anemometry. Particle image velocimetry (PIV) can also be used to find the components of velocity fluctuations that will lead into the turbulence intensity calculations. Table 1 shows the wall shear stress as a function of the free stream velocity and turbulence intensity. The friction velocity $\left(u^* = \sqrt{\tau_w / \rho}\right)$ is calculated by finding the wall shear stress through interpolation from Table 1. The density can be obtained from the pressure and temperature data.

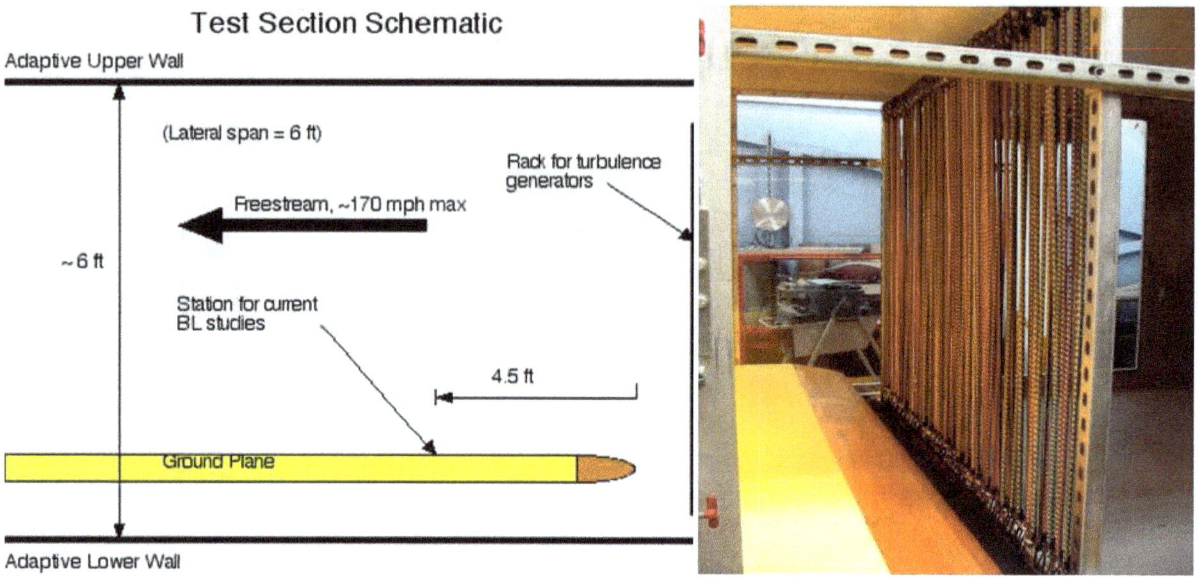

Figure 4. Schematic of Caltech's 6' x 6 ' Wind Tunnel

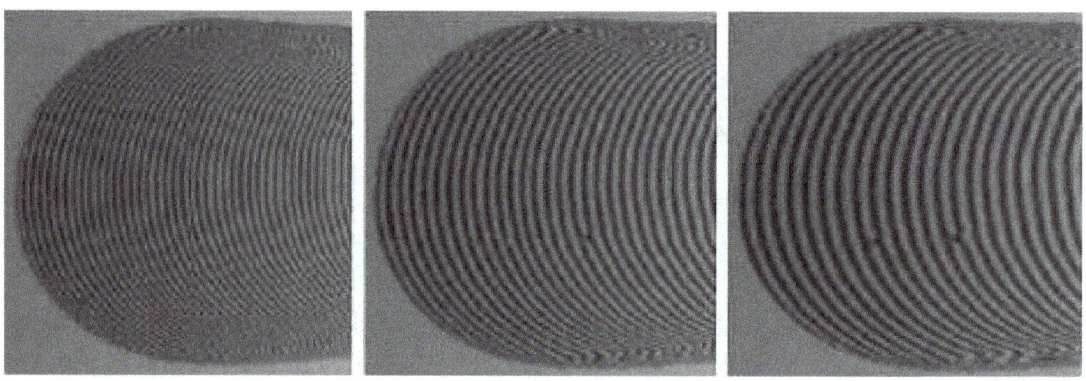

Figure 5. Fringe Spacing Growth Rate for Oil

In a group of wind tunnel experiments, HD droplet evaporation rates were measured for varying droplet volume, mean air speed, and temperature conditions. The turbulence intensity was estimated using the Edgewood Chemical and Biological Center (ECBC) wind tunnel data for the wall shear stress at the velocities that are mentioned hereinto. The wind tunnel tests were conducted for four different free stream velocities, u_∞(m/s)={0.26, 1.77, 3.00, 3.66}, three initial droplet volumes, V(μL)={1, 6, 9}, and five different temperatures, T(°C)={15, 25, 35, 50, 55}. A numerical procedure that solves the governing Equation (17) along with Equation (18) was developed. The scheme is based on the Runge–Kutta fourth order integration algorithm. For the sake of brevity, the model predictions are compared with the experimental data for nine of approximately 60 cases in Figure 6. An excellent agreement is seen for all droplet sizes, air velocities, and temperatures. Detailed comparisons can be found in Navaz et al. [54, 55]

The analysis revealed that there is a difference in how each parameter influences the evaporation rate. Having the analytical formulation validated, the model can be used for parametric studies. To do this, the evaporation rate had to be calculated as the mean flow velocity and turbulence intensities were changed independently. A small (V_0=1 μL) and a large (V_0=9 μL) droplet were exposed to a constant mean boundary layer velocity as the turbulence intensity varied from 0 (laminar) to 6%. Then the mean velocity changed and the same calculations were repeated. Figure 7. (a and b) depict the results of this study and emphasizes the fact that that turbulence intensity is a driving force in convective cooling and the most viable method to account for its contribution is through friction velocity. A similar analysis was performed to examine the significance of the free stream velocity on the evaporation by considering the smallest and largest droplets as shown in Figure 7. (c and d). In this case, a free stream turbulence intensity of 2% with a constant prescribed free stream temperature of 35°C showed that if the free stream velocity is increased by a factor of about 14, the evaporation rate will decrease by a factor of about 4. By the same token, each free stream velocity can be correlated to the evaporation time.

In the last parametric study, the free stream velocity and turbulence intensity were maintained constant for the two droplet sizes and the temperature was varied. It appeared the air temperature (T) had the most significant effect on evaporation as seen from Figure 7. (e and f). For a 14% increase in absolute temperature, the evaporation rate is increased by a factor of 24 due to the vapor pressure being a strong function of temperature (the vapor pressure of HD at 30°C is about 0.25 mmHg).

There are two uncertainties that are encountered in the process of model validation with outdoor data: pertinent free stream velocity and turbulence intensity of the atmospheric flow. In the free atmospheric air flow, the wind velocity varies as a function of height (or vertical direction) and time, i.e., $V(y, t)$. Furthermore, two velocities can be used as the 'free stream velocity' in the prediction of the droplet evaporation rate, one being the instantaneous wind velocity at the specific height $V(y_{spec}, t)$, and the other the height (y) averaged velocity $\bar{V}(t)$. In the outdoor experiments, the instantaneous velocity data were taken at seven or eight vertical locations up to approximately H=5 m high.

The instantaneous velocity profile was used to find the transient average velocity, which is given by:

$$\bar{V}(t) = \frac{1}{H}\int_0^H V(y,t)dy \tag{21}$$

The same procedure was performed for the temperature profile and the graphical representations of these averaged values are embedded in all figures related to the outdoor data.

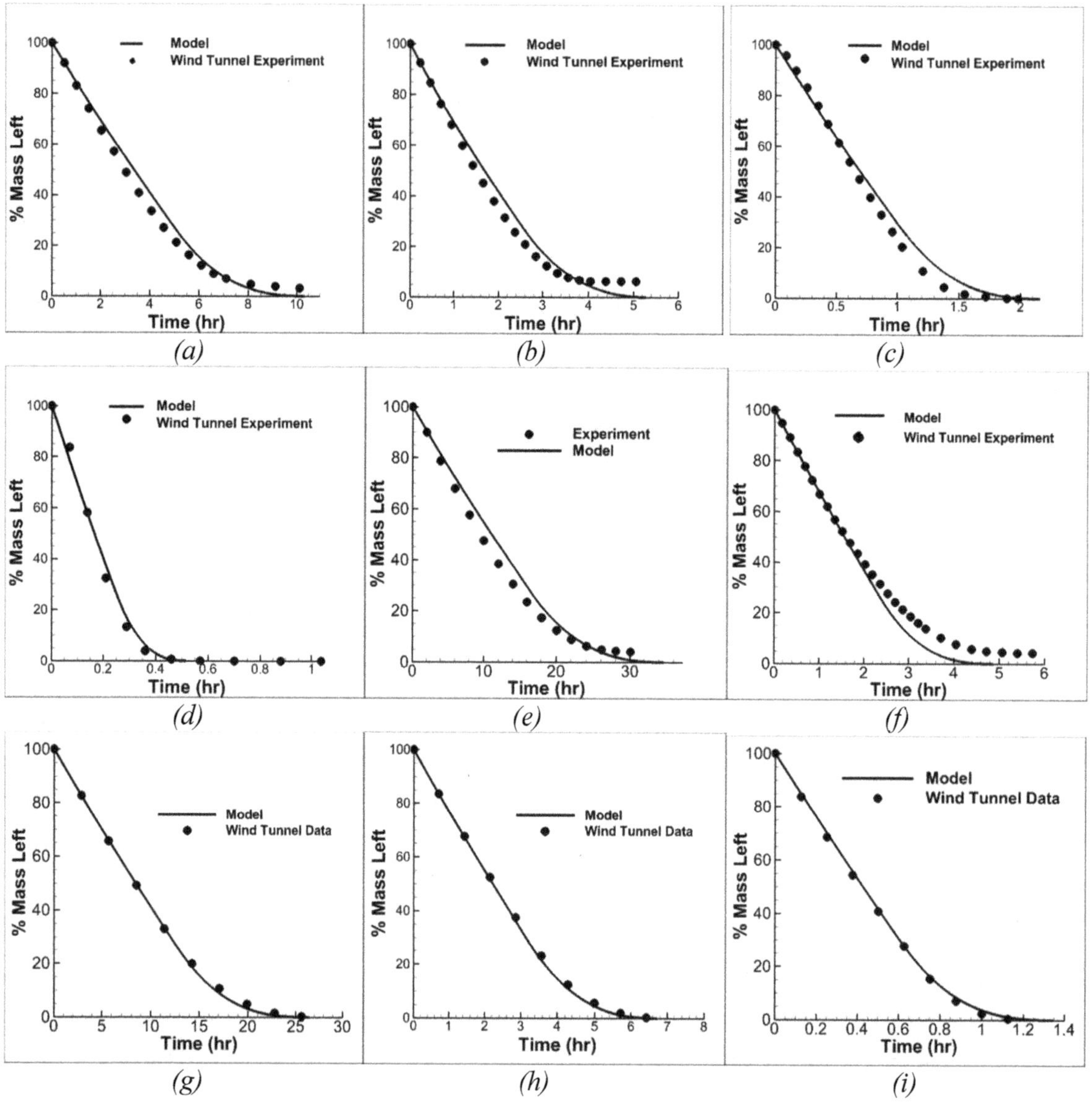

Figure 6. Surface evaporation model prediction compared with wind tunnel experimental data for the following HD droplet sizes, wind speeds, and temperatures: (a) 1 μL, 1.77 m/s, 15°C, (b) 1 μL, 0.26 m/s, 35°C, (c) 1 μL, 1.77 m/s, 35°C, (d) 1 μL, 3.66 m/s, 50°C, (e) 6 μL, 1.77 m/s, 15°C, (f) 6 μL, 3.00 m/s, 35°C, (g) 9 μL, 3.00 m/s, 15°C, (h) 9 μL, 1.77 m/s, 35°C, and (i) 9 μL, 3.00 m/s, 55°C.

The second uncertainty encountered in model application is that the free stream turbulence intensity levels could be a function of time. In the atmospheric flow, one expects the existence of smaller turbulence intensity. In this work, the turbulence intensity for the outdoor experiments was assumed to be about 2% based on the "averaged" value of velocity measurements. Note that the turbulence intensity is an independent variable and needs to be extracted from the actual velocity measurement with high sampling rate.

Numerous model validations with open air test data are compiled by Navaz et al. [54, 55]. Two of these cases are presented here and an excellent agreement between outdoor data and model prediction are observed and summarized in Figure 8.(*a* and *b*).

Free Stream Velocity (m/s)	Turbulence Intensity			
	0%	0.3–0.4%	2.6%	4.1–5.4%
	τ_w (Shear Stress at the wall, Pascal)			
0	0.0	0.0	0.0	0.0
2	0.0037	0.0249	0.0251	0.0253
5	0.0147	0.0626	0.0663	0.0723
10	0.0466	0.2048	0.2102	0.2483
15	0.0765	0.4096	0.4450	0.5115

Table 1. Wall Shear Stress Data as a Function of Turbulence Intensity and Free Stream Velocity

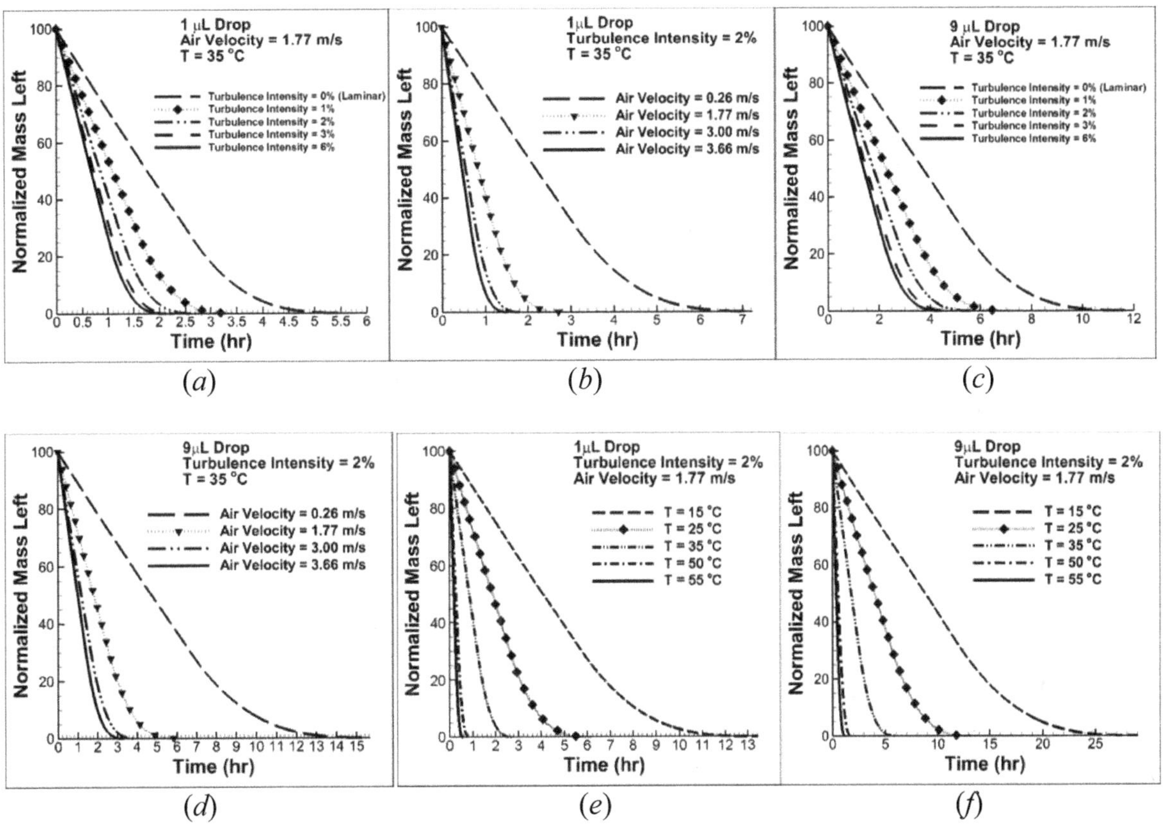

Figure 7. Performing parametric study to explore the effect of turbulence intensity (a and b), free stream velocity (c and d), and air temperature (e and f) on evaporation.

1.4. Porous media models

The porous medium consists of solid inclusions with void spaces in between. The porosity is the ratio of the void volume to the total volume. It is an averaged property of a porous medium for "relatively" homogeneous substrates. For heterogeneous substrates, the porosity becomes a local value expressed as $\phi = \phi(x,y,z)$. Furthermore, in the presence of chemical reaction, the

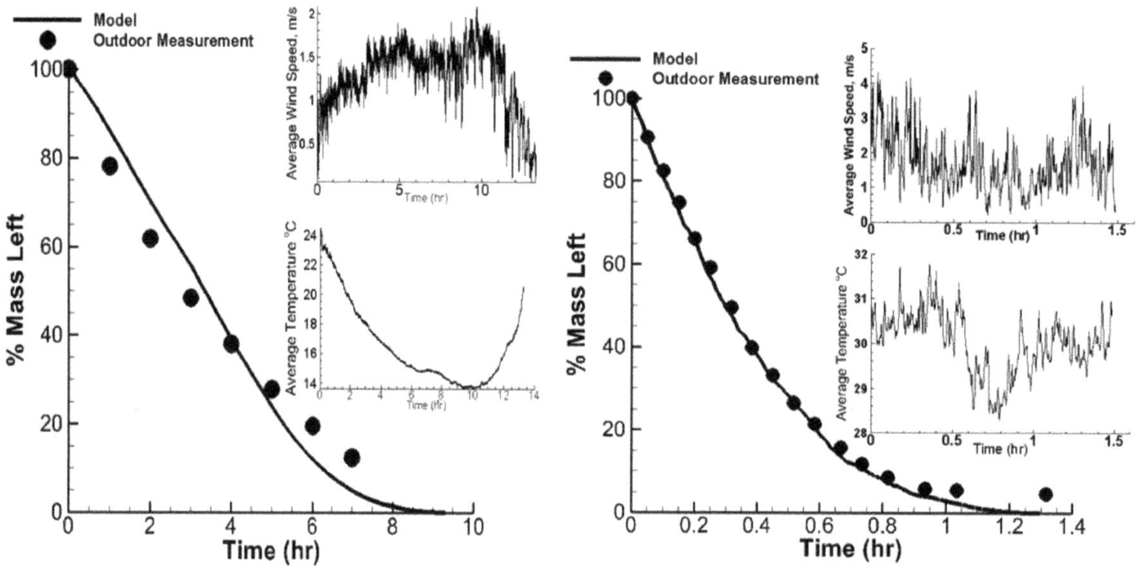

(a) (b)

Figure 8. Model predictions for real-time wind velocity and temperature data as compared with the measurements for 1 μL average HD drop size. The embedded figures represent the instantaneous wind velocity and temperature.

local porosity changes in time as any of the reaction products may appear in solid phase making the porosity a function of time (or local solid phase density), i.e.. $\phi = \phi(x,y,z,t) = \phi(x,y,z,\rho_{sk})$. For the time-dependent porosity, the gradient of the porosity should be included in the governing equations. This will make the equations more complex. However, since the reactions are generally slow, the porosity distribution and change in time is lagged by one time step (taken from the previous time step). This makes the algorithm less complicated and still very accurate.

Another property of the porous medium is the saturation permeability (K in Equation 10) that needs to be known. It is a momentum transport property of the porous medium. It is basically a measure of how "easily" fluid flows through fully saturated porous material. Therefore, this permeability is referred to as single-phase permeability, saturation permeability, or simply permeability. An experiment can be performed with any nonvolatile liquid including water to measure this property. [54, 63].

We have measured the saturation permeability for different porosities using glass beads. This was a necessary step for nonhomogeneous porous media. Saturation permeability is a function of porosity and changes locally as the porosity will change, especially due to those chemical reactions that produce an additional solid phase as a product. This relationship is:

$$K = \exp\left(\log_{10} \varphi(x,y,z,t) - 22.67\right) \tag{22}$$

where K is the saturation permeability in m^2or ft^2 and ϕ is the instantaneous local porosity (0.3, 0.5, etc.). This equation is general and valid for porosities up to 60% or 0.60. This procedure is detailed by Zand et al. [63]

Contrary to the single-phase (fully saturated) flow, multiphase flow can occur when a fluid phase prefers a specific path, thereby leaving some of the porous medium regions filled with the originally present phase, e.g., vapor–air mixture. The measure of how "easily" a phase flows through the porous medium in the multiphase flow is referred to as phase permeability. The relative permeability for each constituent in liquid and gaseous phase ($k_{\ell i}$, k_{gj}) is defined as a ratio of the phase permeability to the saturation permeability and $0 \le k_{\ell i}$ *and* $k_{gj} \le 1$ (zero for immobile phase and one for fully saturated flow). We eliminate the index "*i*" and "*j*" for simplicity, keeping in mind that these properties belong to each constituent in the liquid and/or gas phase. Hence, it is evident that the relative permeability is a function of the phase content or saturation, i.e.,$k_\ell = k_\ell(s_\ell)$ and $k_g = k_g(s_g)$.

A hybrid approach (combined modeling and experimental studies) was used to find expressions for relative permeability. [54, 55] This hybrid approach needs to be performed separately for agents and the substrates of interest.

$$\begin{aligned} k_\ell &= s_\ell^k, \quad k = 2 \\ k_g &= 1 + s_\ell^2 (2s_\ell - 3) \end{aligned} \tag{23}$$

The capillary pressure function (see Equation 10) is not known for sessile droplets. Therefore, it is postulated that it has the following form:

$$\begin{aligned} & p_{ci} = f_i(*) \frac{\sigma_i \cos\theta_i}{\sqrt{K/\varphi}} J(s_{\ell i}), \\ & \text{where } J(s_{\ell i}) = 1.417(1 - s_{\ell i}) - 2.120(1 - s_{\ell i})^2 + 1.263(1 - s_{\ell i})^3 \end{aligned} \tag{24}$$

where σ_i is the surface tension, θ_i is the contact angle inside the pores, K is the saturation permeability, $s_{\ell i}$is the saturation for each constituent, and $f_i(*)$ is a function to be determined. This form is based on the original equation proposed by Leverett [13] and Udell [14]. In Equation (24), a geometrical scale is defined as $\sqrt{K/\varphi}$, which is not necessarily the correct value as capillary pressure can be scaled with a different geometrical scale. Therefore, the capillary pressure is corrected for a potential error in geometrical scale using the function $f_i(*)$.

Initially, the Buckingham theorem was used to identify the basic non-dimensional groups and the computational model developed was utilized to combine these basic **Π** groups into a more meaningful similar solution. A value of $f_i(*)=1$ was assumed and numerical simulations were carried out over a wide range of initial droplet volume and wetted area (maintaining a constant initial contact angle), porous medium saturation permeability, liquid viscosity, and surface tension. For a constant initial contact angle and a specific substrate, a similar solution can be obtained with the following **Π** groups:

$$\Pi_1 = function(\Pi_2) \quad \text{for a constant contact angle}$$

$$\Pi_1 = \frac{t\sigma K^{0.5}}{\mu\left(\frac{V_{on\ the\ surface}}{r^2}\right)^2} \qquad \Pi_2 = \frac{V_{on\ the\ surface}}{r^3} \tag{25}$$

It should be noted again that the properties must be calculated for each constituent in the liquid phase marked by subscript "*i*" earlier. In Equation (25) *r* and *V* represent the wetted radius and the volume of the droplet left on the surface, respectively, and *t* is the time taken by a droplet to disappear from the surface. Figure 9.(*a*) shows a family of similar solution for these non-dimensional groups. Although this curve proves that the proposed non-dimensional groups are correct scalable quantities, it becomes a singular function when $V_{\text{on the surface}} \to 0$. Note that the properties relate to each liquid constituent on the surface only.

The above functions were used in an alternate method to eliminate this singularity. Again, the value *f(*)*=1 was kept, and the porosity and contact angle were changed to generate a family of curves with returning the instantaneous volume and wetted radius to their initial values, i.e., $V_{Initial}$ and $r_{Initial}$. Figure 9.(*b*) is a collection of modified Π_1 group as a function of contact angle for substrate porosity ranging from 0.1 to 0.9. The following procedure can be used to find the unknown *f*(*). The details of this method and all the experimental results and validations are compiled and discussed by Navaz et al. [64]. The process of finding *f*(*) is as follows:

- Measure the initial contact angle, and based on the value of porosity, find the modified Π_1 group representing t_{ref} (reference time).
- Determine the function *f(*)* based on the following equation:

$$\frac{1}{f_i(*)_{Unknown}} = \frac{\text{nondimensional measured time for the droplet to disppear from the surface}}{\text{nondimensional reference time } (\Pi_1 \text{ Read from Figure 9b})} \tag{26}$$

- Using the calculated value from step 2 should yield the correct *f(*)* value that can be used in Equation (24) for simulating any other scenario.

Table 2. shows the results of this study for five different liquids (including HD and VX, known as Mustard and nerve agents, respectively). More validations for this method is compiled by Navaz et al. [54]

Another transport property that needs to be known is the effective diffusion coefficient for the gas phase as indicated in Equations (5) and (6). The evaporation of a liquid inside a porous material generates a vapor phase concentration profile throughout the porous substrate, and the transport is described by a specie diffusion equation. This equation requires the diffusion coefficient as a transport parameter. In a porous medium, the molecular diffusion coefficient needs to be reduced as the vapor transport is affected by a presence of solid and liquid phases.

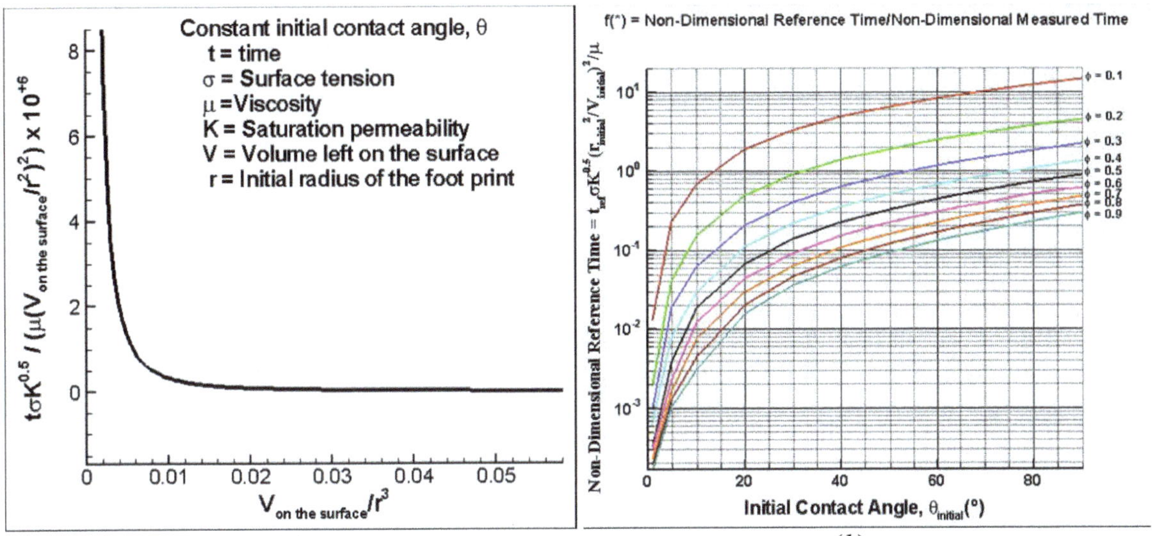

(a) (b)

Figure 9. Nondimensional groups and the similar solutions for the Π groups (a), and the nondimensional reference time as a function of the initial contact angle and porosity (b).

Substance	Experiment (time, s) – Extracted from Video	Model (time, s)
1-2 Propandiol	1.45	1.50
Castor Oil	27.50	28.30
Glycerin	12.30	11.95
VX	1.50	1.45
HD	0.90	0.87

Table 2. Measured and calculated time for the disappearance of a droplet from the surface of a porous material

This results in a definition of an effective diffusion coefficient, which is a function of porosity and saturation, as shown below:

$$D_{eff} = f\left(D_{molecular}, \varphi, s_{\ell}\right) \tag{27}$$

The molecular diffusion coefficient as a function of temperature and molecular structure was taken from Treyball. [65] The function that describes how porosity and saturation attenuate this quantity was derived experimentally and is detailed by Navaz et al. [54, 55]

The effective diffusion coefficient, as given by Equation (27), is measured on a porous substrate bed where the boundary concentrations are given and transport takes place in the principal direction *L*. With reference to Figure 10., if the vapor is traveling from the lower surface to the upper surface (or vice versa), a one-dimensional diffusion problem governs this transport. The measurement technique is based on the gas chromatography and mass spectroscopy (GC/MS) method when the steady state of one-dimensional diffusion is achieved. At steady state, a constant evaporation rate is achieved, and therefore a constant slope mass loss or zero derivative of mass loss in time ($\dot{m}=dm/dt$) should be observed. Then the equation in Figure (23) is used to find D_{eff}.

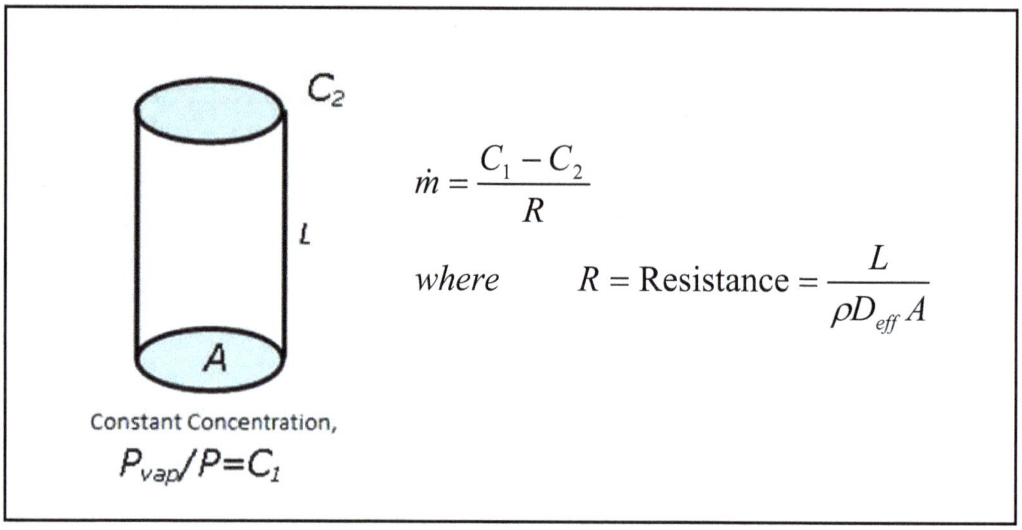

Figure 10. Experimental concept for the effective diffusivity measurements.

After the data reduction, the effective diffusion coefficient (Eq. 28) is obtained.

$$D_{j-mix} = \left\{-0.5855\left(1-s_{\ell i}\right)^3 + 0.4591\left(1-s_{\ell i}\right)^2 + 0.1264\left(1-s_{\ell i}\right)\right\}\varphi(x,y,z,t)D_M \tag{28}$$

where D_M is the molecular diffusion coefficient.

1.5. Contact dynamics and governing equations for the liquid bridge

Many parameters affect the amount or the concentration of transferred liquid into a secondary surface coming into contact with a sessile droplet on a substrate. The approach velocity between two surfaces affects the amount of mass transfer due to the resulting footprint after the contact. The mass transfer between the two surfaces, after the contact, occurs through the area of this footprint. There are basically two pathways by which a contact transfer can occur. These are schematically shown in Figures 29 and 30.

In this document, the bottom and top surfaces are referred to as the primary and contacting or secondary surfaces, respectively. The model is purely driven by the physiochemical properties of the medium, solids, liquids, and gases. When a droplet is placed on a primary surface (Figures 11a and 12a), capillary transport, surface evaporation (due to wind and temperature), and chemical reaction with the surface or other preexisting chemicals in the pores (such as moisture) are simultaneously initiated and the chemical available for transfer

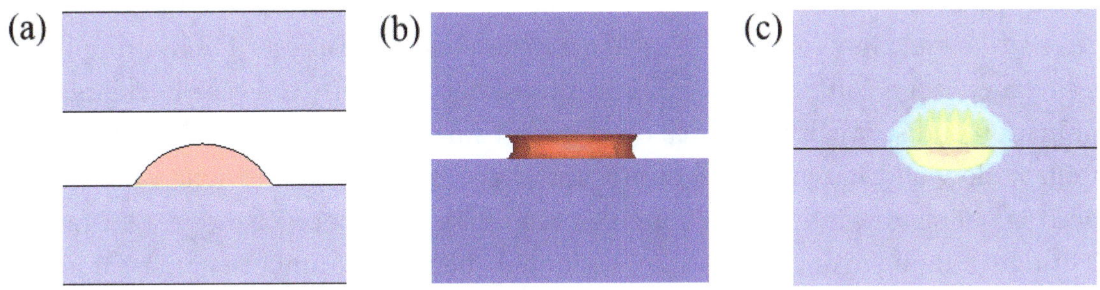

Figure 11. Pathway 1: (a) Spread into the bottom surface, (b) contact dynamics, and (c) contact transfer

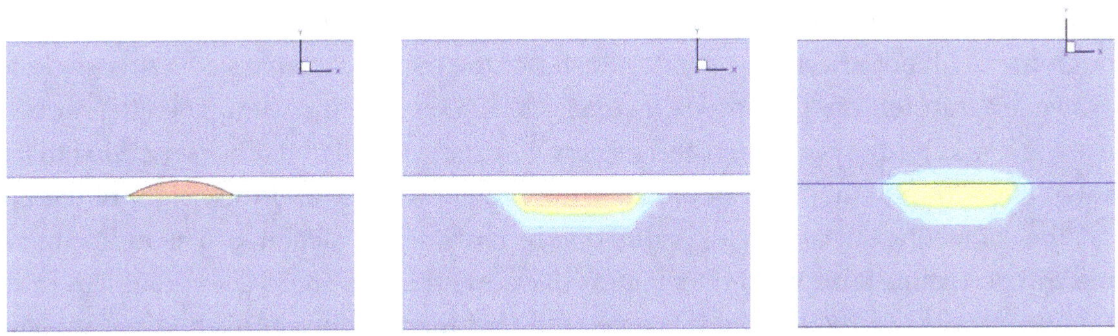

Figure 12. Pathway 2: (a) Spread into the bottom surface, (b) contact dynamics, and (c) contact transfer

through contact is the end product or resultant of all these processes. We refer to this phenomenon as the precontact process. After this point, two possible scenarios could occur: creation of a liquid bridge between the two surfaces and its spread as the two surfaces get closer (Figure 11.b), or complete disappearance of the liquid droplet from the surface of the substrate (Figure 12.b). The last stage or post-contact period is when both surfaces come together and capillary transport occurs at the interface. The mass transfer to the contacting surface always occurs regardless of which pathway is taken. It should be noted when the upper surface comes into contact with or without the formation of the liquid bridge, chemical reaction, and evaporation can also be initiated in the upper surface.

The upper surface moves according to Newton's second law assuming constant acceleration. The y-coordinate of the lower boundary of the upper surface (in motion) can be calculated by the equation of motion:

$$\begin{array}{l} \text{Equation of Motion:} \quad y = \frac{F}{2m}t^2 + V_o t + y_o \\ y = \text{Coordinate system (Vertical)} \qquad V_o = \text{Initial Velocity} \\ y_o = \text{Initial space between the two surfaces} \\ F = \text{Force exerted on the upper surface to move it down,} \\ m = \text{mass of the upper surface} \end{array} \tag{29}$$

The liquid bridge has some free surface. It has been reported by Coblas et al. [66] that the shape of the free surface and its evolution varies as a function of inertial effects. For hydrophilic liquid bridges formed upon contact with small droplets (that is usually the case in chemical agent contamination and spray of pesticides on common surfaces), the liquid bridge is influenced by capillary flow and it generally assumes the shape of a hyperbloid when the velocity of approach is not significantly high (Longley et al.) [67]. The shape of the free surface of the liquid bridge (bulging in or out) has a very minimal effect on the amount of mass transfer into the contacting surfaces. Rather, the footprint area of the liquid bridge defines the surface available for the mass transfer. In the current work, this footprint is assumed to be in the shape of a circle, although the algorithm developed here may be modified to explore other shapes.

The contact angle of this hyperboloid and the surfaces can be found through experiments, although they will not affect the overall transfer of the mass. We have taken this angle to be $\pi/8$. The mass transfer into the porous media is calculated in a time step. Then the remaining mass of the liquid bridge is updated and its new volume is calculated. The separating distance is known from the equation of motion. By knowing the volume and the height of the liquid bridge, the contacting surface area (assumed to be circular) is calculated. The radius increase in time will determine the spread rate. Figure 13. shows that the arc FMG revolves around the *C–C′* axis to produce a hyperboloid. We need to find the volume of this body of revolution analytically. If point *A* is the centroid of the surface engulfed by the arc *FMG*, the volume of the partial torus shaped resulting from this revolution according to Pappus centroid theorem is:

Volume of the partial torus = $V_{pt} = 2\pi A\overline{OA}$ where *A* is the area of the arc. Then the volume of the liquid between the two surfaces will be:

$$V_{Liq} = \pi r^2(t)h(t) - V_{pt} \tag{30}$$

where *t* represents the time. Note that both the footprint radii and the liquid bridge height both are functions of time and are calculated in each time step. If $\hat{\theta} = GBF$, then from basic trigonometry:

$$\begin{aligned} AB &= \frac{4R(t)\sin^3\frac{\theta}{2}}{3(\theta - \sin\theta)} \\ R(t) &= \frac{h(t)}{2\cos\beta} \\ V_{Liq} &= \pi r^2(t)h(t) - 2\pi\left[\frac{R^2(t)}{2}(\theta - \sin\theta)\right]\left[r(t) + \frac{h(t)}{2}\tan\beta - AB\right] \end{aligned} \tag{31}$$

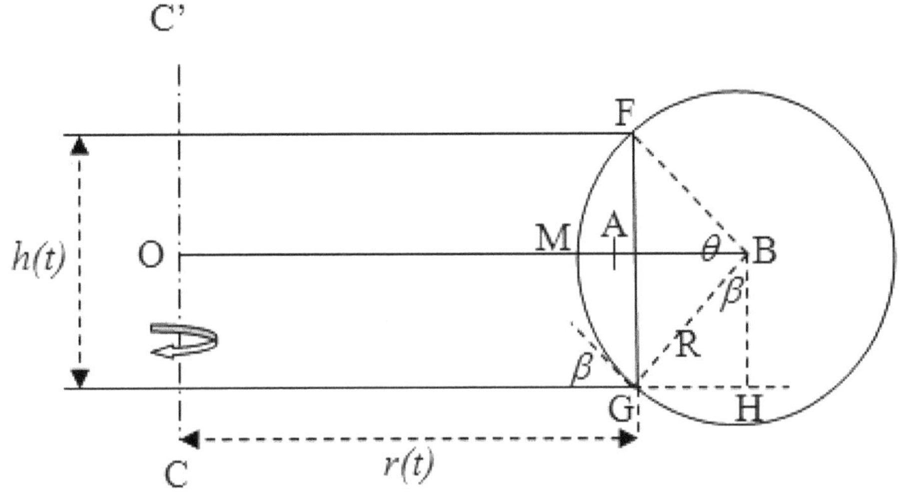

Figure 13. Schematic of the hyperboloid geometry

The equation correlating instantaneous $r(t)$ and $h(t)$ with the liquid volume $\forall_{Liquid\ \ Bridge}(t)$ can be obtained as:

$$r^2(t) - \frac{h(t)G(\beta)}{4\cos^2\beta}r(t) - \frac{h^2(t)G(\beta)\tan\beta}{8\cos^2\beta} + \frac{h^2(t)}{6} - \frac{\forall_{Liquid\ Bridge}}{\pi h(t)} = 0$$
$$G(\beta) = \pi - 2\beta - \sin(\pi - 2\beta) \tag{32}$$

β is the contact angle between the surface and the liquid bridge. The droplet base radius (footprint), $r(t)$, is calculated from the above equation. Both domains, or media, are remeshed and adapted to the new footprint of the liquid bridge. The remeshing is done after each time step.

As the gap between the two surfaces is closing, the height will reduce. At the same time, the liquid is absorbed into the porous media. Therefore, $h(t)$ and $\forall_{Liquid\ \ Bridge}(t)$ are updated in each time step. Then the new footprint radius is obtained by Equation (32). It should be noted that when the height of the liquid bridge becomes less than a minimum threshold, the spread is stopped. This minimum threshold is a function of surface roughness and viscosity of the fluid. However, it is quite possible to assume a value of 10–100 μm for this minimum height. At this point the radius of the footprint remains constant and the absorption is driven solely by capillary pressure causing a reduction in $h(t)$. This value of $h(t)$ can be calculated by solving the above equation again with the "final" liquid footprint radius remaining a constant. This equation is given as:

$$\left(\frac{\pi}{6} - \frac{\pi G(\beta)\tan\beta}{8\cos^2\beta}\right)h^3(t) - \frac{\pi r(t)G(\beta)}{4\cos^2\beta}h^2(t) + \pi r^2(t)h(t) - \forall_{Liquid\ Bridge} = 0 \tag{33}$$

This will be a cubic equation in $h(t)$. As the absorption into the porous media occurs, the volume of the very thin liquid bridge is updated and the new $h(t)$ is calculated. In our model, this switching occurs when $h(t) < 0.08$ mm. This value is a function of surface roughness and can be specified by the user.

1.6. Chemistry model

The chemistry model is based on the Joint Army–Navy–NASA–Air Force (JANNAF) standard methodology. This methodology has three components: a reaction rate processor, species production destruction term calculator, and NASA thermodynamic file (provides specific heat, enthalpy, entropy, and Gibbs free energy) as a function of temperature. [68, 69, 70] This is shown in Equation (34). The outcome of this section will provide the term $\dot{\omega}^{\text{Reaction}}$ in the governing equations.

$$\begin{aligned}
\frac{\overline{c}_p}{\overline{R}} &= \frac{a_1}{T^2} + \frac{a_2}{T} + a_3 + a_4 T + a_5 T^2 + a_6 T^3 + a_7 T^4 \\
\frac{\overline{h}}{\overline{R}T} &= -\frac{a_1}{T^2} + a_2 \frac{\ell n T}{T} + a_3 + \frac{a_4 T}{2} + \frac{a_5 T^2}{3} + \frac{a_6 T^3}{4} + \frac{a_7 T^4}{5} + \frac{a_8}{T} \\
\frac{\overline{s}}{\overline{R}} &= -\frac{a_1}{2T^2} - \frac{a_2}{T} + a_3 \ell n T + a_4 T + \frac{a_5 T^2}{2} + \frac{a_6 T^3}{3} + \frac{a_7 T^4}{4} + a_9 \\
\frac{\overline{g}}{\overline{R}T} &= \frac{\overline{h}}{\overline{R}T} - \frac{\overline{s}}{\overline{R}}
\end{aligned} \tag{34}$$

1.6.1. Bidirectional reactions

A general chemical reaction can be written in terms of its stoichiometric coefficients ν_{ij} and $\nu_{ij}^{'}$ as:

$$\sum_{i=1}^{NSP} \nu_{ij} \overline{M}_i \Leftrightarrow \sum_{i=1}^{NSP} \nu_{ij}^{'} \overline{M}_i \tag{35}$$

where $\overline{M}_i$ represents the i-th chemical species name (i=1, 2,...., NSP) and j represents the j-th reaction (j=1, 2,..., L). These reactions proceed (forward or reverse) according to the law of mass action, which states: "The rate at which an elementary reaction proceeds is proportional to the product of the molar concentrations of the reactant each raised to a power equal to its stoichiometric coefficient in the reaction equation."

Let $[\overline{M}]$ denote the molar concentration of species i. The forward (left to right) reaction rate, $Rate_j\,(L \rightarrow R)$, for reaction j can be written as:

$$Rate_j\left(L \rightarrow R\right) = k_{f_j} \prod_{i=1}^{NSP} \left[\overline{M}_i\right]^{\nu_{ij}} \tag{36}$$

where k_{f_j} is the forward reaction rate constant. For species i, v_{ij} moles on the left side of the reaction becomes v'_{ij} moles on the right side of the reaction. Consequently, the forward reaction for reaction j yields a time rate of change in the molar concentration of species i as follows:

$$(Forward) = \frac{d\left[\bar{M}_i\right]_j}{dt} = \left(v'_{ij} - v_{ij}\right) k_{f_j} \prod_{i=1}^{NSP} \left[\bar{M}_i\right]^{v_{ij}} \tag{37}$$

Similarly, the reverse reaction for reaction j yields:

$$(Reverse) = \frac{d\left[\bar{M}_i\right]_j}{dt} = \left(v_{ij} - v'_{ij}\right) k_{b_j} \prod_{i=1}^{NSP} \left[\bar{M}_i\right]^{v'_{ij}} \tag{38}$$

where k_{b_j} is the backward reaction rate constant. Thus, the net rate of change in the molar concentration of species i for reaction j (denoted by X_{ij}) is as follows:

$$X_{ij} = \frac{d\left[\bar{M}_i\right]_j}{dt} = \left(v'_{ij} - v_{ij}\right) \left[k_{f_j} \prod_{i=1}^{NSP} \left[\bar{M}_i\right]^{v_{ij}} - k_{b_j} \prod_{i=1}^{NSP} \left[\bar{M}_i\right]^{v'_{ij}} \right] \tag{39}$$

The "species production rate" is the time rate of change for the species density. For reaction j the net species production rate for species i is $M_{wi} X_{ij}$ where M_{wi} is the molecular weight of species i. Summing over all reactions gives the net species production rate ω_i for the reaction set

$$\omega_i = M_{wi} \sum_{j=1}^{L} X_{ij} \tag{40}$$

The molar concentration of species i can be written as $\frac{\rho_i}{M_{wi}}$, which is the species i mass density divided by the species i molecular weight.

$$\left[\bar{M}_i\right] = \frac{\rho_i}{M_{wi}} = \frac{\rho C_i}{M_{wi}} \tag{41}$$

It follows that in terms of species i mass fraction Equation (39) becomes:

$$X_{ij} = \left(\nu_{ij}^{'} - \nu_{ij}\right)\left[k_{f_j}\prod_{i=1}^{NSP}\left(\frac{\rho C_i}{M_{wi}}\right)^{\nu_{ij}} - k_{b_j}\prod_{i=1}^{NSP}\left(\frac{\rho C_i}{M_{wi}}\right)^{\nu_{ij}^{'}}\right] \quad \text{or}$$

$$X_{ij} = \left(\nu_{ij}^{'} - \nu_{ij}\right)\left[K_j\prod_{i=1}^{NSP}\left(\frac{\rho C_i}{M_{wi}}\right)^{\nu_{ij}} - \prod_{i=1}^{NSP}\left(\frac{\rho C_i}{M_{wi}}\right)^{\nu_{ij}^{'}}\right]k_j \tag{42}$$

$$K_j = \frac{k_{f_j}}{k_{b_j}} \quad \text{and} \quad k_j = k_{b_j}$$

The reaction rate k_j is from right to left (reverse) in the above equation and is often represented by the Arrhenius form:

$$k_j = a_j T^{-n_j}\exp\left(\frac{-b_j}{\overline{R}T}\right) \tag{43}$$

where a_j is the preexponential coefficient, n_j is the temperature dependence of the preexponential factor, and b_j is the activation energy. The ratio of forward to backward rate, K_j in Equation (42) is related to the equilibrium constant, K_{eql},by the following equation:

$$K_j = \frac{k_{f_j}}{k_{b_j}} = K_{eql}\left(\overline{R}T\right)^{\lambda_j} \tag{44}$$

where

$$\lambda_j = \sum_{i=1}^{NSP}\left(\nu_{ij}^{'} - \nu_{ij}\right) \tag{45}$$

The quantity $|\lambda_j| + 1$ is known as the order of the j-th reaction and the equilibrium constant is:

$$K_{eql} = \exp\left(-\frac{\Delta F}{\overline{R}T}\right), \qquad \Delta F = \sum_{i=1}^{NSP} f_i\nu_{ij} - \sum_{i=1}^{NSP} f_i\nu_{ij}^{'}$$

$$f_i = \text{Gibbs Free Energy} = \text{Chemical Potential} = \overline{h}_i - T\overline{S}_i \tag{46}$$

$$\overline{h}_i = \text{Molar specific enthalpy}$$

$$\overline{S}_i = \text{Molar specific entropy}$$

and

$$X_{ij} = \left(v_{ij}^{'} - v_{ij}\right)\left[K_j \prod_{i=1}^{NSP} \bar{C}_i^{v_{ij}} - \rho^{\lambda_j} \prod_{i=1}^{NSP} \bar{C}_i^{v_{ij}^{'}}\right] k_j \rho^{\sum_{i=1}^{NSP} v_{ij}}$$
$$\bar{C}_i = \frac{C_i}{M_{wi}} (\text{ Molar mass fraction}) \tag{47}$$

1.6.2. Third body dissociation recombination reactions

Reactions involving a third body have a distinct reaction rate for each particular third body. Benson and Fueno [71] have shown theoretically that the temperature dependence of recombination rates is approximately independent of the third body. Available experimental recombination rate data also indicates that the temperature dependence of recombination rates is independent of the third body within the experimental accuracy of the measurements. Assuming that the temperature dependence of recombination rates is independent of the third body, the recombination rate associated with the *k*-th species (third body) can be represented as:

$$k_{kj} = a_{kj} T^{-n_j} \exp\left(\frac{-b_j}{\bar{R}T}\right) \tag{48}$$

where only the constant a_{kj} is different for different species (third bodies). Assuming that the reference species (third body) whose rate is specified in the program input has index $k = k_{ref,}$ we may write:

$$k_{kj} = \frac{a_{kj}}{a_{k_{ref}}} a_{k_{ref}} T^{-n_j} \exp\left(\frac{-b_j}{\bar{R}T}\right) = m_{kj} k_{ref(j)} \quad \text{for } k^{th} \text{ species}$$
$$m_{kj} = \frac{a_{kj}}{a_{k_{ref}}} \tag{49}$$

For the *k*-th species (third body) the stoichiometric coefficient is 1 on both sides of reaction *j* giving:

$$X_{ij}\left(\text{for } 3^{rd} \text{ body } k\right) = \left(v_{ij}^{'} - v_{ij}\right)\left[K_j \frac{\rho C_k}{M_{wk}} \prod_{i=1}^{NSP}\left(\frac{\rho C_i}{M_{wi}}\right)^{v_{ij}} - \frac{\rho C_k}{M_{wk}} \prod_{i=1}^{NSP}\left(\frac{\rho C_i}{M_{wi}}\right)^{v_{ij}^{'}}\right] =$$
$$\left(v_{ij}^{'} - v_{ij}\right)\left[K_j \prod_{i=1}^{NSP}\left(\frac{\rho C_i}{M_{wi}}\right)^{v_{ij}} - \prod_{i=1}^{NSP}\left(\frac{\rho C_i}{M_{wi}}\right)^{v_{ij}^{'}}\right] m_{kj} k_{ref(j)} \tag{50}$$

Summing over all third bodies k we get the total X_{ij}:

$$X_{ij} = \sum_{k=1}^{NSP} X_{ij}\left(\text{for } 3^{rd} \text{ body } k\right) = \left(v'_{ij} - v_{ij}\right)\left[K_j \prod_{i=1}^{NSP}\left(\frac{\rho C_i}{M_{wi}}\right)^{v_{ij}} - \prod_{i=1}^{NSP}\left(\frac{\rho C_i}{M_{wi}}\right)^{v'_{ij}}\right] M_j k_{ref(j)}$$
$$M_j = \rho \sum_{k=1}^{NSP} m_{kj}\left(\frac{C_k}{M_{wk}}\right) = \rho \sum_{k=1}^{NSP} m_{kj}\bar{C}_k = \rho \sum_{k=1}^{NSP} m_{kj}\left[\bar{M}_k\right] \tag{51}$$

This is the standard production rate modified by the factor M_j, which is an effective third body molar concentration. In terms of $\bar{C}_i$ we may rewrite Eq. (51) as:

$$X_{ij} = \sum_{k=1}^{NSP} X_{ij}\left(\text{for } 3^{rd} \text{ body } k\right) = \left(v'_{ij} - v_{ij}\right)\left[K_j \prod_{i=1}^{NSP} \bar{C}_i^{v_{ij}} - \rho^{\lambda_j} \prod_{i=1}^{NSP} \bar{C}_i^{v'_{ij}}\right] M_j k_j \rho^{\sum_{i=1}^{NSP} v_{ij}}$$
$$M_j = \begin{cases} \rho \sum_{i=1}^{NSP} m_{ij}\bar{C}_k & \text{for reactions involving } 3^{rd} \text{ bodies} \\ 1 & \text{for all other reactions} \end{cases} \tag{52}$$

1.6.3. Unidirectional reactions

For unidirectional reactions the reaction only proceeds from left to right and has the assumed form

$$\beta\bar{B} \rightarrow \alpha_1\bar{A}_1 + \alpha_2\bar{A}_2 + ... + \alpha_n\bar{A}_n \tag{53}$$

where the chemical species $\bar{B}$ does not occur on the right side of the equation. Let $\bar{B}$ have species index *K-UNI*. A unidirectional reaction is obtained by setting the backward reaction rate constant to zero, so that for reaction j:

$$X_{ij} = \left(v'_{ij} - v_{ij}\right)\left(\frac{\rho C_{K-UNI}}{M_{w_{K-UNI}}}\right) k_{fj}$$

So that:

$$\frac{d\left[\bar{B}\right]}{dt} = -\beta\rho\bar{C}_{K-UNI} k_{fj} \quad \text{and} \quad \frac{d\left[\bar{A}\right]}{dt} = -\alpha_i\rho\bar{C}_{K-UNI} k_{fj} \tag{54}$$

2. Numerical scheme and algorithm

The governing equations are transformed into the computational domain by rewriting them into a vector form with **U** being the conserved variable vector, **E**, **F**, and **H** are the inviscid flux vectors, and $\mathbf{E}_v$, F_v, and H_v the viscous fluxes in each three directions, respectively. Vector G contains all source terms.

$$\frac{\partial \mathbf{U}}{\partial t}+\frac{\partial \mathbf{E}}{\partial x}+\frac{\partial \mathbf{E}_v}{\partial x}+\frac{\partial \mathbf{F}}{\partial y}+\frac{\partial \mathbf{F}_v}{\partial y}+\frac{\partial \mathbf{H}}{\partial z}+\frac{\partial \mathbf{H}_v}{\partial z}+\mathbf{G}=0 \tag{55}$$

The following transformation is used to cast the conservation equations in the computational domain of ξ, η, and ζ that correspond to x, y, and z.

$$\begin{aligned}\xi &= \xi(x,y,z)\\ \eta &= \eta(x,y,z)\\ \zeta &= \zeta(x,y,z)\end{aligned} \tag{56}$$

According to the chain rule of differentiation:

$$\begin{aligned}\frac{\partial}{\partial x} &= \xi_x\frac{\partial}{\partial \xi}+\eta_x\frac{\partial}{\partial \eta}+\zeta_x\frac{\partial}{\partial \zeta}\\ \frac{\partial}{\partial y} &= \xi_y\frac{\partial}{\partial \xi}+\eta_y\frac{\partial}{\partial \eta}+\zeta_y\frac{\partial}{\partial \zeta}\\ \frac{\partial}{\partial z} &= \xi_z\frac{\partial}{\partial \xi}+\eta_z\frac{\partial}{\partial \eta}+\zeta_z\frac{\partial}{\partial \zeta}\end{aligned} \tag{57}$$

where

$$\begin{aligned}
\xi_x &= \frac{y_\eta z_\zeta - y_\zeta z_\eta}{|\mathbf{n}|}, & \xi_y &= -\frac{x_\eta z_\zeta - x_\zeta z_\eta}{|\mathbf{n}|}, & \xi_z &= -\frac{x_\eta y_\zeta - x_\zeta y_\eta}{|\mathbf{n}|}\\
\eta_x &= -\frac{y_\xi z_\zeta - y_\zeta z_\xi}{|\mathbf{n}|}, & \eta_y &= \frac{x_\xi z_\zeta - x_\zeta z_\xi}{|\mathbf{n}|}, & \eta_z &= -\frac{x_\xi y_\zeta - x_\zeta y_\xi}{|\mathbf{n}|}\\
\zeta_x &= \frac{y_\xi z_\eta - y_\eta z_\xi}{|\mathbf{n}|}, & \zeta_y &= -\frac{x_\xi z_\eta - x_\eta z_\xi}{|\mathbf{n}|}, & \zeta_z &= \frac{x_\xi y_\eta - x_\eta y_\xi}{|\mathbf{n}|}
\end{aligned} \tag{58}$$

With the norm of:

$$|\mathbf{n}| = x_\xi y_\eta z_\zeta - x_\xi y_\zeta z_\eta + x_\eta y_\zeta z_\xi - x_\eta y_\xi z_\zeta + x_\zeta y_\xi z_\eta - x_\zeta y_\eta z_\xi \tag{59}$$

Equation (55) will now be transformed into:

$$\frac{\partial\tilde{\mathbf{U}}}{\partial\tilde{t}}+\frac{\partial\tilde{\mathbf{E}}}{\partial\xi}+\frac{\partial\tilde{\mathbf{F}}}{\partial\eta}+\frac{\partial\tilde{\mathbf{H}}}{\partial\zeta}+\frac{\partial\tilde{\mathbf{E}}_v}{\partial\xi}+\frac{\partial\tilde{\mathbf{F}}_v}{\partial\eta}+\frac{\partial\tilde{\mathbf{H}}_v}{\partial\zeta}+\tilde{\mathbf{G}}=0 \qquad \text{with } \tilde{\mathbf{U}}=\frac{1}{J}\mathbf{U}, \quad \tilde{\mathbf{G}}=\frac{1}{J}\mathbf{G}$$
$$\tilde{\mathbf{E}}=\frac{1}{J}\left(\xi_x\mathbf{E}+\xi_y\mathbf{F}+\xi_z\mathbf{H}\right),\ \tilde{\mathbf{F}}=\frac{1}{J}\left(\eta_x\mathbf{E}+\eta_y\mathbf{F}+\eta_z\mathbf{H}\right),\ \tilde{\mathbf{H}}=\frac{1}{J}\left(\zeta_x\mathbf{E}+\zeta_y\mathbf{F}+\zeta_z\mathbf{H}\right) \tag{60}$$
$$\tilde{\mathbf{E}}_v=\frac{1}{J}\left(\xi_x\mathbf{E}_v+\xi_y\mathbf{F}_v+\xi_z\mathbf{H}_v\right),\ \tilde{\mathbf{F}}_v=\frac{1}{J}\left(\eta_x\mathbf{E}_v+\eta_y\mathbf{F}_v+\eta_z\mathbf{H}_v\right),\ \tilde{\mathbf{H}}_v=\frac{1}{J}\left(\zeta_x\mathbf{Ev}+\zeta_y\mathbf{F}_v+\zeta_z\mathbf{H}_v\right)$$

The Jacobian of the transformation is:

$$J=\xi_x\left(\eta_y\zeta_z-\eta_z\zeta_y\right)-\xi_y\left(\eta_x\zeta_z-\zeta_x\eta_z\right)+\xi_z\left(\eta_x\zeta_y-\zeta_x\eta_y\right) \tag{61}$$

The explicit form of all of the flux vectors are described by Navaz et al. [55,56]. The above sets of equations are solved by finite difference method on a structured mesh. The explicit Runge–Kutta integration algorithm was used to find the distribution of all variables (gas, liquid, and solid concentrations, capillary pressure, velocities, etc.) in time. The program can accept externally generated mesh by commercial software such as GRIDGEN™ [72] or can generate a grid internally for one or two sessile droplet(s) residing on a porous on nonporous substrate. In the presence of a liquid bridge and its spread, a built-in and automated adaptive mesh generator is embedded in the computer model. The mesh or grid point distribution is based on Coon's Patch [73]. First, the mesh for the substrate is generated and then the mesh defining the droplet(s) is overlaid on the substrate geometry. The mesh points defining the droplets are collapsed on the surface of the substrate except in the area(s) that droplet(s) reside on the surface. Figure 14 shows the top view of the adaptive mesh when liquid bridge expands during the contact process. Figures 11 and 12 shows snapshots of the absorption process during the contact. It displays the precontact and postcontact configurations.

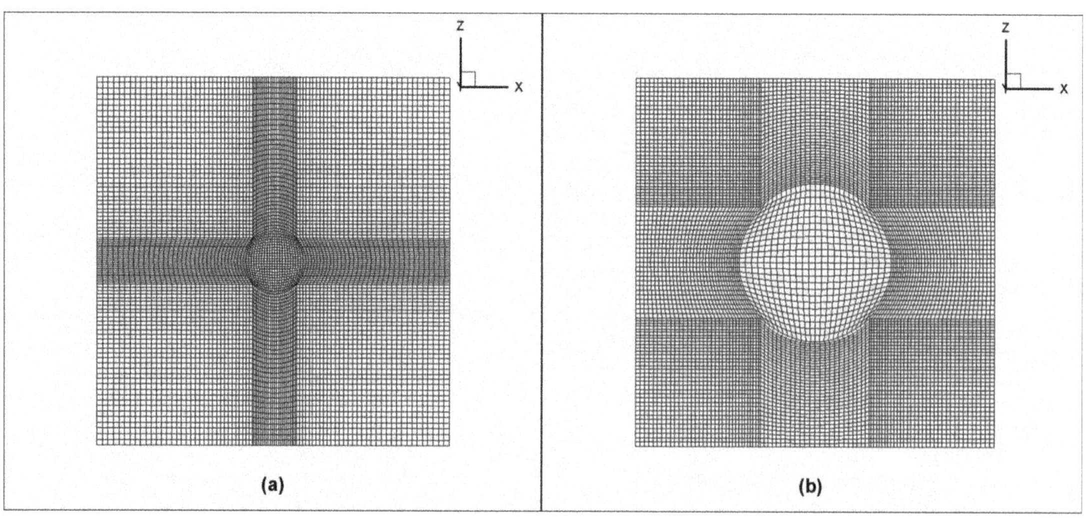

Figure 14. Top view of the adaptive mesh when liquid bridge still expanding, until the conditions for Equation (33) are reached.

The calculations start with the specified initial conditions and then the continuity and momentum equations are numerically integrated in time to find the distribution of liquid inside the pores. The saturation is equal to unity ($s_\ell=1$) at the interface between the droplet or liquid bridge with the porous media. The hydrostatic pressure is also added to the capillary pressure for more accurate calculations. That is, $P_\ell = P - P_{ci} + \rho_\ell g h^*$ where h^* is the local height of the droplet above the surface. Mass is being transported into porous medium according to $(\rho_\ell \tilde{v} \varphi)/J$; where J is the Jacobian for the transformation and $\tilde{v}$ is the contra-variant vertical velocity given by: $\tilde{v} = \eta_x u_\ell + \eta_y v_\ell + \eta_z w_\ell$ with u_ℓ, v_ℓ, w_ℓ being the three components of the velocity and η_x, η_y, η_z being the metrics of the transformation. The mass transfer is calculated in each time step and the instantaneous remaining mass yields the liquid bridge volume. The remaining mass divided by the liquid density will yield the instantaneous volume of the liquid bridge. Since the distance gap or the height of the liquid bridge is known from the equation of motion (Eq. 29), the temporal (or instantaneous) base or footprint radius can be found from Equation(32). When the distance between surfaces become less than a threshold, the spread is stopped, that is, the base radius stays constant at its last value and the capillary transport into both surfaces continues, causing a reduction in the height of the liquid bridge. At this point the instantaneous or temporal height of the liquid bridge can be calculated from Equation (33). This process continues until the amount of liquid between the two surfaces becomes zero.

3. Test cases

The computational model developed is called modeling of chemical hazard—dispersion and contact (MOCHA-DC). Test cases are designed to validate the physiochemical phenomena that are formulated in this chapter. More details are available in Navaz et al. [55, 56]. A test case describing the spread of VX nerve agent into sand is described. The capillary pressure, which is the main driver for the spread and transport of liquid into porous media, was obtained by the method outlined in Section 1.4.

3.1. Capillary transport and secondary evaporation

The topology of the penetration of a 6 μL VX droplet into the UK sand (ϕ = 0.44) after about 120 minutes is shown in Figure 15. VX has a very low vapor pressure at room temperature and so evaporation could be ignored. The shape of the liquid front is obtained experimentally at ECBC and is compared to the MOCHA-DC code prediction (saturation contours) in Figure 15. A comparison for the prediction of the depth of spread in time is observed from Figure 16. as compared with experimental data.

A similar validation was performed for the penetration of 1,2-propandiol, glycerin, and castor oil into a ceramic tile (ϕ = 0.26), coarse sand (ϕ = 0.31), and glass beads (ϕ = 0.28). The penetration was recorded after 2, 3, 5, 10, 20, 30, 60 minutes, for up to a week (for castor oil). The model prediction was excellent and they are detailed by Navaz et al. [55]. A sample case for glycerin spread into sand is shown in Figure 17.

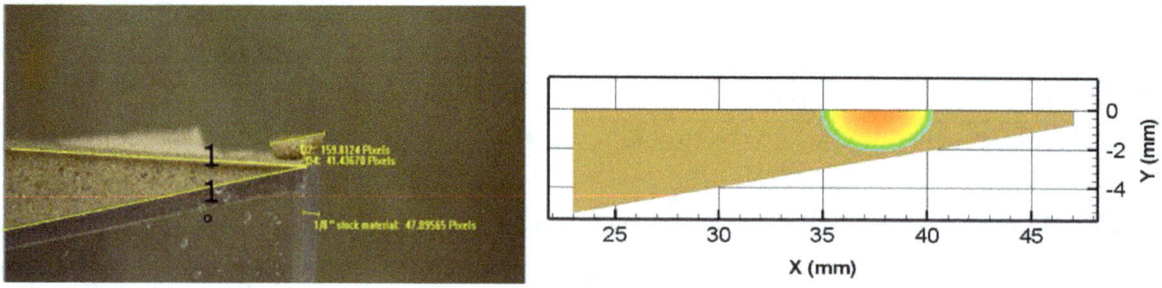

Figure 15. Comparison of COMCAD prediction and experiment for VX on UK sand.

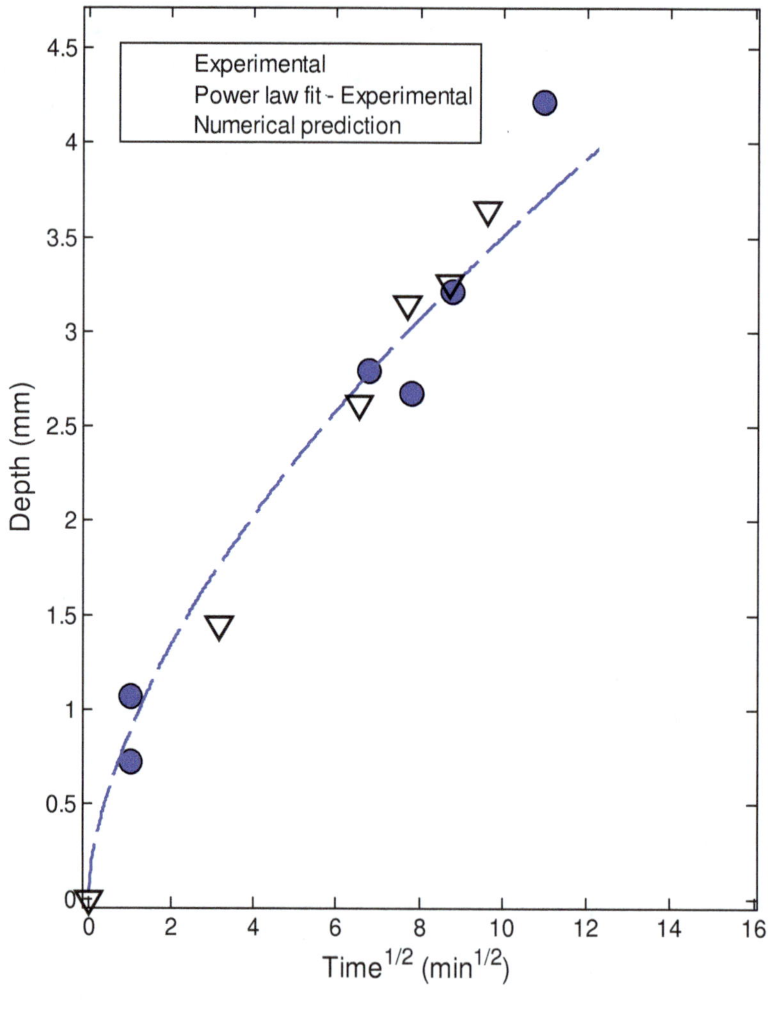

Figure 16. Progression of depth in time for VX spread/transport through UK sand as predicted by the model and experiment

The secondary evaporation of several nerve agents are measured by Navaz et al.[55]. The validation of the model was performed using data from Reis et al. [74] and Mantle et al. [7] They studied the evaporation of water inside porous media using magnetic resonance imaging (MRI). Figure 18. shows an excellent comparison for the evaporation of water in sand and 120- and 400-μm diameter glass beads.

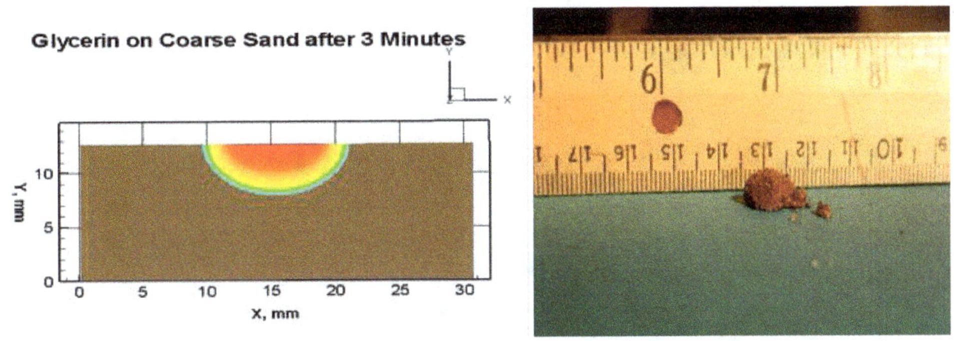

Figure 17. Penetration of glycerin into the coarse sand after 3 minutes (depth ~ 6 mm) and wetted pad from the experimental study

Data collected in collaboration with Kiple Acquisition Science Technology Logistics & Engineering Inc. (KASTLE), B.O.I.S.-Filtry, Brno, Czech Republic and VOP-026 Sternberk, Brno, Czech Republic, was used to validate the model. The data was taken in open air and the model was run under the same conditions (wind speed and temperature) for EDB (ethylene dibromide). An excellent agreement between the model prediction and data exists, as can be seen from Figure 19.

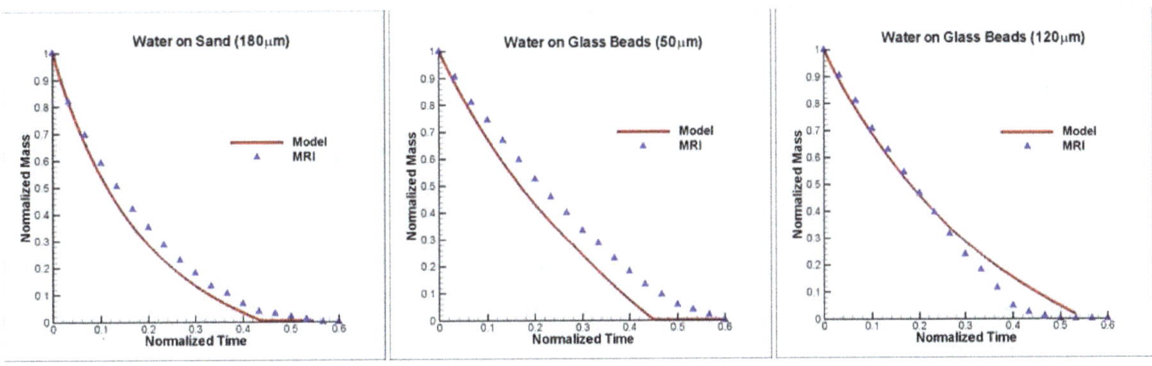

Figure 18. Model validation for the secondary evaporation of water inside sand and glass beads compared with MRI data

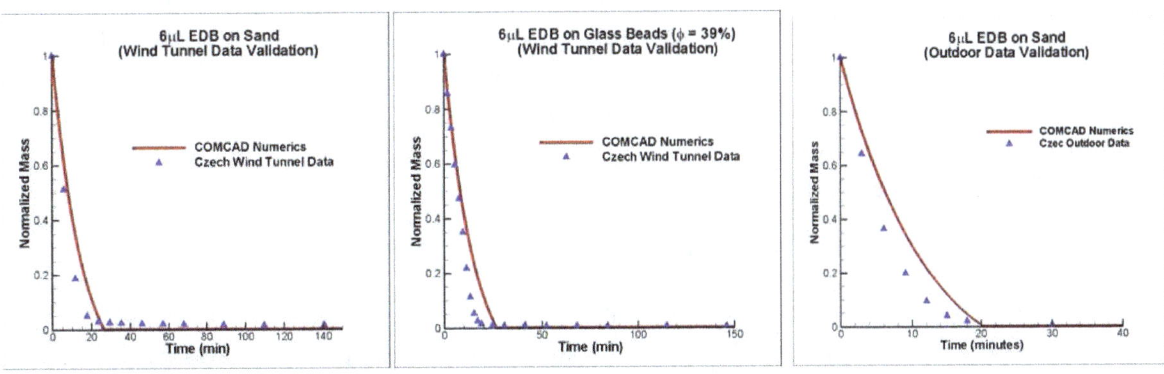

Figure 19. Model/experiment comparisons with Czech Data for 6 μL Ethylene Dibromide (EDB) droplet spread inside porous media (sand and glass beads) with secondary evaporation.

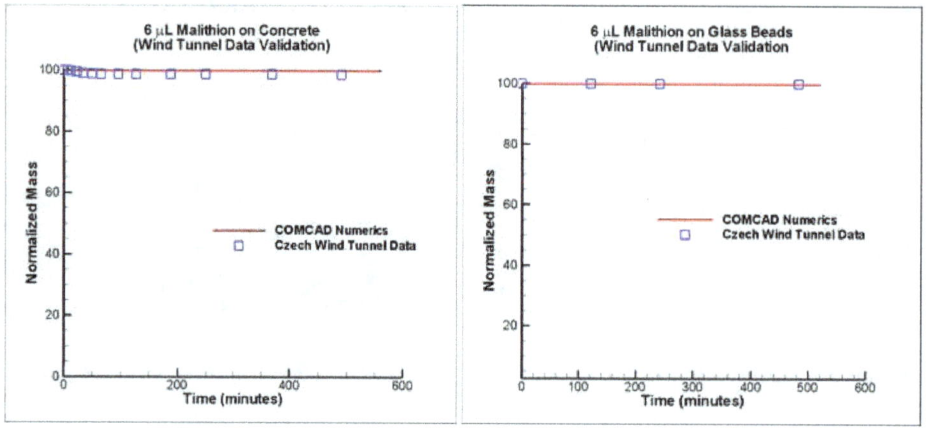

Figure 20. Model/experiment comparisons with Czech Data for 6 μL Malathion (MAL) droplet spread inside porous media with secondary evaporation.

Wind tunnel test were performed for the evaporation of Malathion in dry concrete and glass beads. There was no (or minor) evaporation of Malathion inside dry concrete or glass beads due to the low vapor pressure of the species, as seen from Figure 20. However, it will be shown that in the presence of moisture or water this picture will change.

3.2. Chemical reaction in porous media

In this section series of validation, test cases are presented that involve chemical reaction(s) in different phases.

We chose to simulate two cases for a liquid interacting with a solid. The first case was a 50 μL drop of sulfuric acid placed on a brick of solid sodium sulfide. The second case was a 50 μL drop of sulfuric acid on a mixture of 75% sand and 25% sodium sulfide. In both cases the samples were open to the air to allow the product gas to escape. Mass loss was monitored and rates of product formation/reactant degradation inferred. Earlier experiments documented that during the time scale for these reactions the sulfuric acid did not react with the sand (100% sand case) and adsorption of atmospheric water was insignificant. Therefore, the mass change was due to the reaction alone. The reaction stoichiometric equation is given below. The model simulation of the liquid on the solid brick of sodium sulfide and the sand/sodium sulfide mixture are given in Figures 21a and b. Both results are validated with experimental data.

$$Na_2S_{(s)} + H_2SO_{4\,(aq)} \rightarrow Na_2SO_{4\,(s)} + H_2S_{(g)}$$

(sodium sulfide + sulfuric acid → sodium sulfate + hydrogen sulfide)

In the next test case, glass beads were chosen as the porous substrate and droplets of cyclohexanol and phosphoric acid were deposited about 1 mm apart. The reaction started when the two droplets diffused together due to capillary forces, and the reaction proceeded thereafter. Figure 22. shows the contours of remaining reactants and the amount of product (cyclohexene) as predicted by the MOCHA-DC code from the initiation of the process (diffusion and

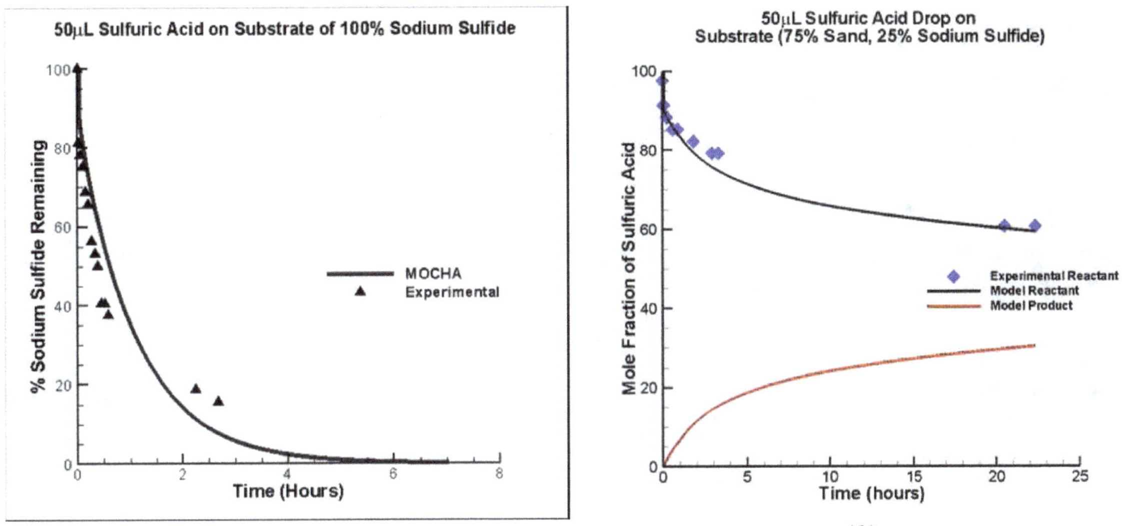

(a) *(b)*

Figure 21. Model/ experiment comparison for (a) 100% sodium sulfate and 50 μL sulfuric acid reaction, and (b) 3:1 mixture of sand: sodium sulfate and 50 μL sulfuric acid droplet

reaction). The overall destruction of cyclohexanol is similar to experimental data and is described in Figure 22..

The reaction of chemical warfare agents with water is of special interest in this section. This is called a hydrolysis reaction. VX and Malathion are among the chemicals that react with water. Water can be present as moisture in sand, concrete, and soil, or can be added to a preexisting chemical agent in a dry porous medium as rain. The model is capable of simulating all possible scenarios. The hydrolysis of VX proceeds as multiple chemical reaction processes that result in several breakdown products. The evolution of these products and the spread of a droplet therefore provide a reasonable validation of the interaction of the reaction and diffusion functions in the code. A simulation of a 6 μL droplet of *O*-ethyl-*s*-[2-*N*,*N*-(diisopropylamino)ethyl] methylphosphonothioate (VX) spread into sand with porosity of 35%. The hydrolysis reaction rates were defined using those for VX in moist sand given in Brevett et al. [75, 76] Brevett et al.'s [75, 76] work indicates that VX degrades into several products in the presence of water. All of these reactions were taken as first order in VX, due to the reported 25-32 fold molar excess of water present in the experiments. The droplet size was taken from the same study. The porous field in the model extended beyond the region of droplet spread. All model boundaries were set as impermeable, as the experimental study involved degradation under sealed conditions. The quantity of VX and the appearance of breakdown products indicated by the models throughout the simulation were consistent with data from the experimental study. VX spread to a radius of 7 mm in the model, similar to the spread reported by Brevett et al. [75, 76] and Wagner et al. [77] The match between the amount of VXH+ in the experiment and in the model during the degradation process results from the model's ability to handle multiple reactions. The rate of production/destruction of each chemical species is a local event, calculated at each node. Species concentration and, therefore, the reaction rate vary across the saturation region within the sand, yielding a complex mixture of breakdown products. If the

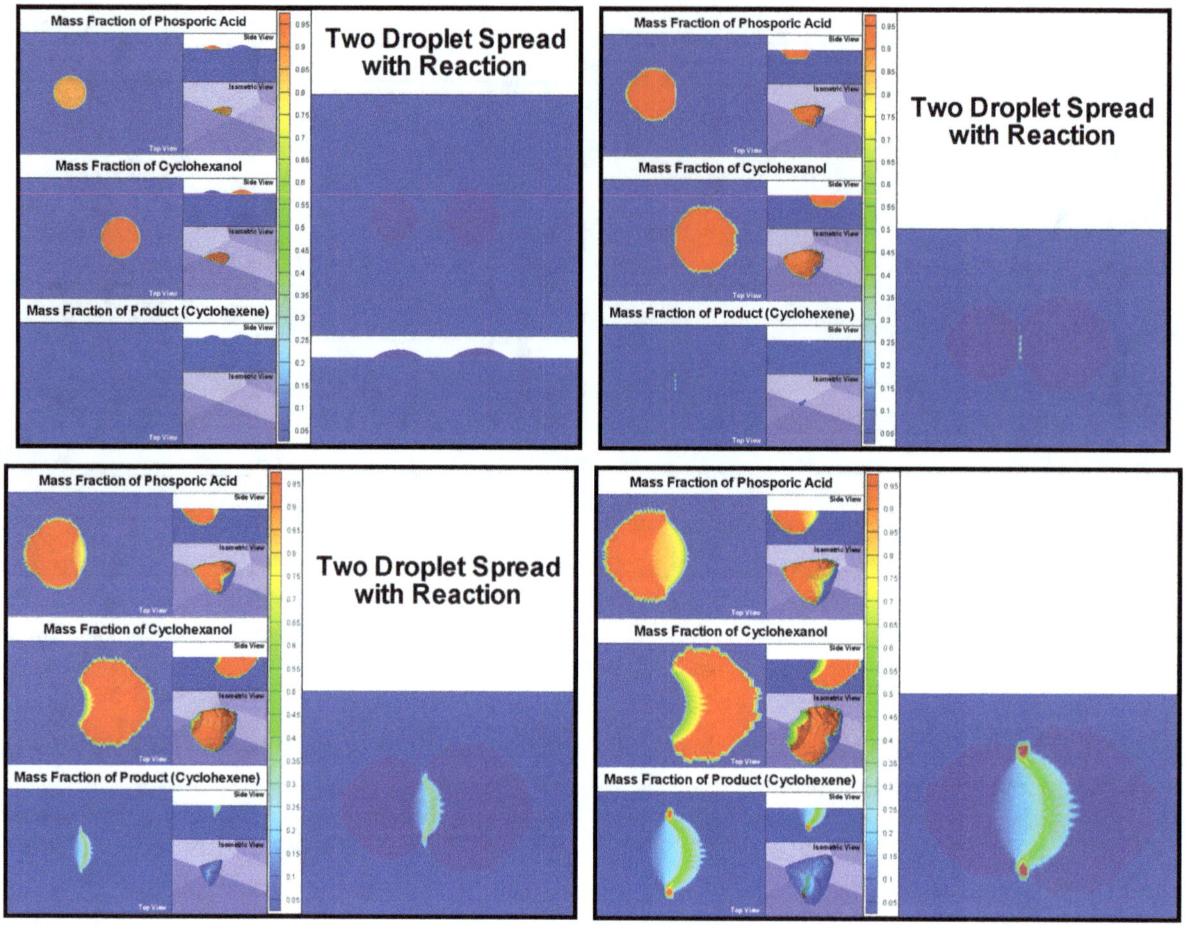

Figure 22. Development of reaction zone between cyclohexanol and phosphoric acid Modeled by the MOCHA-DC Code (from top left to lower right) t=0, t=12 s, t=144 s, t=3600 s.

fluid motion predicted in the model was inappropriate it would lead to unrealistic local concentrations, which would cause the local rates of production/destruction (based on local concentrations in the reaction model) to be inappropriate. The agreement between the model and the experimental data, therefore, demonstrates robustness in both the fluid motion and reaction modeling methods. The result of this study is shown in Figure 24.

3.3. Contact hazard

In this section, some test cases are presented to illustrate the model prediction for a sessile droplet spread dynamics between two moving nonporous or porous surfaces. The presented test cases will examine the gas and liquid mass transfer and the effect of chemical reactions during the droplet spread process.

In the first case, the spread of de-ionized water between a nonporous and porous glass plates as a function of velocity of approach was examined. A linear stage actuator (Thorlabs™ LNR50SEK1) controls the distance between the two parallel plates and a load cell (Interface™ MB-LBF5) is designated for measuring the force acting on the solid body. Two optical windows (Newport™ 20BW40-30) are used as the upper and lower plates as shown in Figure 25.. A

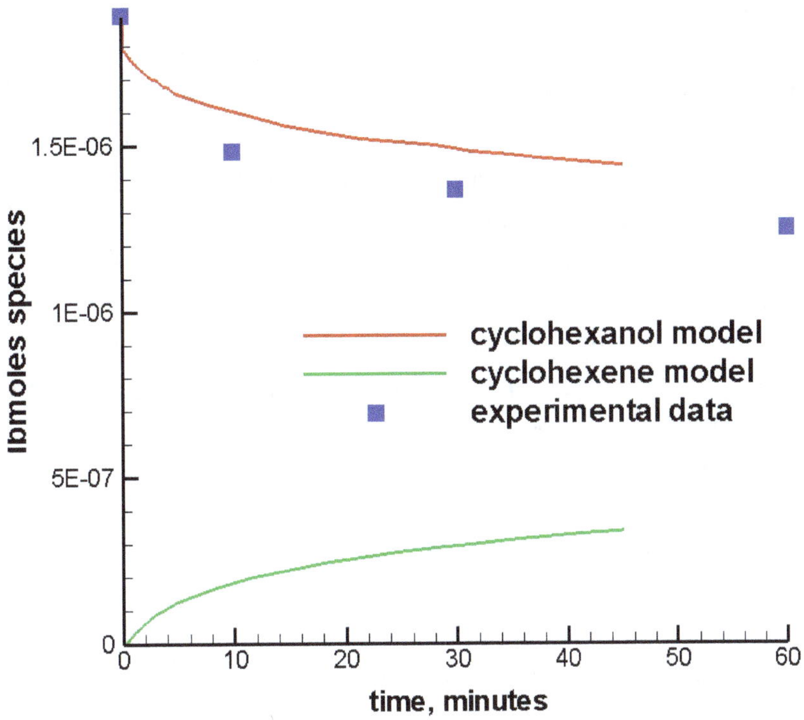

Figure 23. Comparison of destruction of cyclohexanol to experimental data

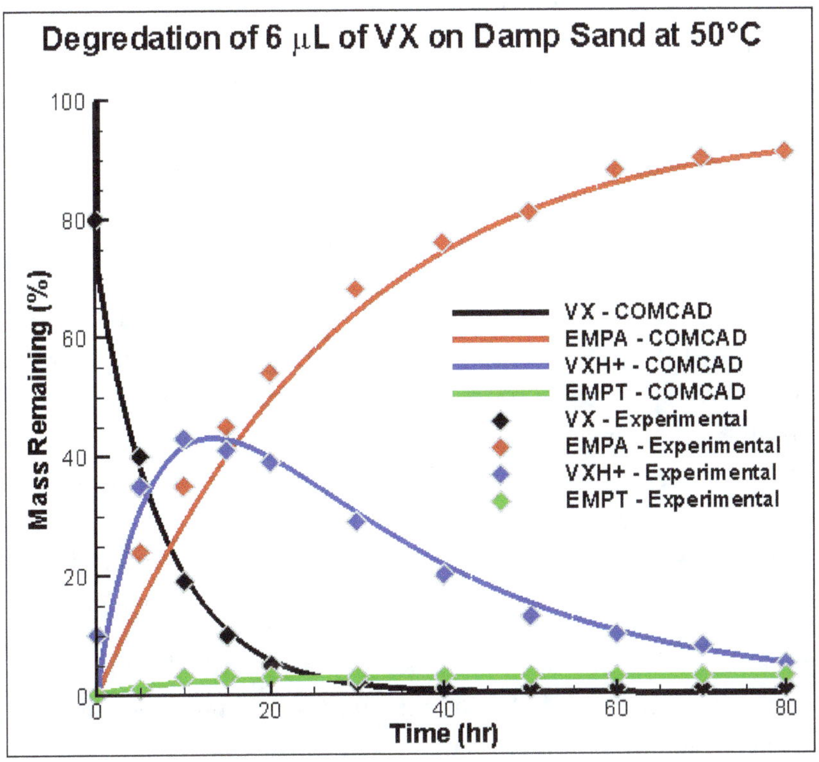

Figure 24. A 6 μL VX droplet undergoes hydrolysis in damp sand at 50°C. Both agent and breakdown product amounts present in the sand, as measured in Brevett et al. [75, 76], are characterized by the model

feedback control loop was created, connecting the load-cell force input, the actuator position, and speed inputs and outputting the upper plate acceleration and speed. This feedback enabled to simulate a freely moving solid body with a given mass, m_1, reacting to external forces and the force applied by the liquid bridge connected to a solid body with mass $m_2 \to \infty$. The plates were cleaned before each experiment using acetone, isopropanol, and nitrogen gas. The advancing and receding wetting angles were estimated from quasi-static force measurements. The experiment utilized a 200 μL deionized (DI) water as a sessile droplet deposited onto a fused silica optical window (an impermeable surface), and a porous glass (Vykor™ 7930) was used as the permeable contacting surface. The physical properties of water are $\mu = 0.001 Pa.s$, $\rho = 1000 kg/m^3$, $\sigma = 0.072 N/m$. The porous glass had a porosity of 28% with permeability of 2.08×10^{-19} m^2/s (values provided by manufacturer). The surfaces were brought into contact with constant relative speeds of 0.5, 1.5, 2.0, and 2.5 μm/s, and the liquid bridge radius was measured in the experiment and compared to that calculated by the model (Figure 26). In all cases, there was a rapid initial change in geometry upon contact of the liquid with both plates, and a transition from sessile droplet geometry to the hyperboloid geometry of the liquid bridge. For smaller approach velocities, the absorption into the porous medium proceeded faster than the radial spreading throughout the liquid penetration process, causing a steady reduction in the footprint radius in time. However, for faster approach speeds (2 μm/s and above) the radius of the liquid bridge was only reducing in the initial stages due to rapid penetration into the porous medium. After some penetration of the liquid into the porous medium, and thus increasing the viscous resistance of the flow, the liquid began to spread radially and the radius of the liquid bridge increased with time. This finding demonstrated that the depth of liquid penetration and the imprint radius of droplet into the porous medium is determined by the speed of the approaching surfaces. We have also modeled two additional approach velocities of 1.0 and 3.0 μm/s to demonstrate the consistency of the results (Figure 26.). The model and experimental results indicated that the spreading increases as the approach velocity increases. This causes a larger footprint to be issued by the process.

An experiment conducted to measure the amount of mass transfer for two surfaces coming into contact with a liquid droplet in between. A 20 μL glycerin droplet was deposited on a porous media composed of play sand (porosity = 35%). Then a cloth at an initial distance of 11 cm with the speed of 35 cm/s was brought into contact with the glycerin. The amount of mass absorbed during the contact process into the cloth was measures and is compared to the model prediction. These tests were conducted with three additional chemicals and the results were compared to the model's response. These conditions are shown in Figure 27.a and the agreement between the model and experimental data is seen to be excellent.

Another experiment was conducted with the same setup with the exception of varying the time of the contact. The purpose of this experiment is to demonstrate that even after the disappearance of the sessile droplet from the surface, mass transfer to a secondary porous material will take place. That is to say, even though a hazardous liquid substance may not be visible on a surface of a porous material, it can still pose a threat through contact. However, kitchen tile (porosity of 24%) with a lower porosity was selected for this experiment to have more control over the disappearance time of the sessile droplet from the surface. A wafer,

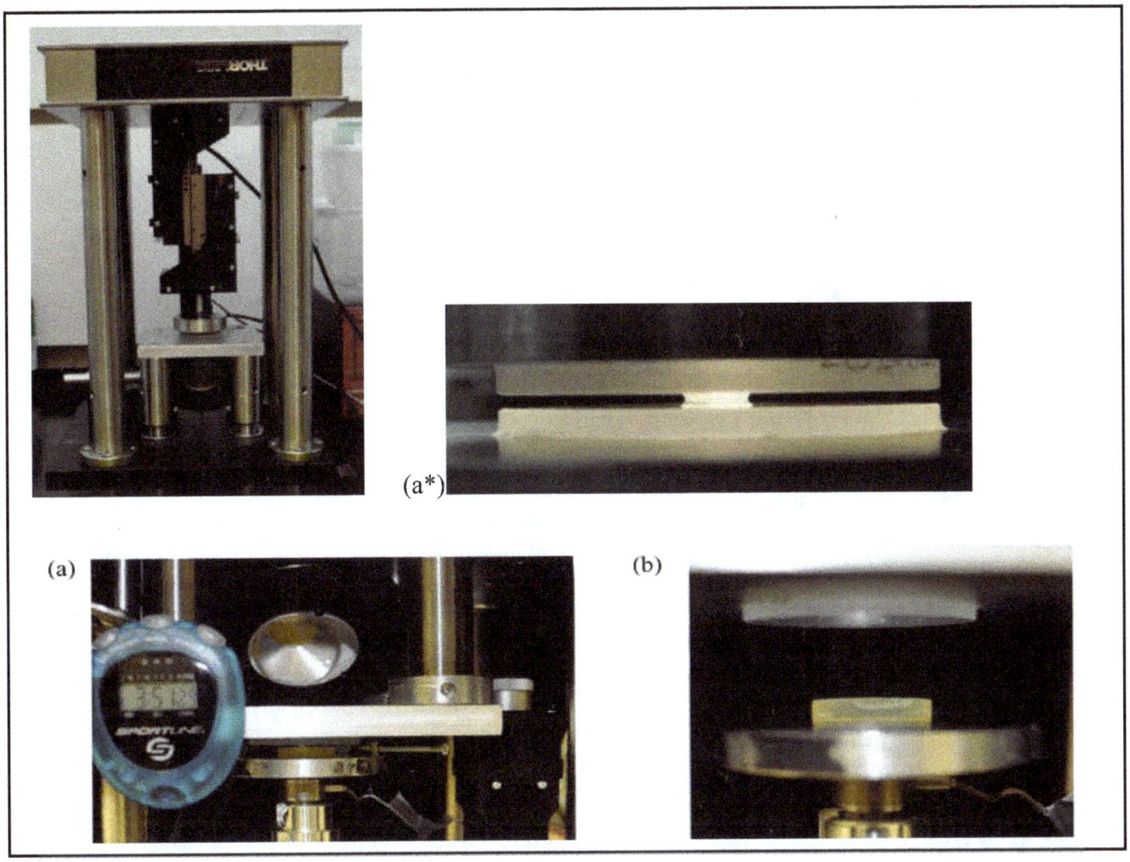

Figure 25. The experimental setup: two fused-silica plates connected by a DI water liquid bridge. (a) Plates in fixture and (*a**) a closeup view of the liquid bridge (*b*) shows the porous glass used in the experiment

fabricated in our laboratory from filter paper pulp, with a porosity of 60%, served as the secondary surface. Five separate experiments (three repetitions for each experiment) were conducted with a contact time to be, respectively, 1, 10, 20, 30, and 40 minutes after the droplet is placed on the kitchen tile. Every experiment was repeated three times at an initial gap of 2.5 cm between the two surfaces and a downward force of 1 N. The two surfaces were brought into contact. The amount of glycerin transferred into the secondary surface (wafer) was measured and compared favorably with the model predictions (Figure 27.b). These comparisons indicate that the model was fairly accurate in predicting the amount of mass being absorbed into a contacting porous surface.

In order to provide data for cases where contact between two surfaces results in a chemical reaction in one or both of the contacting surfaces, experiments were conducted measuring the rate of formation of product where one reactant resides on one surface and the other on the second, contacting, surface. Since there are many permutations for these types of experiments, we began our investigation for a case where the primary surface is nonporous, a glass slide, and the contacting surface is porous, an adsorbent pad. The reaction needed to be fast enough so that the product formation could be measured accurately and in a relatively short time span to avoid complications associated with evaporation of chemicals. The reaction chosen was addition of bromine to alkenes, since the progress of this reaction could be monitored both

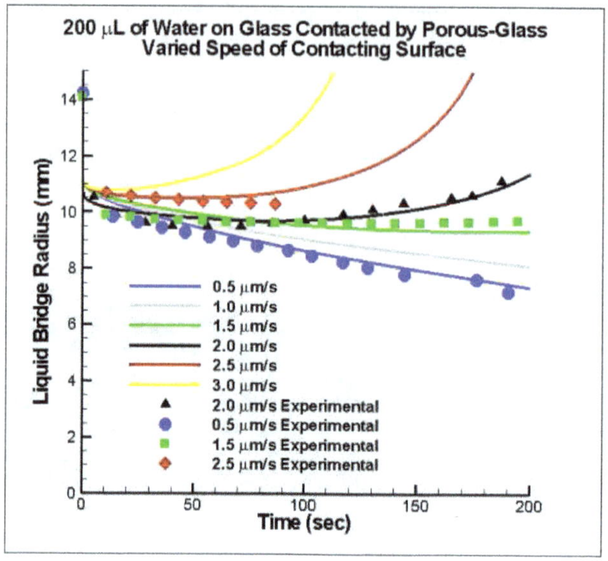

Figure 26. 200 μL of water on glass contacted by porous glass. The approach velocities of 0.5, 1.5, 2.0, and 2.5 μm/s are compared with the model. Two additional velocities of 1.0 and 3.0 μm/s are also modeled. For lower velocities the absorption into porous glass is at a faster pace than the spread speed, causing a continuous reduction in radius in time. As the approach velocity increases, the spread will proceed at a faster pace than the absorption, causing a continuous increase of radius in time. It appears that at velocity of 1.5 μm/s both processes take place at the same "speed"

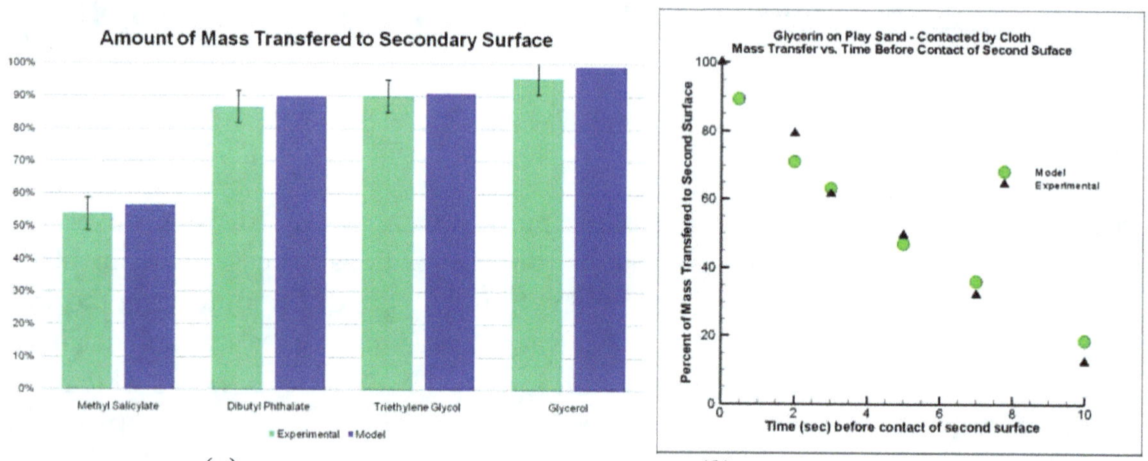

(*a*) (*b*)

Figure 27. (*a*) Mass transferred to contacting surface as a function of the type of chemical and comparison of the model prediction with experimental data for each case, and (*b*) mass transfer of glycerin, that was initially deposited on sand, to a contacting cloth, where the amount transferred was a function of the time elapsed between deposition and the contact, the figure also shows the model prediction for each case.

quantitatively and qualitatively. A 20 μL of styrene solution was placed on the surface of a glass slide and an adsorbent pad saturated with 1 mL of bromine in carbon tetrachloride solution was placed on top of the slide. The contact was stopped at various time intervals. The reaction was quenched by addition of excess cyclohexene to the pad, after removal from the slide, and adsorbent pad and the glass slide were each extracted with 2 mL of dichloromethane. The solutions were then analyzed by GC/MS and GC/FID. No traces of styrene or the addition product were observed on the glass slide even for the smallest time step, indicating complete adsorption into the pad. Figure 28. shows the model/experiment comparison of the product form according to:

The styrene–bromine reaction is as follows: $C_8H_8(L)+Br_2(L) = C_8H_8Br_2(L)$

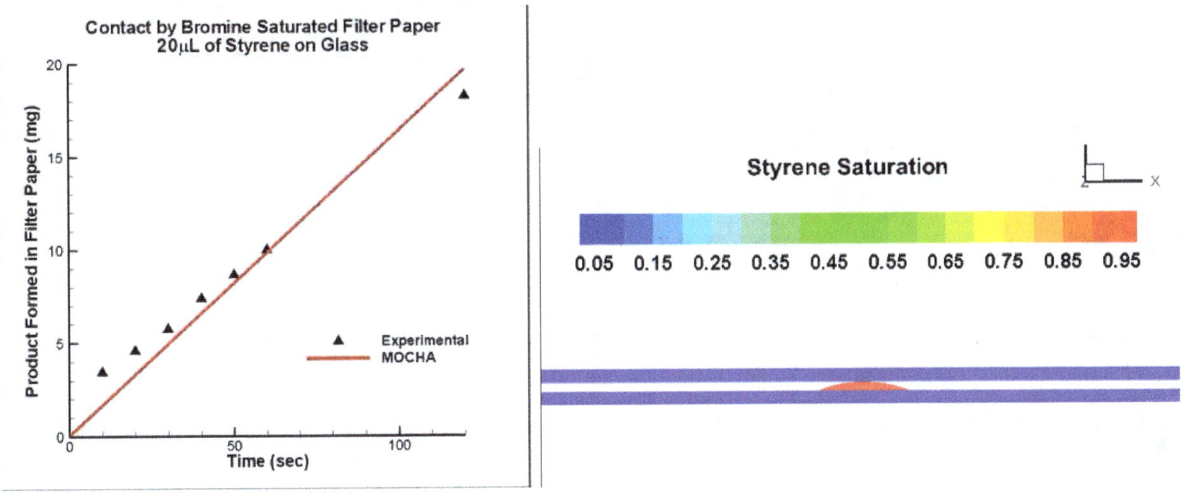

Figure 28. Product formation in time for the styrene and bromine reaction from the experiment and as predicted by MOCHA-DC and the model showing the styrene droplet just prior to contact with the bromine saturated upper layer

The second experiment is similar to the previous experiment but the glass substrate is replaced with permeable sand. A 60 μL styrene droplet was placed on sand and after completely disappearing from the surface was contacted by filtered paper saturated by bromine. The mass exchange occurs at the interface such that the reaction takes place in both media. The results are shown in Table 3. The comparison is quite satisfactory and the model accurately predicts the mass transfer between the two contacting surfaces and the amount of product in both zones.

	Styrene Sand (mg)	Styrene Pad (mg)	Product Sand (mg)	Product Pad (mg)
MOCHA-DC	44.50	7.12	4.90	1.75
Data	43.74	7.40	5.13	1.56

Table 3. Data and model reactant and product mass comparison in two porous media

In another test case, a 6 μL VX droplet is situated on moist sand (5% saturation). The reaction occurs until a secondary moist porous medium (cloth) is brought into contact with the first surface after about 0.5 hour. The remaining VX at the surface of the sand starts to react with water content in or from the contacting surface. The reaction in both zones continues for about 3 hours. Figure 29. shows the amount of VX left and all products during this process in real time. We do not have any experimental data to compare with these results, however; although there are no data available for this case, it is a very challenging case to run and it demonstrates the robustness and generality of the code.

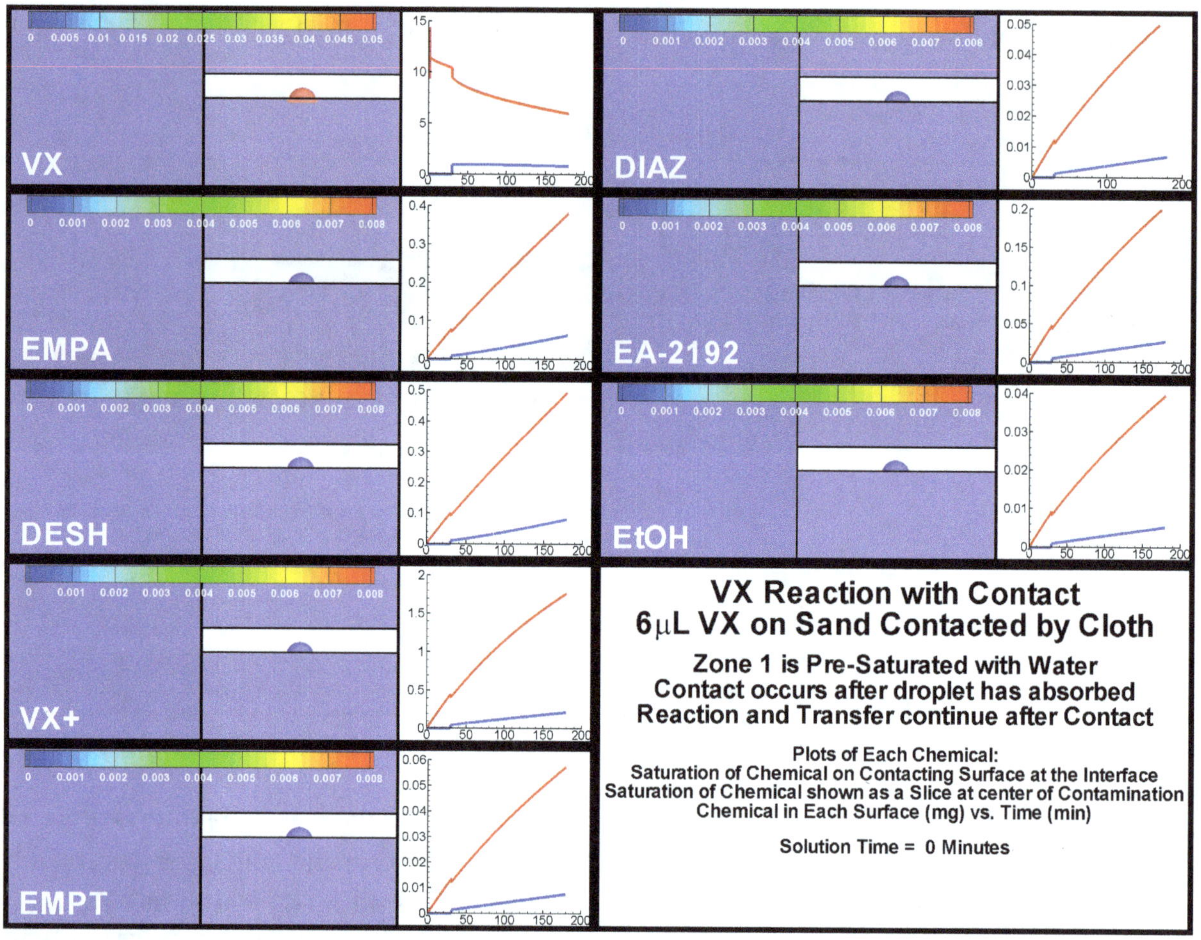

Figure 29. VX on wet sand and contact with another wet porous surface

In the following test case, liquid and vapor transport through a dual layer fabric is examined when each of the layers has a different permeability. The layer with small permeability (pores of 0.1 μm) is defined hereafter as the outer later and the layer with the larger permeability (pores of 5 μm) is defined as the inner later. The first experiment started with a 40 μL de-ionized water droplet on the inner-layer. Figure 30.a shows the saturation distribution and its migration to the outer layer. In the second experiment the droplet was placed on the outer layer and it did not penetrate to the inner layer as seen in Figure 30.b.

Computer simulations were performed with the same conditions. The evaporation module was activated to monitor the vapor concentration as well as liquid saturation. The MOCHA-DC prediction is shown in Figure 31. for the case of a droplet deposited on the low-porosity outer layer. The model indicates the liquid spreads laterally, instead of continuing to spread vertically, when it reaches the interface with the high-porosity inner layer. Due to the smaller pores, the capillary pressure at the outer layer is greater than that of the inner layer. It is seen that this spread pattern virtually eliminates the possibility of liquid passing through the higher porosity inner layer. Since the double layer material is thin, the aspect ratio of vertical direction in these plots is taken to be eight times greater than other directions to make the visuals clearer.

On the other hand, when the droplet is placed on the inner layer as seen from Figures 30a and 32, its preferred direction for movement is vertical when it reaches the interface between the layers, rather than spreading horizontally. In the top image of Figure 32., the droplet is initially placed on the inner layer and gradually is transported vertically, due to capillary pressure, toward the outer layer. The liquid keeps moving upward toward the outer layer saturating the upper portion of the fabric. The fluid therefore travels through both the layers. This behavior would facilitate the transport of fluid, such as sweat, through a garment constructed of such a material. As inner layer fluids reach the outer layer, they might evaporate in the surrounding environment. Conversely, when fluids contact the outer layer they do not reach the inner layer surface, thus preventing contamination of the inner surface. This result suggests it may be possible to eliminate the likelihood of hazardous liquid contact with skin if such a fabric were used for protective clothing.

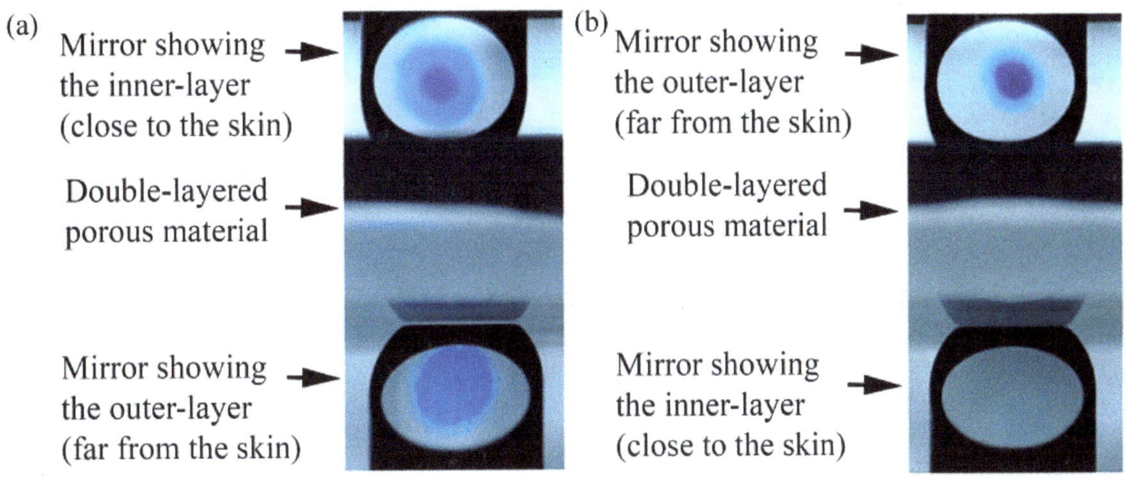

Figure 30. Sessile droplet positioned on the inner layer (a) and the outer layer (b) on a double-layered fabric. The outer layer pore size is order of magnitude smaller compared with the inner layer. (Adapted from Gat et al. 2013.)

The vapor phase also followed the same pattern. Although a small amount of vapor was predicted at the inner layer, it was significantly smaller than the amount of vapor present in the outer layer. Figure 33. shows the distribution of mole fraction of water vapor inside both layers. The vapor inside the bottom layer is one to two orders of magnitude less than the upper layer, implying that the capillary pressure difference between the two layers can be utilized as a tool to prevent volatile chemicals from reaching skin.

Given that the pore size influences capillary pressure, a range of porous glasses were coupled to study the transport behavior. Figure 34. demonstrates the experimental and model prediction for all combinations of the two pore sizes for the evaporation of deionized water placed on the bottom surface of the dual layer.

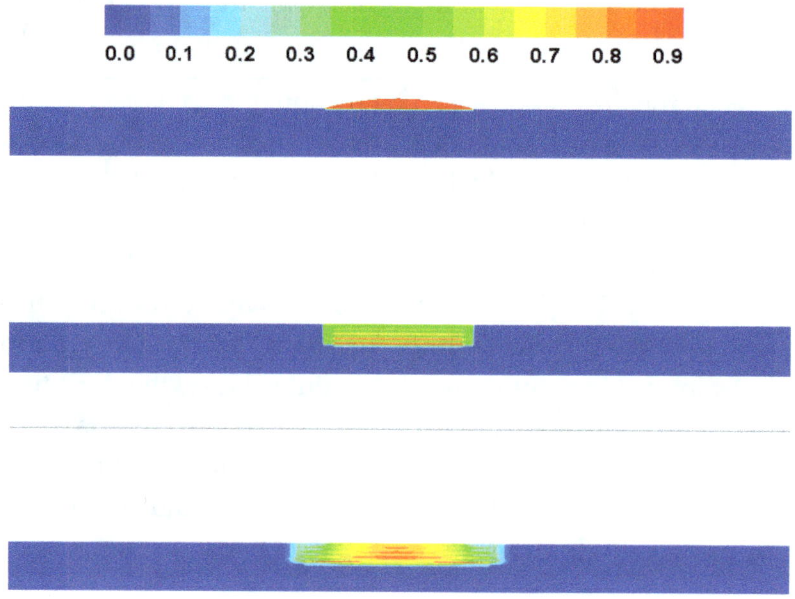

Figure 31. Liquid transport and spread from the upper layer to the lower layer in the model. This time series shows the saturation contours in the center plane of the domain as the droplet is absorbed and moves and spreads laterally, but not into the lower layer of the dual layer fabric. The saturation within the lower layer remains zero.

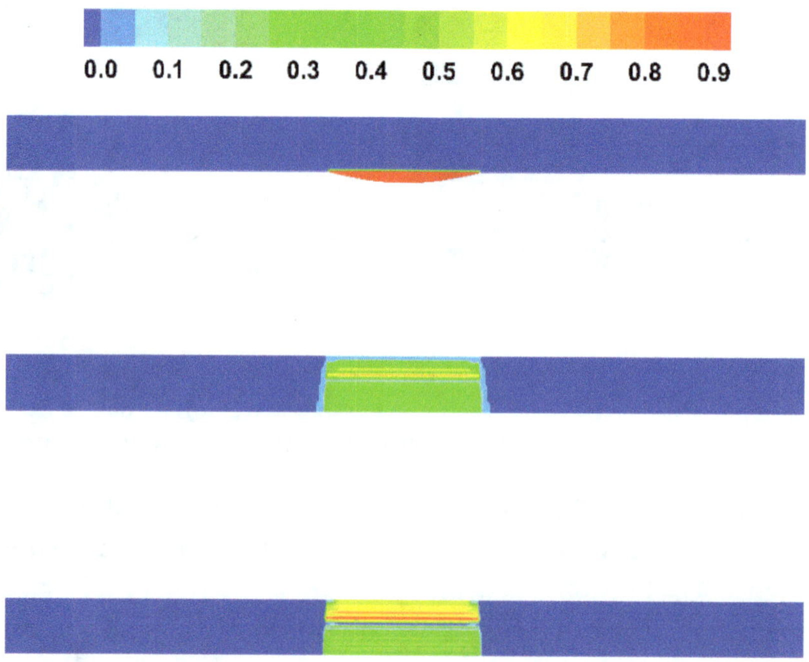

Figure 32. Liquid transport and spread from the lower layer to the upper layer as progressed in time. The figures show the saturation contours at the center plane of the domain. The top picture is the initial condition and as the droplet is absorbed it moves toward the upper layer rather immediately and then spread laterally.

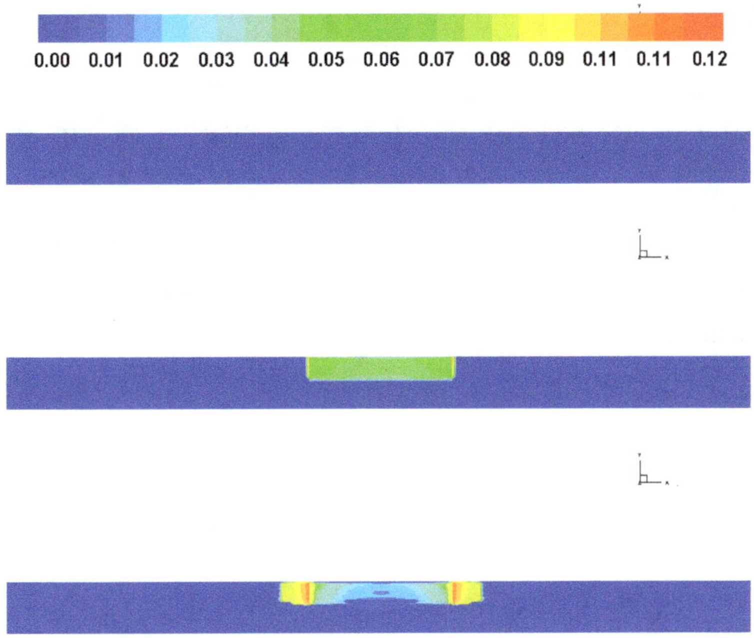

Figure 33. Image of the vapor transport and spread in the upper and lower layers when a droplet is placed on the upper surface. The vapor phase demonstrates a behavior similar to that of the liquid phase. The droplet is 40 μL water. The evaporation occurs somewhat faster at the edges of the wetted region. The mole fraction of the vapor phase is shown.

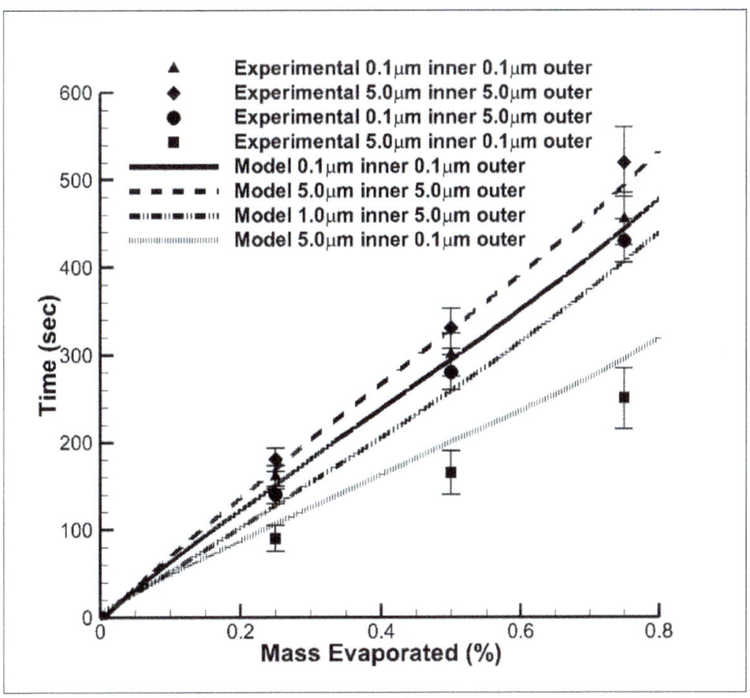

Figure 34. Evaporation vs. time for 200 μL ethanol droplet, positioned on the inner layer, into various combinations of nylon membranes (Scientific Tisch, TM) with 5 and 0.1 μm average pore size. Both model and experiment predict smaller evaporation time for the smaller inner pore size. For all cases the droplet was placed on the lower surface.

When the outer (upper) layer's pore size is an order of magnitude smaller than the inner (lower) layer's, the time for evaporation of a droplet on the lower surface was reduced by about $\approx 30-40\%$. This suggests that such a combination can be used to increase the liquid evaporation rate from the inner side of a protective fabric so as to enhance the pass through of perspiration. For the smallest outer layer pore size, the model predicts somewhat slower evaporation. This could be due to the mesh size, which may need to be refined for faster processes. However, since the results are fairly close to the experimental values, we decided not to repeat this calculation and emphasize the fact that there are a number of values to the dual layer fabric design that make it suitable for a variety of applications.

4. Nomenclature

4.1. General

A_s *Surface area*

a Pre-exponential coefficient

b Activation energy

C Mass fraction

c_p *Specific heat at constant pressure*

D Diffusion coefficient

E, F, H*Flux vectors in x, y, and z directions*

F Force

f Gibb's Free Energy

f(*) Unknown quantity for capillary pressure

G*Source term vector*

g Gravitational acceleration

h Height of apex of a sessile droplet

$\bar{h}$ *Molar enthalpy*

h Local height of the droplet*

H Enthalpy

h_{fg} *Latent heat of evaporation*

H Maximum vertical distance for boundary layer measurements

J Jacobian for the transformation

K Saturated permeability for isotropic medium

K_{eql} *Equilibrium reaction rate*

k Reaction rate

k_{eff} *Effective conductivity*

k_g, k_ℓ*Relative permeability in gas or liquid phases*

M Molecular weight

Molar concentration

m Mass

n Index of refraction

P Pressure

P_v *Vapor pressure*

$\dot{\mathbf{q}}$*Heat flux*

r Wetted area radius

R or R_s *Radius of curvature*

Re Reynolds number, u^*R/v

S Entropy

s Saturation, fraction of the pore volume occupied by a phase (or state)

Sc Schmidt number

t Time

T Temperature

$\mathbf{U}$*Conserved variables vector*

u^* *Friction velocity*

V Velocity

v and v′ Stoichiometric coefficients

$\tilde{v}$*Contra-variant vertical velocity*

u, v, w Velocity components

x, y, z Cartesian coordinate system

X Net rate of change in molar concentration

Greek Symbols

α *Thermal diffusivity*

β *Angle defined in*Figure 12.

ϕ *Porosity*

λ or λ_r *h/r*

λ_j or λ_{ij}*Reaction order*

μ *Viscosity*

ρ *Density*

$\dot{\rho}$*Rate of phase change*

σ *Surface tension*

τ *Shear stress*

$\forall$ *Volume*

ω *Species production/destruction rate*

Contact angle

$_a$ *Contact angle at which the evaporation topology changes*

$_i$ *Contact angle inside the pores*

ξ, η, ζ *Computational domain coordinates*

Subscripts

A Air

b Backward

c Capillary

eql Equilibrium

f Forward

fg Fluid to gas (evaporation)

g Related to the gaseous phase

i Liquid phase constituent

j Gas phase constituent

k Solid phase constituent

ℓ, liq Related to the liquid phase

m, n, ℓRefer to xyz for shear stress tensor

M Related to mass transfer

mix Mixture of liquid and gas phase

n Temperature dependence

o Initial

pt Partial torus

r Relative

s Surface of the droplet, or related to the solid phase

$s-\ell$ *Solid to liquid phase change*

$\ell-g$ *Liquid to gas phase change*

$s-g$ *Solid to gas phase change*

T Related to heat transfer

vap Vapor phase

$\forall$ *Volume*

w At the wall

∞ *Far field – Free stream*

Superscripts

~ *Vector*

UNI Unidirectional

per unit time

Acknowledgements

This project was collectively supported by The Air Force Research Laboratory, Human Effectiveness Directorate, Biosciences and Protection Division, Wright-Patterson AFB, US Army's Edgewood Chemical and Biological Center under contract, and The Defense Threat Reduction Agency (DTRA) under the contract HDTRA1-10-C-0064. The authors wish to thank all the above agencies and our CORs Drs. William Ginely, Sari Paikoff, and Brian Pate for their support and guiding this effort. Special thank also goes to Mr. Joseph B. Kiple of KASTLE Corporation for his involvement with the open air testing and sharing his experiences and vision with the entire team and Dr. D'Onofrio of ECBC for providing some of test data. We also appreciate Dr. Miroslav Skoumal and his team in Czech Republic for participating and sharing their state-of-the art open air testing facilities and their laboratories. A final word of thanks goes to the many Kettering University students who helped with the experimental and numerical studies.

Author details

Navaz Homayun[1*], Zand Ali[1], Gat Amir[2] and Atkinson Theresa[1]

*Address all correspondence to: hnavaz@kettering.edu

1 Kettering University, Flint, USA

2 Technion – Israel Institute of Technology, Haifa, Israel

References

[1] Davis, S. H., Hocking, L. M., (1999), "Spreading and imbibition of viscous liquid on a porous base," Phys. Fluids, 11, 48-57

[2] tarov, V. M., (2004), "Surfactant solutions and porous substrates: spreading and imbibition," Adv. Colloid Interface Sci., 111, 3–27

[3] Alleborn, N., Raszíller, H., Anthonissen, K., and Lievens, O. (2003), "Spreading and sorption of a droplet on a porous substrate," Fifth Europian Coating Symposium, Fribourg, Switzerland, pp. 246–251

[4] Alleborn N. and Raszíller, H. (2004), "Spreading and sorption of a droplet on a porous substrate," Chem. Eng. Sci., 59, 2071–2088

[5] Denesuk, M., Smith, G. L., Zelinski, B. J. J., Kreidl, N. J., and Uhlmann, D. R., (1993), "Capillary penetration of liquid droplets into porous materials," J. Colloid Interface Sci., 158, 114–120

[6] Holman, R. K., Cima, M. J., Uhland, S. A., and Sachs, E., (2002), "Spreading and infiltration of inkjet-printed polymer solution droplets on a porous substrate",J. Colloid Interface Sci., 249, 432–440

[7] Mantle M. D., Reis N. C., Griffiths R. F., and Gladden, L. F. (2003), "MRI studies of the evaporation of a single liquid droplet from porous surfaces," Magnetic Res. Imag., 21, 293–297.

[8] Bear, J. (1988), Dynamics of Fluids in Porous Media, Dover

[9] Vafai, K. (2000), Handbook of Porous Media, Marcel Dekker, Chapter 17

[10] Chen, Q. and Balcom, B. J. (2005), "Measurement of rock core capillary pressure curves using a single-speed centrifuge and one dimensional magnetic resonance imaging," Journal of Chemical Physics, Vol. 122, 214720

[11] Brooks, R. H. and Corey, A. T. (1964), "Hydraulic Properties of Porous Media," Colorado State University Hydrology Paper No. 3

[12] Van Genuchten, M. T. (1980), "A closed-form equation for predicting the hydraulic conductivity of unsaturated soils," Soil Sci. Soc. Am. J., 44, 892–898

[13] Leverett, M. C. (1941), "Capillary behavior in porous solids," AIME Trans., 142, 152–169

[14] Udell, K. S. (1985), "Heat transfer in porous media considering phase change and capillarity—the heat pipe effect," Int. J. Heat Mass Transfer, 28, 485–495

[15] Starov, V. M., Kostvintsev, S. R., Sobolev, V. D., Velarde, M. G., and Zhdanov, S. A., (2002), "Spreading of liquid drops over dry porous layers: Complete wetting case," J. Colloid Interface Sci., 252, 397–408

[16] Lenormand, R., Touboul, E., and C. Zarcone, (1988), "Numerical models and experiments on immiscible displacement in porous media," J. Fluid Mech., 189, 165–187

[17] Valavanides, M. S., and Payatakes, A. C. (2001), "True-to-mechanism model of steady-state two-phase flow in porous media, using decomposition into prototype flows," Adv. Water Resour., 24, 385–407

[18] Fatt, I. (1956), "The network model of porous media III. Dynamic properties of networks with tube radius distribution," Trans AIME, 207, 164–181

[19] Prat, M. (1993), "Percolation model of drying under isothermal conditions in porous media," Int. J. Multiphase Flow, 19,691–704

[20] Constantinides, G. N. and Payatakes, A. C. (1996), "Network simulation of steady-state two-phase flow in consolidated media," AIChE J., 42, 369–382

[21] Markicevic, B., D'Onofrio, T. G., and Navaz, H. K. (2010), "On spread extent of sessile droplet into porous medium: Numerical solution and comparisons with experiments,", Phys. Fluids, 22, Art. No. 012103

[22] Markicevic, B. and Navaz, H. K. (2010), "Primary and secondary infiltration of wetting liquid sessile droplet into porous media," Trans. Porous Media, 85, 953–974

[23] Markicevic, B. and Navaz, H. K. (2010), "The influence of capillary flow on the fate of evaporating wetted imprint of the sessile droplet in porous medium," Phys. Fluids, 22, Art. No. 122103

[24] Lago, M., and Araujo, M. (2001), "Capillary rise in porous media," J. Colloid Interface Sci., 234, 35–43

[25] Markicevic, B. and Navaz, H. K. (2009), "Numerical solution of wetting fluid spread into porous media", Int. J. Num. Meth. Heat Fluid Flow., 19, 521–534

[26] Popov, Y. O. (2005), "Evaporation deposition pattern: Spatial dimensions of the deposit," Phys. Rev., E 71, 036313-1–036313-17

[27] Hu, H. and R. G. Larson, (2002), "Evaporation of a sessile droplet on a substrate," J. Phys. Chem., B, 106, 1334–1344

[28] Baines, W. D. and James, D. F. (1994), "Evaporation of a droplet on a surface," Ind. Eng. Chem. Res., 33, 411–416

[29] Winter, S., Karlsson, E., and Nyholm, S. (1990). Models for the Evaporation of Chemical Warfare Agents and Other Tracer Chemicals on the Ground. A Literature Review, Department of NBC Defense Report, S-901 82 Umea, Sweden, Art Hin Fokke, Oeseburg: TNO, Prins Maurits Laboratorium, The Netherland..

[30] Navaz, H. K., Dang, A. L., and Rangel, R. H. (1992), "Numerical analysis of bipropellant combustion in liquid thrust chambers by an eulerian-eulerian approach," AIAA Paper No. 92-3735.

[31] Ward, P., Collings, and Hay, N. (1985), "The comparison of simple models of turbulent droplet diffusion suitability for use in computation of spray flames," J. Engng Gas Turbines Power, Transactions of ASME, 107, 690–694

[32] Metzger, M. M. and Klewicki, J. C. (2001),"A comparative study of near wall turbulence in high and low Reynolds number boundary layers," Phys. Fluids, 13(3), 692–701

[33] Klinger, B. A., Huang, B., Kirtman, B., Schop, P., and Wang, J. (2004), Monthly Climatology of Friction Velocity Cubed from SSM/I Wind Data, COLA Technical Report

[34] Weber, R. O. (1999), "Remarks on the definition and estimation of friction velocity," Boundary-Layer Meteorology 93,197–209

[35] Frandsen, S. and Madsen, P. H. (2005), Spatially Average Of Turbulence Intensity Inside Large Wind Turbine Arrays, Riso National Laboratory, 4000 Roskilde, Denmark

[36] D'Asaro, E. A. and Geoffrey, G. T. (1997), "Turbulence intensity measurements in a wind-driven mixed layer," J. Phys. Oceanogr., 27,2009–2022

[37] Zarrouk, S. J. (2004), Simulation of Complex Multi-Phase, Multi-Component Reacting Flows in Porous Media, Ph.D. thesis, The University of Auckland

[38] Zarrouk, S. J. (2008), Reacting Flows in Porous Media: Complex Multi-Phase, Multi-Component Simulation, VDM Verlag

[39] Xu, T., Sonnenthal, E. L., Spycher, N., and Pruess, K. (2004), TOUGHREACT User's Guide: A simulation program for non-isothermal multiphase reactive geochemical transport in variably saturated geologic media. Report LBNL-55460, Lawrence Berkeley National Laboratory, Berkeley, CA

[40] Lichtner, P.C. (1985), "Continuum model for simultaneous chemical reactions and mass transport in hydrothermal systems," Geochem. Cosmochem. Acta49, 779–800

[41] Steefel, C. I. (2001), GIMRT, Version 1.2: Software for modeling multicomponent, multidimensional reactive transport, User's Guide. Report UCRL-MA-143182, Lawrence Livermore National Laboratory, Livermore, California

[42] Butt, H.-J. and Kappl, M. (2008), "Normal capillary forces," Adv. Colloid Interface Sci.146, 48

[43] Carter, W. C. (1988), "The forces and behavior of fluids constrained by solids," Acta Metall.36, 2283

[44] Mazzon, D. N., Tardos, G. I., and Pfeffe, R. (1987), "The behavior of liquid bridges between two relatively moving particles," Powder Technol.51, 71

[45] Dodds, S., Carvalho, M. D. S., and Kumar S. (2009), "Stretching and slipping of liquid bridges near plates and cavities," Phys. Fluids,21, 092103

[46] O. Pitois, O., Moucheront, P., and Chateau, X. (2000), "Liquid bridge between two-moving spheres: An experimental study of viscosity effects," J. Colloid Interface Sci. 231, 26

[47] Meurisse, M.-H. and Querry, M. (2006), "Squeeze effects in a flat liquid bridgebetw-een parallel solid surfaces," J. Tribol.128, 575

[48] De Sauza, E. J., Gao, L., McCarthy, T. J., Arzt, E., and Crosby, A. J. (2008), "Effectof contact angle hysteresis on the measurement of capillary forces," Langmuir24, 1391

[49] Ngo, M.A., O'Malley, M., and Maibach, H. I. J. (2010), "Percutaneous absorption and exposure assessment of pesticides," Appl. Toxicol., 30(2),91–114

[50] Kilpatrick, W. T., Ling, E. E., Hin, A. R. T., and Brevett, C. A. S., Fagan, M. W., and Murdock, W. P., Jr. (2004), "Testing requirements for predictive model development using environmental wind tunnels," (HEDs)," AFRL-HE-WP-TR-2004-DRAFT, Air Force ResearchLaboratory Report

[51] Savage, J. J. (2006). "Agent Fate Program Overview," 2006 Scientific Conference on Chemical Biological Defense Research, Hunt Valley, MD

[52] Savage, J. J. (2006), "Agent Fate Program Overview," 74th Military Operations Research Society Symposium (MORSS), Colorado Springs, CO

[53] Munro, N. B., Talmage, S. S., Griffin, G. D., Waters, L. C., Watson, A. P., King, J. F., and Hauschild, V. (1999),"The sources, fate, and toxicity of chemical warfare agent degradation products, "Environ. Health Perspec., 107, 933

[54] Navaz, H. K., Zand, A., Markicevic, B., Herman, M., Atkinson, T., Li, Nowakowski, A., Gheresus II, P., Rothstein, M., Kiple, J., and Middleton, V. (2011) Agent Fate Model Integration, Final Report, Prepared for Edgewood Chemical and Biological Center (ECBC)

[55] Navaz, H. K., Zand, A., Atkinson, T., Nowakowski, A., Kiple, J., Kamensky, K., and Jovic, Z. (2013), Agent Fate Modeling-COMCAD Theory and Analysis, Code Structure, and Users Guide (Volumes I, II, and III), Final Report prepared for Edgewood Chemical and Biological Center (ECBC), January 7

[56] Navaz, H. K., Zand, A., Dang, A. L., Gat, A., Gharib, M., Atkinson, T., Nowakowski, Kamensky, K., and Jovic, Z. (2014), Predictive Model for Assessment of Chemicals on and in Surfaces vs. Chemicals Available for Contact and Transport—Volume I: Theory and Analysis,Final Report prepared for Defense Threat Reduction Agency (DTRA)

[57] Navaz, H. K., Chan, E., and Markicevic, B. (2008), "Convective evaporation of sessile droplets in a turbulent flow—comparison with wind tunnel data," International Journal of Thermal Sciences 47, 963–971

[58] Navaz, H. K., and Dang, A. D. (1994),"The Development of the Liquid Thrust Chamber Performance (LTCP) Code for Turbulent Two-Phase Flow Combustion of Dense Sprays," Technical Report prepared for NASA/MSFC, Contract No. NAS8-38798

[59] Kuo, K. K. (1986), Principles of Combustion, Wiley, Chapters 6 and 8

[60] Chiu, H. H. and Zhou, X. Q., (1983) "GEMCHIP-CODE: Spray Combustion Process in Air Breathing Propulsion Application," University of Illinois at Chicago

[61] Chiu, H. H. (1989), "Theory of Droplet: I Renormalization Laws of Droplet Vaporization in Non-Dilute Sprays," Constitutive Relationships and Models in Continuum Theories of Multiphase Flows, NASA Conference Publication 3047

[62] Nikolopoulos, N., Theodorakakos, A., and Bergeles, G., (2007), "A numerical investigation of the evaporation process of a liquid droplet impinging onto a hot substrate", Int. J. Heat Mass Transfer, 50, 303–319

[63] Zand, A., Sikorski, Y., Sanders, M., and Navaz, H. K. (2007) "A simple laboratory experiment for the measurement of single phase permeability," J. Physical Nat Sci, 1(2), pp 1-10

[64] Navaz, H. K., Markicevic, B., Zand, A., Sikorski, Y., Chan, E., Sanders, M., and T. D'Onofrio (2008), "Sessile Droplet spread into porous substrates—Determination of capillary pressure using a continuum approach," J. Colloid Interface Sci., 325, 440–446

[65] Treyball, R. E. (1980), Mass-Transfer Operation, McGraw Hill.

[66] Coblaş, D., Broboană, D., Bălan, C., and Hejjam, M. (2013) " Numerical Simulation of Constant Velocity Squeeze Flow," U. P. B. Bul., Series D, Vol. 75, Issue 2

[67] Longley, L. E., Dooley, E., Givler, D. M., Napier, III, W. J., Chaudhury, M. K., and Daniel, S. (2012) "Drop motion induced by repeated stretching and relaxation on a gradient surface with hysteresis," Langmuir 28 (39), pp 13912–13918

[68] Gordon, S., and McBride, B. (Oct. 1994) Computer Program for Calculation of Complex Chemical Equilibrium Composition and Applications, NASA Reference Publication 1311

[69] McBride, B. and Gordon, S. (Nov. 1992) Computer Program for Calculating and Fitting Thermodynamic Functions, NASA Reference Publication 1271

[70] McBride, B., Gordon, S., and Reno, M. (Oct. 1993) Coefficients for Calculating Thermodynamic and Transport Properties of Individual Species, NASA Technical Memorandum 4513

[71] Benson, S. W., and Fueno, T. (1962) "Mechanism of atom recombination by consecutive vibrational deactivation," J. Chem. Phys., 36, 1597

[72] GRIDGENTM, (2009), Reliable CFD Meshing, Pointwise Corporation, Fortworth, TX

[73] Coons, S. A. (1966), "Computer graphic and innovative engineering design—super-Sculptor," Datamation,12(5) 32–34

[74] Reis, N., and Griffiths, R. (2003), "Investigation of the evaporation of the embedded liquid droplets from the porous surfaces using magnetic resonance imaging," Int J Heat Mass Transfer

[75] Brevett, C. A. S., Kenneth B., Sumpter, J. P., and Robert G. (2009), "Evaporation and degradation of VX on silica sand," J. Phys. Chem.,113(16), pp 6622-6633.

[76] Brevett, C, Sumpter, K, Pence, J. et al. (2009) "Evaporation and degradation of VX on silica sand", J. Phys. Chem., 113(16), 6622–6633

[77] Wagner, G. W., O'Connor, R. J., Edwards, J. L., and Brevett, C. A. (2004) "Effect of drop size on the degradation of VX in concrete," Langmuir, 20, 7146-7150

3

Effect of Certain Ethylene Oxide Heterogeneous Heterobifunctional Acyclic Oligomers (HEHAO) on Wetting

Calin Jianu

Abstract

The effect of ethylene oxide (EO) heterogeneous heterobifunctional acyclic oligomers (HEHAO) on wetting represents a continuation of the review of the research partly published by the author on ethylene oxide (EO) homogeneous heterobifunctional acyclic oligomers (HOHAO). Heterogeneous polyoxyethylene chains obtained by the anionic polymerization of EO ("anionic ring-opening polymerization") with a certain polydispersity degree are hydrophilic, flexible (specific spatial conformation), biocompatible "bridges," without toxicity. After the purification of polyethoxylated higher alcohols (technical products), heteroderivatization performed by an adapted classical reaction scheme led to the colloidally evaluated HEHAO series.

Keywords: Polyoxyalkylene ether acid with higher alkyl group, alkyl polyethoxy carboxylate surfactants, carboxy propylated nonionic alkyl surfactants, wetting, ethylene oxide heterobifunctional acyclic oligomers

1. Introduction

Ethylene oxide heterogeneous heterobifunctional acyclic oligomers (HEHAO) represent a class of relatively recent (ca. nine decades) "niche" surface-active compounds, if we refer only to molecular architectures considered in this chapter for colloidal study. The material is a continuation of the research partly published by the author and collaborators in the period

1968–2014 on ethylene oxide homogeneous heterobifunctional acyclic oligomers (HOHAO) (Figure 1a). The difference between the two types of acyclic oligomers lies in the fact that HOHAO [1] have in their structure "homogeneous" polyoxyethylene chains (with strictly determined oligomerization degrees (n)), built by the adapted Williamson method [2–4], while HEHAO (Figure 1b) have "heterogeneous" polyoxyethylene chains [1] and oligomerization degrees (n*), respectively, in a statistical distribution dependent on the operating parameters accessed.

a) $R_1{-}O{-}(CH_2CH_2O)_n{-}R_2$ (HOHAO)

n = *homogeneous* oligomerization degree (strictly monitorized value) (3,6,9,12,16,20)
R_1, R_2 = anionic, aliphatic, aromatic or mixed derivation terminals

b) $R_1{-}O{-}(CH_2CH_2O)_{n^*}{-}R_2$ (HEHAO)

n* = *heterogeneous* oligomerization degree (narrow molecular weight distribution- NMWD) (3,6,9,12,16,20)
R_1, R_2 = anionic, aliphatic, aromatic or mixed derivation terminals

Figure 1. General structure of heterobifunctional acyclic oligomers of ethylene oxide

Interest in polyoxyethylene chains (PEO) was generated due to their conformational and colloidal properties. The reactivity of the three-atom ether heterocycle (oxirane), also founded on the "ring tension theory," favors the nucleophilic attack of organic compounds with hydroxyl, thiol, primary and/or secondary amine, etc., function, with breaking of the C–O bond of the oxirane ring [1,5–7].

A narrow-range ethoxylated alcohol, also called "packed ethoxylated" alcohol, has a distribution curve that is narrower than the equivalent standard alcohol ethoxylate and a considerably lower content of unreacted alcohol. Reactions of ethylene oxide with fatty alcohols proceed in the presence of sodium metal, sodium hydroxide, and boron trifluoride and are used for the synthesis of surfactants (including HEHAO) [1,4,5,7].

To facilitate text comments on HEHAO compounds, their chemical names have been abbreviated. The main organic functions are symbolized by the initials of their chemical names (*e.g.*, PN–propionitrile, PD–propionamide [3], PC–β-substituted propionic acids [3], EP–primary ethylamine), the main hydrocarbon-chain homologues by the initials of their common names (*e.g.*, LM–lauryl/myristyl, CS–cetyl/stearyl), followed in parentheses by the ratio (7/3) signifying their relative distribution. The hydrocarbon chains attached later by synthesis are symbolized by the number of carbon atoms contained, for example, EC-1.1.16 [3] in the cationic structure, where we find two methyl groups indicated by 1.1 and a hexadecyl chain.

The actual distribution of PEO chain homologues, and hence of the oligomerization degrees ($n_{average}$), changes in the series $n_{average}$=3–18 from advanced symmetry for $n_{average}\leq$8–9 to pronounced asymmetry for $n_{average}\leq$9–18 [1,4,5,7].

Heterogeneous polyoxyethylene chains obtained by the anionic polymerization of ethylene oxide ("anionic ring-opening polymerization") with a polydispersity degree M_w/M_n<1.1 are hydrophilic, flexible (specific spatial conformation), biocompatible "bridges" [1].

For heterogeneous polyoxyethylene (PEO) chains, the statistical distribution quantitatively expressed through equations: Natta, Weibull/Nycander/Gold, Natta/Mantica, Poisson, etc., is accepted.

During our research, the results obtained contributed decisively to the confirmation of direct participation of PEO chains in nucleophilic addition reactions (cyanoethylation, amidoethylation) [1,3,4] of polyethoxylated higher alcohols purified of free higher alcohols, polyethylene glycols (PEG_n) and water, when processing yields under similar conditions increase proportionally with the oligomerization degree (n*) of the PEO chain [1,3,4] (for further details, see section 4).

An overview of the theoretical and practical colloidal competences regarding several categories of classical ionic (anionic, cationic) and nonionic surface-active structures suggested the idea of making a new class of hybrid surface-active compounds of the nonionic-ionic HEHAO type with synergistic effects. For this purpose, a classical reaction scheme was followed adapted to the HEHAO series [3,4].

The literature reports similar structures of the mainly nonionic–ionic type with a length of the actual polyoxyethylene chain (n*=3). In this chapter, the range of HEHAO surface-active structures is extended, using the reaction scheme not reported previously and higher polyoxyethylene chains (n*=3–20) (Table 1) [8–20].

In the literature, 1939 is accepted as the year of the first reporting of β-alkyl (alkylaryl)-X (oxygen, sulfur, nitrogen) polyethyleneoxy (n*=3–50) functional derivatives of saturated acids, through the patent granted to H. Haussmann at I.G. Farben Industrie.

Traditional soap was gradually replaced in sanitation recipes with similar polyethyleneoxy carboxylic surface-active structures, more soluble and unaffected by the hardness of washing waters. For their synthesis, new methods have been accessed, reported primarily in the patent literature [21-32,47,48].

2. Surface-active nonionic–ionic (anionic and cationic) colloids in the HEHAO category

The research conducted focused exclusively on nonionic–ionic (anionic and cationic) surface-active structures, where the "bridge" between the heterogeneous polyoxyethylene chain (PEO) and the polar (hydrophilic) anionic or cationic group is represented by the hydrophobic moiety R_2 = ethyl/propyl.

In these molecular architectures, the fragment R_1 (the determinant hydrophobic group) [33] is the lauryl/myristyl (7/3) (LM) and cetyl/stearyl (7/3) (CS) radical, respectively, and the fragment F_2 (R_2') as the hydrophilic ionic polar group includes functional groups [33].

In the reaction scheme accessed [5], heterogeneously polyethoxylated (n*=3, 6, 9, 12, 16, 20) $C_{12}H_{25}/C_{14}H_{29}$ (7/3) (LM) higher alcohols and heterogeneously polyethoxylated (n*=3, 6, 9, 12,

16, 20) $C_{16}H_{33}/C_{18}H_{37}$ (7/3) (CS) higher alcohols, respectively, were technical products obtained by the base-catalyzed reaction of the higher alcohols above with ethylene oxide [3,40].

They are mixtures of statistically distributed PEO chain homologues along with non-ethoxylated higher alcohols, polyethylene glycols as by-products, and accidental traces of water. These impurities possessing hydroxyl groups can unwantedly participate simultaneously in the cyanoethylation or amidoethylation processes of heterogeneously (n*) polyethoxylated higher alcohols, generating impurities and serious purification difficulties [4]: additional consumption of acrylic monomer (AN), low useful nucleophilic addition yields, and finally limited purity of the HEHAO obtained.

To eliminate these impediments, the raw materials (technical products) were purified [4] by repeated liquid/liquid extractions in various solvent systems. In the synthesis of HEHAO were accessed purified technical products [heterogeneous polyethoxylated (n*) higher alcohols].

Preliminary cyanoethylation attempts of higher alcohols ($C_{12}H_{25}$–$C_{18}H_{37}$) and of polyethoxylated (n*=3–20) higher alcohols ($C_{12}H_{25}$–$C_{18}H_{37}$) [3,4] revealed that the oligomerization processes of the acrylic monomer, especially at temperatures exceeding 40°C, have a significant share. Nucleophilic addition yields decrease considerably and the reaction tends to proceed vehemently, highly exothermically, with the formation of brown-reddish resins (PPN), soluble in polar solvents (water, acetone, etc.), or yellow solid products $R(EO)_{n*}PPN$ soluble in medium-, low-polarity solvents [4].

Initially we worked without an acrylic monomer oligomerization inhibitor, in order to follow the evolution of cyanoethylation and oligomerization yields. Afterward under the optimum parameters established was introduced as an oligomerization inhibitor $FeSO_4$ [3–5].

The content of oligomers increases with temperature and excess monomer. Compared with cetyl/stearyl alcohol it is noted that cyanoethylation yields are higher at the same temperature value, even below 45°C. This behavior suggests that the cyanoethylation reaction has a reversible character, the polyethoxylated chain favors the addition, and the hydrocarbon chain with its increase, reduces the cyanoethylation yields.

In the research, the formation of AN oligomers was avoided by introducing anhydrous ferrous sulfate ($FeSO_4$) as an inhibitor of the polymerization of the acrylic monomer. For 1% addition of oligomerization inhibitor, the cyanoethylation yield of lauryl/myristyl alcohol (7/3) increases by more than 10%, without the formation of oligomers. Similar results are obtained in the series of polyethoxylated (n*=3–20) higher alcohols ($C_{12}H_{25}$–$C_{18}H_{37}$).

By gradually adding polyethyleneglycols (PEGn*) (n*=9) in the cyanoethylation phase, it was found that:

- -the increase of the oligomer content, because alkaline polyethyleneglycolates (n*=9), as bidentate nucleophiles, have higher basicity than alkaline higher alkoxides;
- -the increase of the content of β-alkyl-(LM)-propionitriles, proportional to the quantity of polyethyleneglycols added due to inter-phase catalysis [4].

The polyethoxylated $C_{12}H_{25}/C_{14}H_{29}$ (7/3) (LM) (n*=3, 6, 9, 12, 16, 20) and $C_{16}H_{33}/C_{18}H_{37}$ (7/3) (CS) (n*=3, 6, 9, 12, 16, 20) higher alcohols can be considered as monoderivatized polyethyleneglycols (primary PEGylation fragment). There are many possibilities for modifying the determinant hydrophobic fragment between the $C_{12}H_{25}/C_{14}H_{29}$ and $C_{16}H_{33}/C_{18}H_{37}$ [40].

The inclusion of β-alkyl ($C_{12}H_{25}/C_{14}H_{29}$) polyethyleneoxy (n*) derivatives [3,4] in the category of surface-active agents (HEHAO) involves the nomination of the hydrophilic (lipophobic)/ hydrophobic (lipophilic) polar/nonpolar (amphiphilic) domain, which next determines their colloidal character.

Due to their composition, specific primary and secondary (conformational) structures, HEHAO have a pronounced heterogeneous character. The idea of their basic colloidal characterization becomes indicative, being the resultant of the colloidal manifestation of 24 unitary homologous structures (4 hydrophobic R series, each with 6 hydrophilic PEO series) [1,3,4,40].

Synergistic cumulation in the same surface-active structural architecture of the ionic (anionic, cationic) functional group (hydrophilic) with one or more polyoxyethylene chains (hydrophilic) with different oligomerization degrees (n*) led to the HEHAO [1,4].

3. Structure–hydrophilicity (wetting competences) correlations in the HEHAO homologous series studied

The main colloidal characteristics of the surface-active compounds (technical products) depend on their structure and heterogeneous composition [3,40]. Laboratory evaluations performed on the homologous series (n*=0–20) of the two sets of HEHAO purified prior to the synthesis fall into this casuistry.

Numerous industrial applications of HEHAO, due primarily to their varied composition and structure, are based on several colloidal phenomena: wetting, colloidal (micellar) solubility, adsorption, chemical reactions at the interface (chemosorption) and actual penetration into the dirt or support. Determining the value of use of HEHAO is most often more important than the classical chemical analysis, if we consider that most marketed products are mixtures with a wide distribution of hydrocarbon-chain R and polyoxyethylene-chain PEO homologues.

Knowing the main surface-active properties [1], are suggested directions for the application of the HEHAO synthesized, dependent on the structural elements of the respective homologous series (the hydrophilic–hydrophobic index, HLB). The use of the most suitable surface-active evaluation methods can generate some justified comments, both by the reproducibility of recording the basic colloidal phenomenon, and by the lack of widely recognized, accepted working practices. Therefore, the values commented have an indicative character, even though they represent the result of the statistical processing of a large number of experimental measurements. Were preferred standardized protocols supplemented with ISO recommendations, where applicable. For the estimation of the wetting effect [49], the data obtained correctly characterize the physico-chemical aspect, but less so the mechanical aspect of the phenomenon, which differs most of the times through the wide variety of equipment employed in the processing, but also by the nature of the supports or the difference between the artificial and the natural polluted support. Laboratory measurements allowed the establishment of structure–surface-activity correlations, following that the absolute, comprehensive assessment on the scope to involve the investigation under conditions of actual equipment, support, float, etc. The main surface-active properties previously considered [3,40] were: surface tension (σ) and critical micelle concentration (CMC), foaming with its three aspects [foamability (F), foam stability (FS), foam density (FD)], wetting effect (U), the softening–antistatic effect (LA) and the washing effect (WE), respectively [1,3,4,7,18,19,33].

The HLB index, the balance between the hydrophilic (lipophobic) and hydrophobic (lipophilic) properties of the HEHAO, and in particular its controlled modification by structural elements allowed to obtain a wide range of HEHAO with directed potential of use.

Surface tension is dependent on the structural characteristics of the studied heterogeneous HEHAO [1,40]. Increasing the oligomerization degree of the PEO chain ($n>6$) induces an increased capacity of surface tension reduction, simultaneously with decreasing the effectiveness of adsorption at the water/air interface [1,40].

The evaluation of wetting competences in the homologous HEHAO series mentioned (Tables 1–3) enabled the formulation (by extrapolation) of structure–wetting competences correlations, knowing a priori the basic colloidal characteristics (surface tension, critical micelle concentration, conformational behavior) [1,3,4,36–38,41–46]. Based on these in this chapter could be proposed and subsequently argued mechanisms of action, suggestively presented schematically (section 4).

Wetting, as an interface phenomenon indicating the magnitude and rapidity of establishing contact between a support and the aqueous float, depends on the ability of the surface-active structure to reduce under dynamic conditions the surface tension at the support–float interface, the molecule's mobility in the float and its capacity for adsorption at the interface. Since the adsorption capacity is also a function of the nature of the support (dielectric constant), are distinguished hydrophilic interfaces (dielectric constant lower than that of water, *e.g.*, cellulose, cotton) easily wetted by water and hydrophobic interfaces (dielectric constant higher than water, *e.g.*, synthetic fibers) little wetted by water or polar fluids. Along with the nature of the substrate, in the adsorption process can be distinguished two more parallel, concurrent phenomena, dependent on superficial structure and the permittivity of the environment: desorption (water adsorption at the interface) and the adsorption proper. The desorption is accompanied by a decrease of the water concentration in the adsorption layer and the local concentration of the colloidal solution is the more advanced as the hydrophilicity of support is higher.

Interpreting the values obtained (Table 1), the following observations can be formulated: at room temperature, on hydrophilic surfaces as such or mixed (B, B–FR, B–PES), due to high desorption and adsorption at the interface, the wetting–lifting effect is pronounced (the wetting power values are of the order of seconds); under the same environmental conditions, on hydrophobic surfaces (PES, PNA, PES-Ac) the lower desorption and adsorption capacity

No.	R	N*	Nature of solution	Float concentration (g/L)	Wetting time (sec.)				Textile support
					Untreated support*	Treated support			
						20°C	40°C	60°C	
1	12–14	3	HCOOH	2	148	5.8	5.6	3.5	PES
2	16–18	12	HCOOH	5[x]	148	5.5	5.4	3.3	
3	12–14	6	AcOH	0.5	148	6.6	5.4	4.3	
4	12–14	9	AcOH	2	148	6.8	5.6	4.4	
5	16–18	3	AcOH	5	184	4.3	3.1	0.8	PES-AC
6	12–14	3	HCOOH	5[x]	184	4.5	3.4	1.2	
7	12–14	6	HCOOH	2	184	4.6	2.4	1.3	
8	12–14	9	AcOH	0.5	184	4.9	2.7	1.5	
9	12–14	3	HCOOH	2	60	2.3	2.2	1.0	B
10	16–18	12	AcOH	5	60	1.4	1.1	0.8	
11	12–14	3	HCOOH	0.5	60	1.8	1.1	0.9	
12	12–14	6	HCOOH	5[x]	60	1.4	1.2	1.0	
13	12–14	12	HCOOH	2[x]	78	3.9	2.7	1.5	B-PES
14	16–18	3	HCOOH	5[x]	78	4.2	3.0	2.9	
15	16–18	6	HCOOH	0.5	78	4.1	4.0	2.8	
16	16–18	9	HCOOH	0.5	78	4.9	3.8	2.7	
17	16–18	12	HCOOH	5[x]	4200	11.2	11.0	10.8	B-FR
18	16–18	3	AcOH	2	4200	11.8	11.5	11.3	
19	16–18	6	AcOH	0.5	4200	20.6	18.4	11.4	
20	16–18	9	AcOH	5	4200	11.5	11.3	11.1	
21	16–18	12	HCOOH	5	116	12.3	12.1	10.7	PNA
22	16–18	9	AcOH	5[x]	116	12.2	12.0	10.6	
23	16–18	3	AcOH	2	116	12.0	9.9	9.7	
24	12–14	9	AcOH	2	116	12.1	11.9	10.8	

*float hardness 5 dG (German degrees)

**not cleaned of softening/antistatic and/or mercerization products

***PES – polyester; PES–AC – polyester acetate; B – cotton; B–PES – cotton polyester; B–FR – cotton frotir; PNA – polyacrylonitrile.

Table 1. Wetting effect of β-alkyl polyethyleneoxy propionamides in formic and acetic solutions (Ac) [R-O-$(CH_2CH_2O)_{n*}$-$CH_2CH_2CONH_2$]

determines the decrease of the wetting effect. Increasing the temperature of the float reduces both interface phenomena, but increases the mobility of HEHAO structures and their ability to reduce surface tension [3,4], so that overall the wetting effect is optimized; concentrations of HEHAO in the treatment floats below CMC provide a good wetting effect, especially at elevated temperatures, since at 20°C the mobility of β-substituted propionamides is greatly reduced.

β-Alkyl-polyethyleneoxy-propionamides in acetic solutions exhibit a greater wetting effect than in formic solutions, due to the difference in solubility in the two organic acids (solvation phenomena). Because β-substituted propionamides in acetic solutions have a better capacity of surface tension reduction [3,4], the wetting effect is more pronounced. Increasing the length of the polyoxyethylene chain increases the hydrophilicity, but lowers the wetting capacity by increasing the hydration degree and reducing the mobility in the float. Increasing the hydrocarbon chain provides a good wetting effect by the advanced reduction of surface tension [4]. The hardness of treatment floats in the case of polyoxyethylene chains larger than 12 oxyethylene units (EO) does not affect the wetting effect both in the case of the formic and acetic solution, due to the coordination phenomena of alkaline earth cations (Table 1).

The analysis of wetting characteristics (Table 2) in the category of nonionic soaps leads to observations similar to those made for β-substituted propionamides. Are noted the optimum wetting properties of the series of anionic–cationic structures of the β-lauryl/myristyl (7/3) polyethyleneoxy (n*=3) ethylammonium β-alkyl ($C_{12}H_{25}$–$C_{18}H_{37}$) polyethyleneoxy (n*=3–20) propionates [HLM–$(EO)_3^-EP^+$] type, probably due to the optimum capacity of surface tension reduction correlated with the high degree of "packing" at the interface (good entropic factor). Compared to the acetic or formic solutions of β-substituted propionamides, nonionic soaps are more efficient in the wetting operation in the same conditions of environment and structure, a fact that recommends them in washing (cleaning) operations.

The analysis of the wetting–lifting effect in these surface-active structures (Table 3) reveals structure–effect parallelisms similar to those observed previously, namely that shifting the polar hydrophilic group toward the center of the hydrocarbon chain R together with increasing the length of the latter, through the unfavorable entropic factor, determines the decrease of the wetting effect parallel with the surface tension reduction capacity.

Increasing the length of the polyoxyethylene chain increases the hydrophilicity, the hydration degree, and drastically reduces the mobility in the float. Along with the unfavorable entropic factor, it negatively affects the capacity under dynamic conditions of reducing the tension at the support–float separation interface. Overall, though the quaternary salts presented are better wetting agents than the nonionic–anionic structures analyzed, due to the affinity of the cationic polar group toward supports with anionic character, independent of their degree of hydrophobicity (cellulose, cotton, natural, or synthetic fibers), which secure the cationic structure at the support–float interface and facilitates the reduction of surface tensions.

No.	R	n*	Nature of cation	Float concentration (g/L)	Wetting time (sec.) Untreated support**	Treated support 20°C	40°C	60°C	Textile support
1	12–14	3	Na	5	148	3.2	2.6	1.5	PES
2	16–18	12		2^x	148	3.3	2.4	1.1	
3	12–14	6	Na	0.5	148	3.6	2.8	1.9	
4	12–14	3	HLM-$(EO)_3$-EP^+	2	148	1	1	1	
5	12–14	0	Na	0.5	184	2.8	1.8	0.9	PES-AC
6	16–18	12	Na	2	184	2.3	1.7	0.8	
7	12–14	20	Na	2	184	3.8	2.2	1.3	
8	12–14	3	HLM-$(EO)_3$-EP^+	2	184	1	1	1	
9	16–18	0	Na	2	60	1.1	1	1	B
10	16–18	9	Na	2	60	1.9	1.1	1	
11	12–14	9	HLM-$(EO)_3$-EP^+	2	60	1	1	1	
12	12–14	20	Na	5	60	2.3	1.3	1	
13	12–14	20	Na	5	78	2.7	1.8	1.2	B-PES
14	16–18	0	Na	2	78	2.5	1.6	1.1	
15	16–18	0	Na	0.5	78	2.3	1.4	1	
16	12–14	12	Na	0.5	78	2.9	1.7	1.2	
17	16–18	9	Na	2	4200	9.3	5.3	3.3	B-FR
18	16–18	9	Na	2	4200	10.4	5.9	5.5	
19	16–18	0	Na	2	4200	7.5	6.0	4.8	
20	12–14	20	HLM-$(EO)_3$-EP^+	2	4200	1.6	1	1	
21	12–14	3	Na	5	116	3.3	2.2	1.8	PNA
22	16–18	3	Na	2	116	10.5	7.5	5.8	
23	16–18	6	Na	2	116	12.3	7.8	6.2	
24	12–14	6	Na	0.5	116	4.5	2.3	1.4	

^{x}float hardness 5 dG (German degrees)

**not cleaned of softening/antistatic and/or mercerization products

***PES – polyester; PES–AC – polyester acetate; B – cotton; B–PES – cotton polyester; B–FR – cotton frotir; PNA – polyacrylonitrile

Table 2. Wetting effect of alkali and ammonium β-alkyl polyethyleneoxy propionates $[R–O–(CH_2CH_2O)_{n*}–CH_2CH_2COO^-X^+]$

No.	R	n*	R_1	Float concentration (g/L)	Wetting time (sec.)				Textile support
					Untreated support**	Treated support			
						20°C	40°C	60°C	
1	12–14	3	1	2	148	2.5	2.1	1.5	PES
2	12–14	9	4	5^x	148	3.0	2.5	1.2	
3	12–14	12	8	2	148	2.7	1.4	1.2	
4	16–18	20	10	0.5^x	148	4.0	2.8	1.7	
5	12–14	3	4	0.5^x	184	1.4	1.2	1	PES-AC
6	12–14	6	2	2	184	2.2	1.8	1.1	
7	12–14	12	2	5	184	1.8	1.1	1	
8	16–18	6	2	2	184	2.3	1.7	1	
9	12–14	12	10	5^x	60	1.6	1.4	1.2	B
10	16–18	9	4	5	60	1.5	1.3	1.2	
11	16–18	12	16	0.5^x	60	1.5	1.2	1.1	
12	16–18	12	20	1	60	1.7	1.5	1.2	
13	12–14	3	1	0.5^x	78	1.5	1.0	1	B-PES
14	16–18	12	4	2	78	1.2	1	1	
15	16–18	20	2	5x	78	1.0	1	1	
16	16–18	9	2	0.5	78	1.5	1.1	1	
17	12–14	3	8	0.5	4200	1.6	1.3	1.1	B-FR
18	12–14	9	2	5^x	4200	1.2	1.0	1	
19	16–18	6	4	2^x	4200	1.7	1.5	1.3	
20	16–18	9	1	0.5	4200	1.5	1.4	1.2	
21	12–14	3	1	2	116	1.9	1.6	1.2	PNA
22	12–14	9	8	5^x	116	1.7	1.5	1.0	
23	16–18	9	12	2	116	1.9	1.7	1.3	
24	16–18	12	20	5^x	116	1.8	1.6	1.4	

*float hardness 5 dG (German degrees)

**not cleaned of softening/antistatic and/or mercerization products

***PES – polyester; PES–AC – polyester acetate; B – cotton; B–PES – cotton polyester; B–FR – cotton frotir; PNA – polyacrylonitrile.

Table 3. Wetting effect of *N,N*-dimethyl-*N*-alkyl (R_1)-*N*-β-alkyl (R) polyethyleneoxyethylammonium chlorides

4. Structure–hydrophilicity correlations (wetting competences) in the homologous HEHAO series studied

To understand and coherently interpret the main experimental data (Tables 1–3), but also to argue the structure–colloidal (wetting) competences correlations previously formulated, it is necessary to consider several structural and colloidal features of the HEHAO evaluated.

4.1. The conformation of polyoxyethylene (PEO) chains and its consequences for the colloidal behavior of HEHAO [40]

The main features of these spatial architectures with consequences for the study of heterogeneous HEHAO are: dimensional flexibility; transfer mobility; the existence of "meander," "zig-zag," "helix" conformations with variable geometries; free C–C/C–O coaxial rotation, and absence of "ring tensions" specific to rigid structures [1,3,4,40].

Subsequent experimental facts [1,40] also confirmed for the first time the approximate dimensions of the coordination "helical cavity" (ca. 6 EO units/sodium cation and ca. 7 EO units/potassium cation, respectively).

4.2. Adsorption of HEHAO at the aqueous solution–air interface; formation of micelles

HEHAO are lyophilic association colloids which group instantly in aqueous floats as macromolecular associations (lyophilic micelles) at the value corresponding to the critical micelle concentration (expressed as mol/L 10^{-5}).

Surface tension as a form of free energy (expressed in N/m or dyne/cm) independent of the shape of the interface separating two phases in a system, is a function of temperature, time, and the structural characteristics of the HEHAO considered. Because the interface equilibrium is established within a short period, the existence of static and dynamic surface tension is accepted [1]. The latter manifests itself in aqueous floats of HEHAO and subsequently acquires great practical significance in the process of their actual use. Two aspects can be distinguished:

- The capacity of HEHAO to reduce the surface tension, expressed as the concentration required to achieve a certain effect of reducing the surface tension
- The effectiveness of HEHAO expressed by the minimum value which is capable of reducing the surface tension

In the case of heterogeneous HEHAO, the diversity of composition, structure, and conformations does not allow for such an approach, but only for the indicative evaluation of the following surface-active parameters: HLB balance, surface tension (σ_{cmc}), and critical micelle concentration (CMC).

In the homologous series of heterogeneous hydrophobic (R) and hydrophilic (PEO) chains of HEHAO, subsequently, structure–surface activity correlations can also be indicatively formulated.

In aqueous solutions, cationic structures form colloidal associations (normal micelles) and in organic solvents (reverse micelles), with a pronounced acidic character of the "cavity" of the micellar site [3,4] (Figure 2). The difference is due to the fact that in water, cationic colloidal electrolytes can dissociate as solvated ion pairs, while in non-aqueous solvents, with low dielectric constant, the solvation phenomena decrease in proportion to its value, and the cationic colloidal electrolytes exist as unsolvated ion pairs.

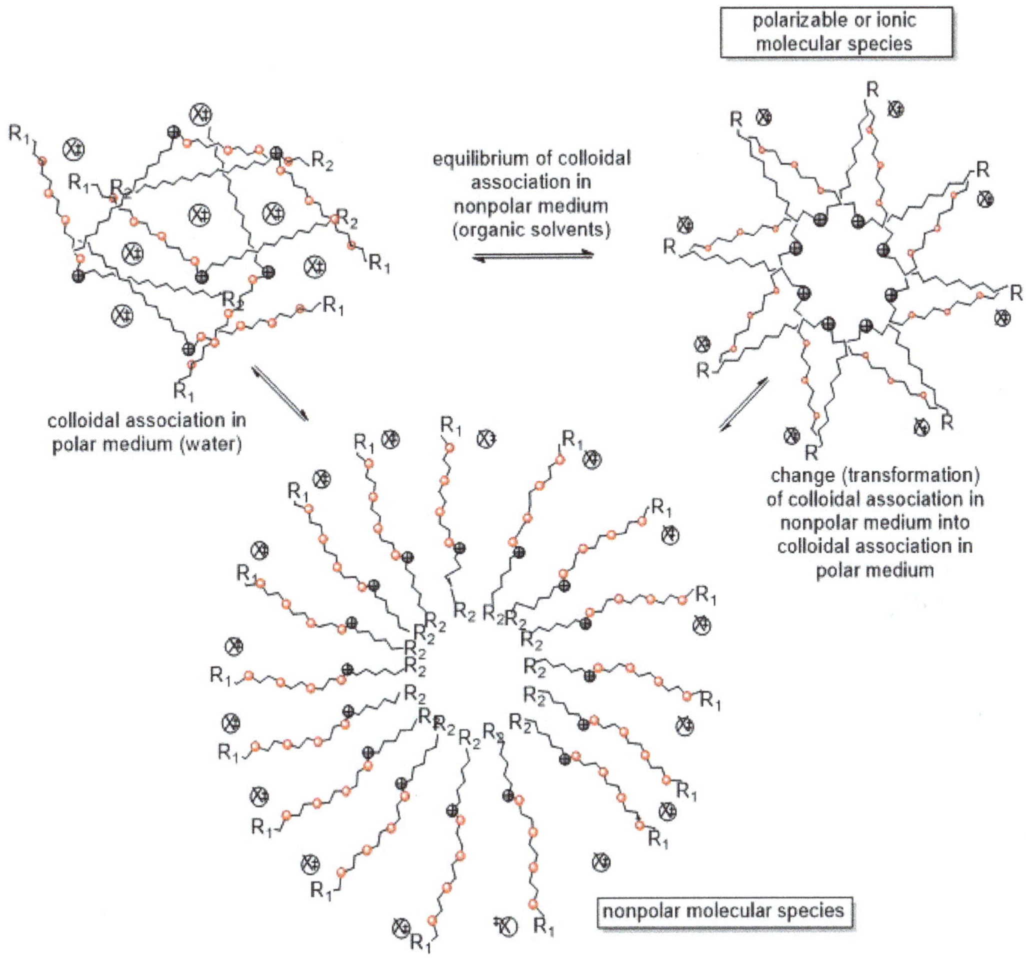

Figure 2. Scheme of the association mechanism of cationic surface-active agents and of micellar solubilization in polar (water) and nonpolar (organic solvents) medium

Modification of the hydrophobic chain influences the mutual steric or electrostatic repulsions through geometric coordinates and steric hindrance [1,4,40]. Corroborating the aforementioned conformational data of PEO chains in the literature with the structure of various HEHAO, in Figure 4 are presented several concrete instances of association at the hydrophobic or hydrophilic separation interface (polar and nonpolar media) of HEHAO. Micellar architectures with polar and nonpolar cavities are thus structured as "solubilization," sequestration, and transfer spaces of potential entities in the sanitation practice (Figure 3).

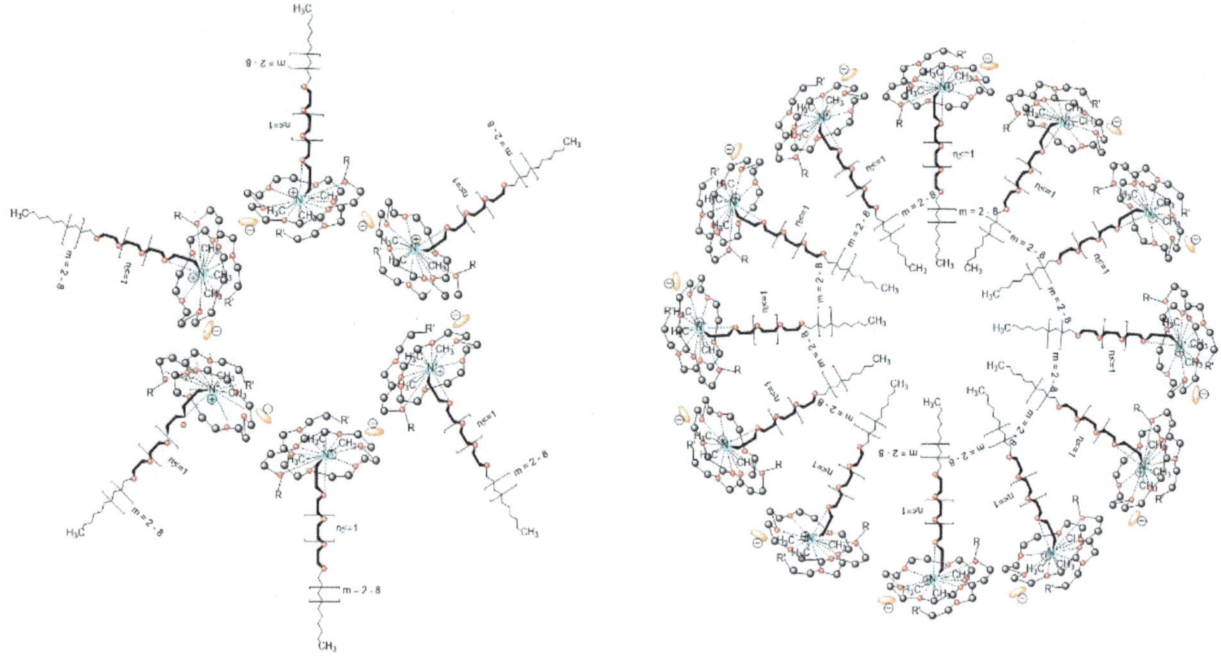

Figure 3. Mixed HEHAO "micellar architectures" with conformational (a), polar micellar (b), and nonpolar micellar (c) solubilization "cavities" (sites)

4.3. Adsorption of mixtures of heterogeneous HEHAO homologues at the aqueous solution–air interface; formation of mixed micelles

Mixtures of two or more different homologues in the HEHAO series show positive or negative "synergistic" interdependences [1,4,7,40]. The overall interfacial properties of the mixture of homologues can be more pronounced than those of a "homogeneous" (unitary) HEHAO structure.

4.4. Cumulated consequences of structural and colloidal characteristics of HEHAO

The major consequences of colloidal and conformational behavior of HEHAO studied for the wetting operation can be summarized thus:

- The competence to sequester (coordinate) small molecular entities with mainly cationic character and variable dimensions depending on the "conformational flexibility" of the heterogeneous polyoxyethylene (PEO) chains
- The possibility of orientation and directed adsorption at the separation interface of two phases depending on their nature (specificity) and hydrophilic and/or hydrophobic character (micelle, support, etc.) (Figure 5)
- The possibility of penetration, displacement, micellar solubilization, and transfer of entities from the interface into its adjacent environment (Figure 6)

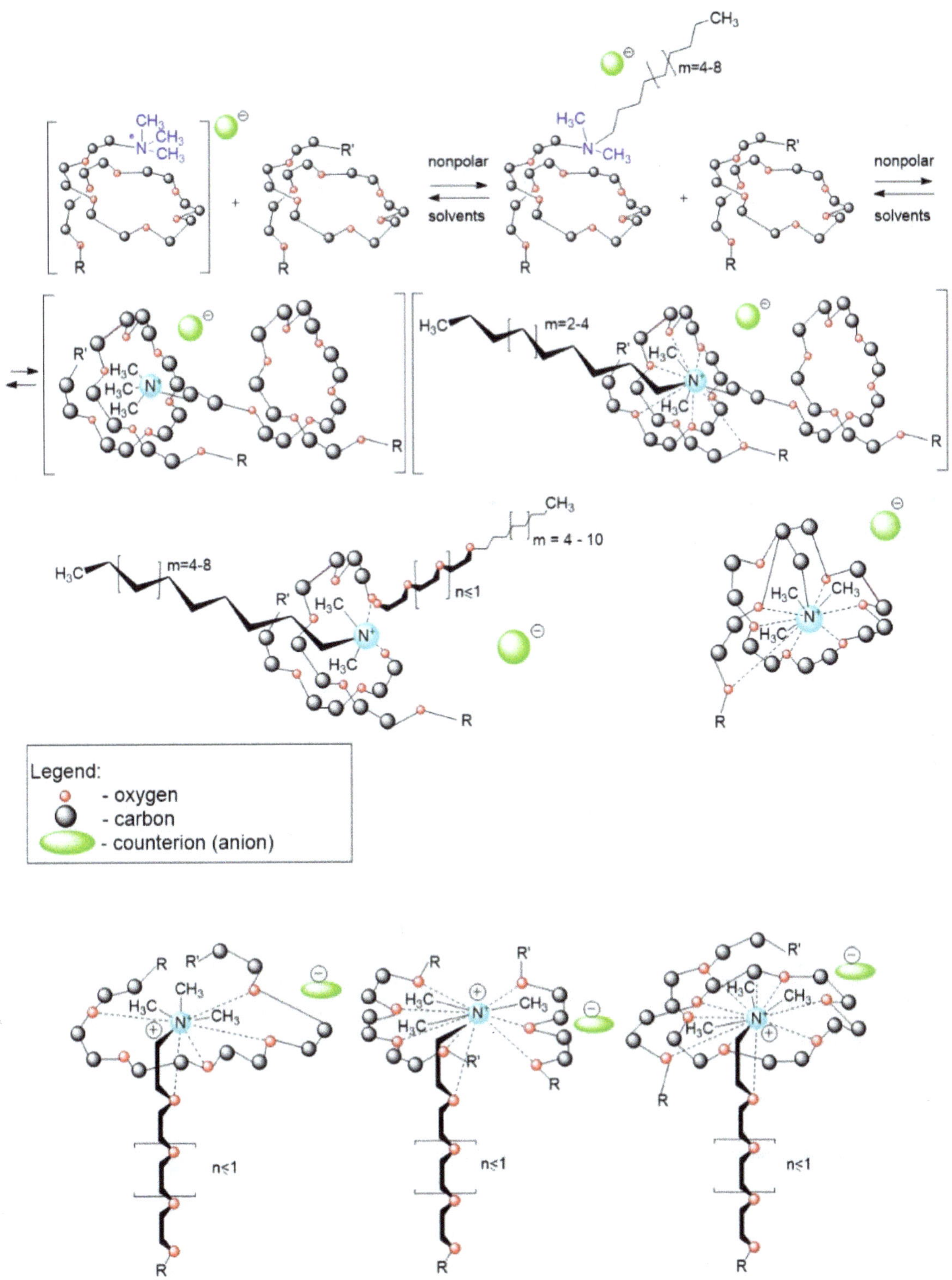

Figure 4. "Micellar architectures" of mixed HEHAO with conformational sequestration (solubilization) "cavities" (sites) [3,4]

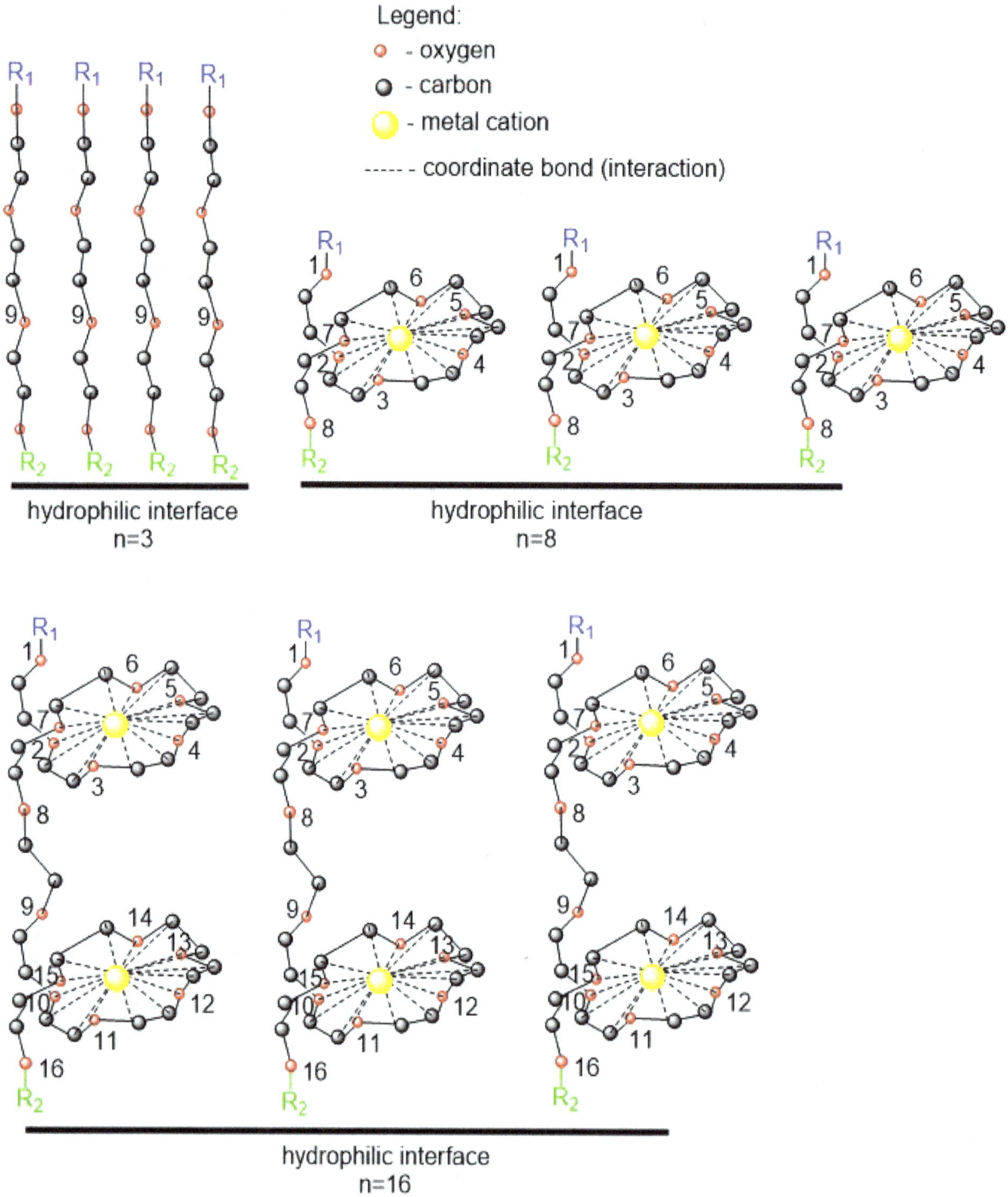

Figure 5. Schematic representation of the orientation and adsorption processes at the hydrophilic interface cumulated with the "encapsulation" of metal cations in the matrix of monoderivatized polyoxyethylene (PEO) (n*=3, 8, 16) chains [4]

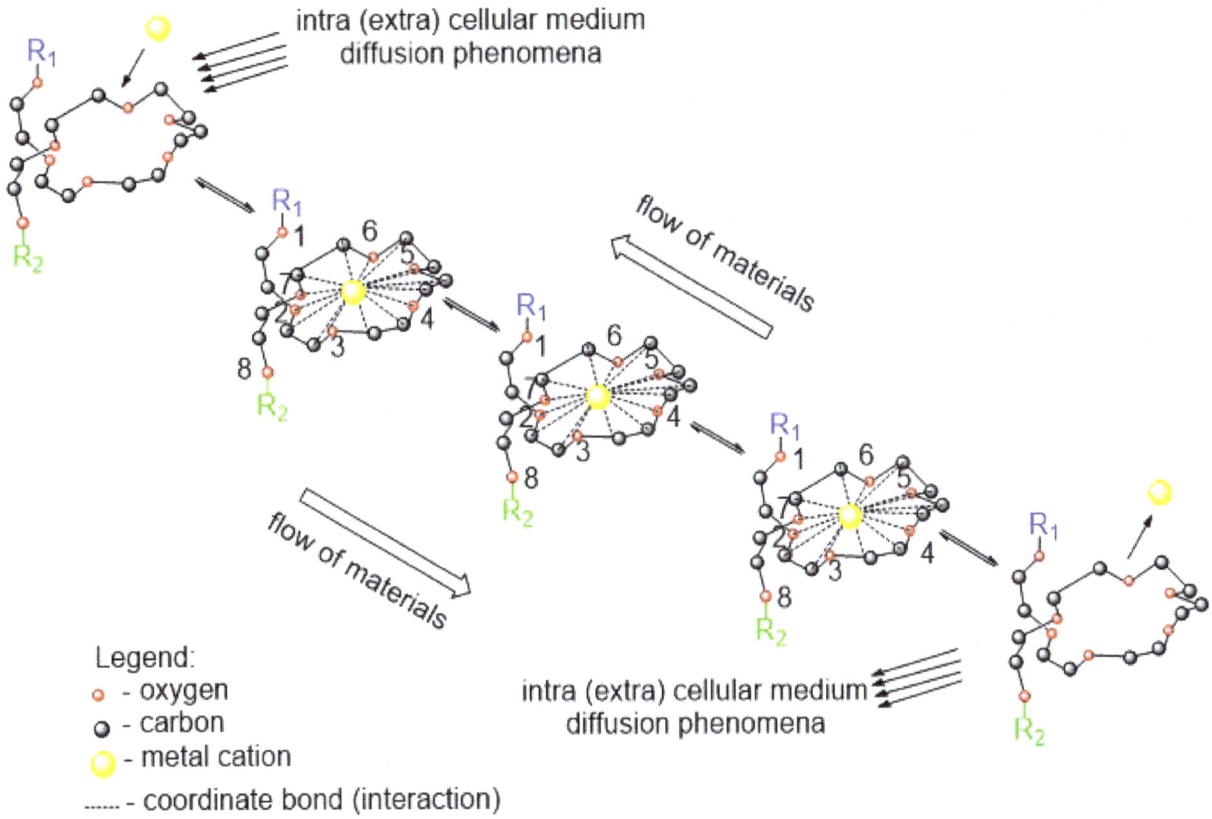

Figure 6. Schematic representation of the process of coordination and interphase transfer of some entities in the conformational "cavity" of HEHAO polyoxyethylene (PEO) (n*=8) chains [4]

5. Conclusions

The present review of the more significant issues relating to the obtaining and characterization of HEHAO in the category of β-alkyl ($C_{12}H_{25}$–$C_{18}H_{37}$) polyethyleneoxy (n*=3–20) ethyl/propyl derivatives (Table 1) represents a novelty in the field of HEHAO. The material prepared allowed an outline of an overall picture of the colloidal (surface-active) potential with possibilities of further controlled directing of their value of use in an extended HLB range (lifting–antistatization, emulsification, micellar solubilization, coordination, interphase transfer, washing, etc.). The beneficial synergistic cumulation of heterogeneous polyoxyethylene (PEO) (n*=3–20) (nonionic hydrophilic) chains with polar ionic groups (anionic and/or cationic) allowed the preliminary confirmation of this latent colloidal potential.

The binary and/or ternary directed (compatible) association in the perspective of various HEHAO structures (including β-alkyl ($C_{12}H_{25}$–$C_{18}H_{37}$) polyethyleneoxy (n*=3–20) propionitriles in acid and/or alkaline media) further extends the area of the application fields.

The biodegradability and advanced purity, along with the reduced toxicity of the raw materials and final molecular architectures (absence of by-products), further recommends the continuation and furthering of the research.

Acknowledgements

My entire gratitude to my mentor, Prof. Univ. Emeritus Ionel Jianu, Ph.D., the founder of Agrifood School of Timişoara, in whose team I started and trained as a researcher and discovered the hitherto unsuspected potential of homogeneous and heterogeneous polyoxyethylene chains.

Author details

Calin Jianu*

Address all correspondence to: calin.jianu@gmail.com

Department of Food Science, Faculty of Food Processing Technology, Banat's University of Agricultural Sciences and Veterinary Medicine, Timişoara, Romania

References

[1] Jianu C. Ethylene oxide homogeneous heterobifunctional acyclic oligomers. In: Claire Lesieur (ed.), *Oligomerization of Chemical and Biological Compounds*. Intech, 2014, pp. 103-132.

[2] Williamson, A. On Etherification. *J Chem Soc* 1852, 4: 229-239.

[3] Jianu I. Aminoethers surfaceactive agents. Cationic surfaceactive agents. PhD Thesis. Polytehnic University; 1984 (in Romanian).

[4] Jianu C. Research concerning the potential of some metal ion complexing additives to improve food value in horticultural raw matter. PhD Thesis. Banat's University of Agricultural Sciences and Veterinary Medicine; 2006 (in Romanian).

[5] Edwards CL. Polyoxyethylene alcohols. In: van Os NM (ed.), *Nonionic Surfactants: Organic Chemistry; Surfactant Science Series*. Volume 72. Marcel Dekker, Inc: New York, USA; 1998, pp. 87-121.

[6] Edwards CL. Chemistry and handling of ethylene oxide. In: van Os NM (ed.), *Nonionic Surfactants: Organic Chemistry Science Series (Chapter 1)*. Marcel Dekker, Inc: New York, USA; 1998, pp. 2-4.

[7] Schick, M. *Nonionic Surfactants* (ed.). Marcel Dekker, Inc: New York, USA; 1967, pp. 80-82; 230.

[8] Sansebastiano G, Zoni R, Bigliardi L. Cleaning and disinfection procedures in the food industry general aspects and practical applications. In: McElhatton A, Marshall

RJ (eds.) *Food Safety A Practical and Case Study Approach.* Springer New York; 2007, pp. 253-280.

[9] Holah JT. Cleaning and disinfection. In: Lelieved HLM, Mostert MA, Holah J, White B. (eds.), *Hygiene in Food Processing.* Cambridge: Woodhead Publishing; 2003, pp. 235-278.

[10] Jianu I. Process for obtaining polyethoxylate soaps, RO Patent 85437 (A2), International Classification: C11D1/02; C11D1/02; (IPC1-7): C11D1/02.

[11] Drugarin C, Jianu I. Process for the preparation of surfaceactive agents, RO Patent 77367 (A2), International Classification: C07D233/08; C07D233/00; (IPC1-7): C07D233/08.

[12] Drugarin C, Jianu I. Process for the preparation of new cationic agents, RO Patent 82439* (A3), International Classification: (IPC1-7): C07C87/30.

[13] Jianu I. Beitrage zum kinetischen und thermodynamischen studium der cyano-ethylierungs reaktionen höcherer alkohole (C_{10}-C_{18}). Proceedings World Surfactants Congress. Surfactants in our World Today and Tomorrow; 1984, II: 188-196 (in German).

[14] Drugarin C, Jianu I. Imidazoline derivatives and process for the preparation thereof, RO Patent 82335*(A2), C.A. 102, 26882p, 1985, International Classification: C07D233/20; C07D233/00; (IPC1-7): C07D233/20.

[15] Weinheimer RM, Varineau PT. Polyoxyethylene alkylphenols. In: van Os NM. (ed.) *Nonionic Surfactants: Organic Chemistry.* Marcel Dekker Inc; 1998, pp. 120.

[16] Drugarin C, Jianu I. The hydrolysis β – [(p) – Alkylphenyloxy polyethyleneoxy] – propionitrils under phase transfer catalysis – Part I: The synthesis of β - [(p) – Alkylphenyloxy polyethyleneoxy] – propionitrils. *Tenside Detergents* 1983; 20: 128-129.

[17] Drugarin C, Jianu I. The hydrolysis β-[(p)-Alkylphenyloxy polyethyleneoxy]-propionitriles under phase transfer catalysis – Part II: The synthesis of β-[(p)-Alkylphenyloxy polyethyleneoxy]–propionitriles. *Tenside Detergents* 1983: 20: 130-131.

[18] Drugarin C, Jianu I. Correlation between molecular structure and surface activity, Part I: The foaming power of the class of β-Alkyloxyethyl and β-Alkylpoly(ethylenoxy] – ethyl-alkyl-dimethylammonium halides. *Tenside Detergents* 1983; 20: 76-77.

[19] Drugarin C., Jianu I., Correlation between molecular structure and surface activity, Part II. The foaming power of the class of β-Alkyloxyethyl and β-Alkylpoly(ethylenoxy]-ethyl-alkyl-dimethylammonium halides. *Tenside Detergents* 1983; 20: 78-79.

[20] Drugarin C, Jianu I. Process for obtaining synthetic triglycerides, RO Patent 82041(A2), C.A. 103, 55720k, 1985, International Classification: C07D67/22; C07C69/003; C07C67/00; C07C69/00 (IPC1-7): C07D67/22; C07C69/003.

[21] Haussmann H, Scheufeer W, Kaupp J. Polyoxyalkylene ether acid compounds containing a higher aliphatic group, 1939. Patent Number US 2183853/1939.

[22] Isa H, Inoniya T, Karube K, Takemoto T, Nagayama M. Verfahren zur Herstellung von carboxylaten: 1973. DE 2.303.829 (A1).

[23] Rodney MW, Thomas AC. Light-duty liquid or gel dishwashing detergent composition containing an alkyl ethoxy carboxylate surfactant, 1993. Patent Number US 5230823/1993.

[24] Kon-Chang L. Preparation of carboxypropylate non-ionic surfactants; 1987. Patent Number US 4692551/1987.

[25] Abe Y. Long-chain fatty acids containing ether linkage. I. The antimicrobial and fungicidal activities of some new p-alkyloxypropionic acids and their methyl esters. *Lipids* 1966, 1(2):141-145.

[26] Abe Y, Osanai S. Long-chain carboxylic acids containing ether linkage. III. The antibacterial and antifungal activities of the amine salts of some β-alkoxypropionic acids. *J Am Oil Chem Soc* 1969, 46(7}365-367.

[27] Abe Y, Osanai S, Matsumura S. Long-chain carboxylic acids containing ether linkage. IV. The antibacterial activities of O-{2-Alkylaminoethyl)-3-oxypropionic and N-(2-Alkyloxyethyl)-3-aminopropionic Acids. *J Am Oil Chem Soc* 1972, 49(6)357-360.

[28] Abe Y, Osanai S, Matsushita T. Long-chain carboxylic acids containing ether linkage. V. N-methyl substituted glycine-type amphoterics containing alkoxy radical. *J Am Oil Chem Soc* 1974, 51(9):385-388.

[29] Schafer W, Schafer R, Schafer D, Aalbers JG, Smid JK. Novel polyether carboxylic acid derivates, as well as their user, 1989. Patent Number US 4818440/1989.

[30] Cripe TA, Ofosu-Asante K. Alkaline light duty diswashing detergent composition containig an alkyl ethoxy carboxylate surfactant, magnesium ions, chelator and buffer, 1994. Patent Number US 5376310/1994.

[31] Ofosu-Asante K. Liquid or gel diswashing detergent containing alkyl ethoxy carboxylate divalent or ions arid alkylpolyethoxy polycarboxylate. 1998. Patent Number US 5739092/1998.

[32] Man VFP. Alkaline cleaners based on alcohol ethoxy carboxylates; 2001. Patent Number US 6274541 61/2001.

[33] SR ISO/TR 896:1995 Surface active agents. Scientific classification (identical with the International ISO/TR 893:1977). Romanian Standards Association (ASRO); http://magazin.asro.ro/.

[34] Jianu C. Synthesis of nonionic-anionic colloidal systems based on alkaline and ammonium beta-nonylphenol polyethyleneoxy (n=3-20) propionates/ dodecyl-benzene-sulfonates with prospects for food hygiene. *Chem Central J* 2012, 6:95, doi: 10.1186/1752-153X-6-95.

[35] Jianu C, Nedelea G, Lazureanu A, Jianu I, Adamescu A. Diversifying and functional featuring of some PEG coacervates as ecological nanocomponents in the soil-plant-

man (food consumer) circuit of some nutrients. *Bull Univ Agric Sci Vet Med* 2004, 60:507.

[36] Jianu C, Butnariu M, Cocan I, Rinovetz A, Bujanca G, Jianu I. Hydrolysis of β-alkyl ($C_{12}H_{25}/C_{18}H_{37}$) polyethyleneoxy (n = 3-20) propionitriles in micellar catalysis conditions (I). *J Agroalim Proc Technol* 2009, 15(1):72-78

[37] Jianu C, Trașcă T, Riviș A, Mișcă C, Chiș M, Jianu I. Hydrolysis of β-alkyl (C12H25/Ci8H37) polyethyleneoxy (n = 3-20) propionitriles in phase transfer catalysis conditions (II). *J Agroalim Proc Technol* 2009, 15(1):79-87.

[38] Jianu C, Jianu I. Colloidal competences of new tailor-made lipids. *J Food Agric Environ* 2010, 8(3&4):148-155.

[39] Jianu C. Colloidal competences of some food cleaning agents based on alkaline and ammonium nonionic soaps. *J Food Agric Environ* 2012, 10(1):10-15.

[40] Riviş Adrian. Basic surfaceactive properties in the homologous series of beta-alkyl ($C_{12}H_{25}/C_{18}H_{37}$) polyethyleneoxy (n=0-20) propionamides. *Chem Central J* 2013, 7:31, doi:10.1186/1752-153X-7-31.

[41] Jianu C, Riviş A, Traşcă T, Jianu Adelina, Bujanca G, Controlled of colloidal parameters through gradual ternary association of new "homogeneous" nonionic structured lipids, J. Agroaliment. Proc. Technol., 2014, 20(4):354-362.

[42] Jianu C, Traşcă T, Riviş A, Jianu Adelina, Costescu Corina, Cazacu Mihaela. Colloidal characteristics in gradual binary associated systems of new "homogeneous" polyoxyethylene structured lipids. *J Agroaliment Proc Technol* 2014, 20(4):383-390.

[43] Yanagida S, Takahashi K. Okahara M. Metal-ion complexation of noncyclic poly(oxyethylene) derivates I. Solvent extraction of alkali and alkaline earth metal thiocyanates and iodines. *Bull Chem Soc Jpn* 1977, 50(6):1386-1390.

[44] Harvey RD, Barlow DJ, Drake AF, Kudsiova L, Lawrence MJ, Brain A, Heenan RK. The effect of electrolyte on the encapsulation efficiency of vesicles formed by the nonionic surfactants, 2C18E12. *J Colloid Interface Sci* 2007, 315:648-661.

[45] Yokoyama Y, Hirajima R, Morigaki K, Yamaguchi Y, Ueda K. Alkali-cation affinities of polyoxyethylene dodecylethers and helical conformation of their cationized molecules studied by electrospray mass spectrometry. *J Am Soc Mass Spectrom* 2007, 18:1914-1920.

[46] Jackson AT, Scrivens JH, Williams JP, Baker ES, Gidden J, Bowers MT. Microstructural and conformational studies of polyether copolymers. *Int J Mass Spectrom* 2004, 238:287-297.

[47] Man VFP. Alkaline cleaners based on alcohol ethoxy carboxylates; 2002. Patent Number US 6479453 82/2002.

[48] Leinweber D. Dahlmann U, Kupfer R. Process for the solvent-free preparation of ether carboxylic acids having a low residual salt content, 2007. Patent Number US 7208118 B2/2007.

[49] SR EN 1772/2009, Surface active agents. Determination of wetting power by immersion (identical with the European Standard EN 1772:2000).

4

Re-derivation of Young's Equation, Wenzel Equation, and Cassie-Baxter Equation Based on Energy Minimization

Kwangseok Seo, Minyoung Kim and Do Hyun Kim

Abstract

Recently, Young's equation, the Wenzel equation, and the Cassie-Baxter equation have been widely used with active research on superhydrophobic surfaces. However, experiments showed that the Wenzel equation and the Cassie-Baxter equation were not derived correctly. They should be reviewed on a firm physical ground. In this study, these equations are re-derived from a thermodynamic point of view by employing energy minimization and variational approach. The derivations provide a deeper understanding of these equations and the behavior of a contact angle. Also, in applying these equations, the limitations and considerations are discussed. It is expected that this study will provide a theoretical basis for the careful use of these equations on rough or chemically heterogeneous surfaces.

Keywords: Young's equation, Wenzel equation, Cassie-Baxter equation, contact angle, energy minimization, variational method

1. Introduction

The easiest way to determine the wetting property is to drop a liquid drop on the surface. The drop on the surface forms a unique contact angle depending on the wetting property. By measuring the contact angle, it is easy to examine the surface wettability. Young's equation on the ideal surface, the Wenzel equation on the surface with roughness, and the Cassie-Baxter equation on the surface with chemical heterogeneity have been widely used for the analysis of the contact angle. Although these equations were not derived correctly, they have been used without consideration of the limitations. Application of these equations to surfaces such as a surface with large contact angle hysteresis that do not meet the conditions for these equations can give errors inherently.

In this chapter, Young's equation, the Wenzel equation, and the Cassie-Baxter equation will be re-derived by energy minimization and variational approach. From analyses of the derivations, properties of a contact angle will be reviewed. Also, the limitations and the considerations will be discussed in applying these equations to various surfaces. We expect that this study will help in the understanding of the nature of the contact angle and its application.

1.1. Young's equation, Wenzel equation, and Cassie-Baxter equation

It is possible to quantify the wettability of a surface by simply measuring the contact angle of a drop resting on a surface. Young's equation has been used as a basic model. Application of this equation is limited to an ideal surface that is rigid, perfectly flat, insoluble, non-reactive, and chemically homogenous. The surface is assumed to have no contact angle hysteresis. On the surface, a contact angle of liquid drop can be described by the following Young's equation:

$$\gamma_{sl} + \gamma \cos\theta = \gamma_{so} \tag{1}$$

where γ, γ_{sl}, and γ_{so} are liquid/gas surface tension, solid/liquid interfacial energy, and solid/gas surface energy, respectively. The apparent contact angle (θ) is an equilibrium contact angle (θ_Y). However, since all the real surfaces are not ideal, models were developed to describe the contact angles on the real surfaces.

There are two models to describe the contact angle on a real surface, i.e. the Wenzel model and the Cassie-Baxter model. Contrary to the ideal surface, the real surface can have chemical heterogeneity and surface roughness. The Wenzel model considers the rough surface but with chemical homogeneity [1]. The Cassie-Baxter model considers the flat surface but with chemical heterogeneity [2].

In the Wenzel model, the surface roughness r is defined as the ratio of the actual area to the projected area of the surface. The Wenzel equation can be written as:

$$\cos\theta^* = r \cos\theta_Y \tag{2}$$

where θ^* is the apparent contact angle and θ_Y is the equilibrium contact angle from Young's equation on an ideal solid without roughness.

In the Cassie-Baxter model, f_1 and f_2 are the area fractions of solid and air under a drop on the substrate. The Cassie-Baxter equation can be written as:

$$\cos\theta^* = f_1 \cos\theta_Y - f_2 \tag{3}$$

where θ^* is the apparent contact angle and θ_Y is the equilibrium contact angles on the solid.

From the Wenzel model, it can be deduced that the surface roughness amplifies the wettability of the original surface. Hydrophilic surface becomes more hydrophilic and hydrophobic surface more hydrophobic. In the Cassie-Baxter model, the area fractions under the drop is important in that the larger the area fraction of air, the higher the contact angle. Although these two models were proposed half a century ago, these equations have been widely used recently with active research on superhydrophobic surface [3-6].

1.2. The fallacy of the Wenzel model and the Cassie-Baxter model

In the Wenzel model and the Cassie-Baxter model, the contact angles were obtained from the non-smooth or chemically heterogeneous state of the surface under the drop. However, Gao and McCarthy demonstrated the fallacy of these models experimentally [7]. They prepared a surface with a hydrophilic spot on a hydrophobic surface, as shown in Fig. 1a. Fig. 1b shows a smooth hydrophobic surface with a superhydrophobic spot. D and d are mean diameters of the drop and the spot.

With various diameters of the drops and the spots, advancing and receding contact angles were measured. They proved that the state of internal surface inside the triple line does not affect the contact angles experimentally and the contact angles are determined only by the state of the surface at triple contact line. It means that the previous Wenzel model and Cassie-Baxter model should be revised for rigid physical meaning [7]. Since then, an active discussion on them has been made [8-12]. Also, these models have been derived in a more rigorous way. We have summarized the derivations of these models studied to date in Table 1. All the derivations verify that a contact angle is determined at the triple line regardless of the external fields. Experiments also confirmed these findings [13-15]. Here, we will introduce the derivations by energy minimization using simple mathematics or calculus of variations.

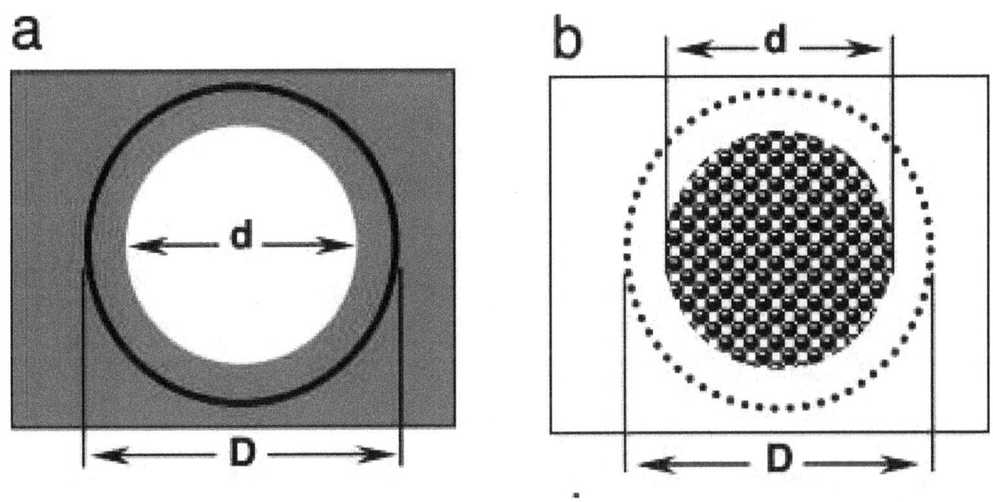

Figure 1. Depictions of (a) a hydrophilic spot on a hydrophobic surface and (b) a superhydrophobic spot on a smooth hydrophobic surface. Reprinted with permission from reference [7].

Derivation method	External field	Region to determine a contact angle	Reference
Homogenization approach	N/A	At triple line	Xu and Wang, 2010 [16]
Fundamental calculus	N/A	At triple line	Seo et al., 2013 [17]
Fundamental calculus	N/A	At triple line	Whyman et al., 2008 [18]
Variational approach	Gravity	At triple line	Bormashenko, 2009 [19]
Variational approach	Electric field	At triple line	Bormashenko, 2012 [20]

Table 1. Derivations for the Wenzel model and the Cassie-Baxter model.

2. Derivation with simple mathematics

For the derivation of Young's equation in a rigorous way, the following assumptions will be used. First, the surface is ideal and it has no contact angle hysteresis. Thus, the contact line can freely move around. Second, the drop is in zero gravity and the shape of the drop is always a section of sphere, i.e., spherical cap.

As shown in Fig. 2, when the shape of the drop is deformed by spreading or contracting, the solid/liquid interfacial area varies with a contact angle that is a one-to-one function of the interfacial area. By the free movement of the contact line on an ideal surface, the drop can change freely its shape in order to satisfy the minimum energy state of the system. When the drop is at the equilibrium state, there will be no residual force at the contact line. At this point, the contact line and the shape of the drop will be fixed.

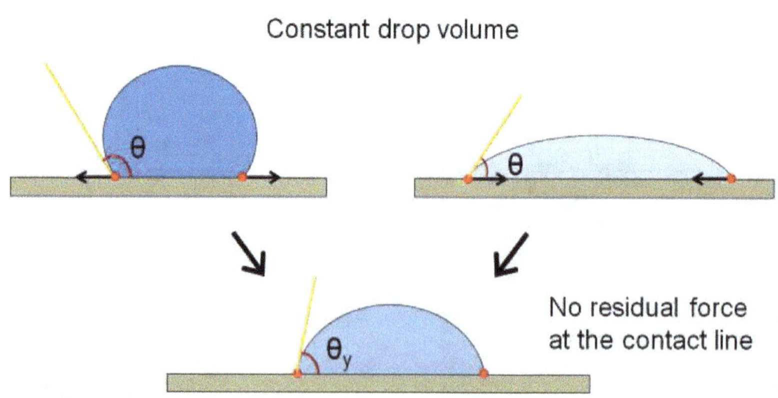

Figure 2. Formation of the contact angle of a drop on an ideal surface. Reprinted with permission from reference [17].

2.1. Derivation of Young's equation

With a thermodynamic approach, Young's equation can be derived with simple mathematics. Fig. 3 shows a drop on an ideal surface. The volume of the spherical cap is $V = \frac{\pi h}{6}(3r^2 + h^2)$ or

$V=\frac{\pi R^3}{3}(1-\cos\theta)^2(2+\cos\theta)$. The surface area of the cap is $A=2\pi Rh$ or $A=2\pi R^2(1-\cos\theta)$. The total energy of the system can be written as

$$E=\pi r^2\left(\gamma_{sl}-\gamma_{so}\right)+2\pi Rh\cdot\gamma=\pi(R\sin\theta)^2\left(\gamma_{sl}-\gamma_{so}\right)+2\pi R^2\left(1-\cos\theta\right)\gamma \tag{4}$$

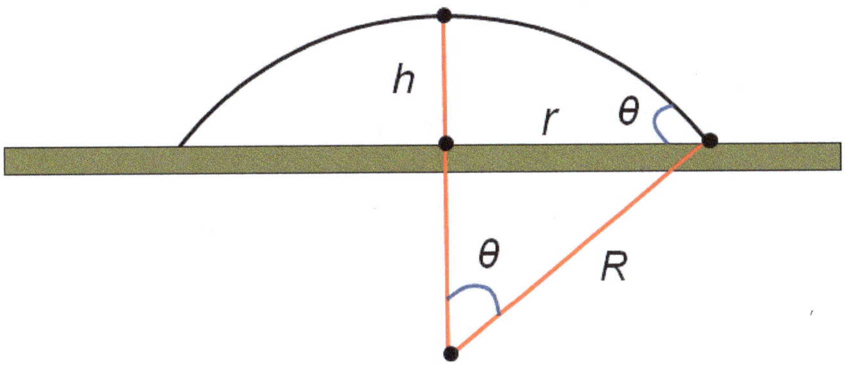

Figure 3. Cross-section of the drop on an ideal surface. Reprinted with permission from reference [17].

The variation of the energy is written as

$$dE=2\pi R\cdot\left[\left(\gamma_{sl}-\gamma_{so}\right)\cdot\left(\sin^2\theta dR+R\sin\theta\cos\theta d\theta\right)+\gamma\cdot\left(2\left(1-\cos\theta\right)dR+R\sin\theta d\theta\right)\right] \tag{5}$$

The variation of the energy is equal to zero at the equilibrium state ($dE=0$). Dividing both sides by dR is written as

$$\frac{dE}{dR}=2\pi R\cdot\left[\left(\gamma_{sl}-\gamma_{so}\right)\cdot\left(\sin^2\theta+R\sin\theta\cos\theta\frac{d\theta}{dR}\right)+\gamma\cdot\left(2\left(1-\cos\theta\right)+R\sin\theta\frac{d\theta}{dR}\right)\right]=0 \tag{6}$$

$\frac{dE}{dR}=0$ means that there is no energy variation by an infinitesimal change of the shape of the drop. At this point, the shape of the drop satisfies the minimum energy state. Here, in order to induce $\frac{d\theta}{dR}$, the constant-volume condition of the drop is used ($dV=0$).

The volume of the drop is given by

$$V=\frac{\pi R^3}{3}\left(1-\cos\theta\right)^2\left(2+\cos\theta\right) \tag{7}$$

$\frac{d\theta}{dR}$ can be obtained from the condition of the constant volume.

$$\frac{d\theta}{dR} = -\frac{(1-\cos\theta)(2+\cos\theta)}{R\sin\theta(1+\cos\theta)} \tag{8}$$

Substituting Eq. (8) into Eq. (6) gives

$$\begin{aligned}&(\gamma_{sl}-\gamma_{so})\cdot\left(\sin^2\theta+\cos\theta\left(-\frac{(1-\cos\theta)(2+\cos\theta)}{(1+\cos\theta)}\right)\right)\\&+\gamma\cdot\left(2(1-\cos\theta)-\frac{(1-\cos\theta)(2+\cos\theta)}{(1+\cos\theta)}\right)=0\end{aligned} \tag{9}$$

Rearranging the above equation gives rise to the Young's equation.

$$\gamma_{sl} + \gamma\cos\theta = \gamma_{so} \tag{10}$$

2.2. Derivation of the Wenzel equation

Fig. 4 shows a drop in the Wenzel state. The radius of the drop is a. The radius of the smooth region of the substrate under the drop is b. When b is equal to zero, the substrate becomes a uniform rough surface. The contact line of the drop is assumed to be located on the rough region of the substrate.

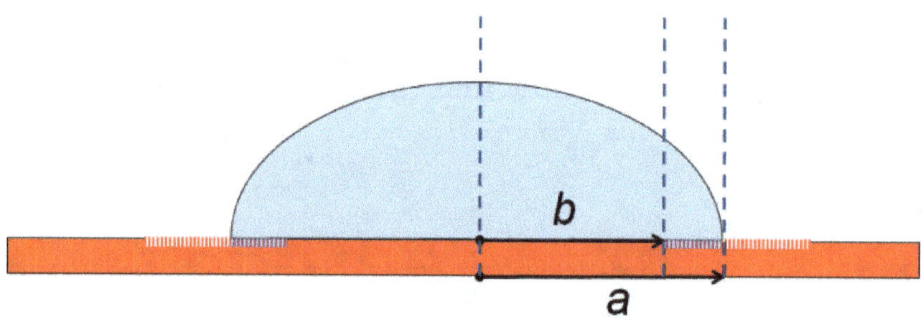

Figure 4. Schematic of the drop in the Wenzel state. The radius of the drop is a. The smooth region is b. Reprinted with permission from reference [17].

From the figure, the total energy of the system can be written as

$$E = \pi b^2(\gamma_{sl}-\gamma_{so}) + K\pi(a^2-b^2)(\gamma_{sl}-\gamma_{so}) + A_s\gamma \tag{11}$$

K is the surface roughness factor. A_s is the gas/liquid interfacial area. The variation of the energy is written as

$$dE = \mathrm{d}\left[\pi b^2(\gamma_{sl}-\gamma_{so})\right] + d\left[K\pi(a^2-b^2)(\gamma_{sl}-\gamma_{so})\right] + d\left[A_s\gamma\right] \tag{12}$$

Rearranging above equation, it will be written as

$$dE = K\pi\left(\gamma_{sl} - \gamma_{so}\right)d\left[a^2\right] + \gamma d\left[A_s\right] \tag{13}$$

It should be noted that the terms related to b^2 were eliminated because they are constant. Thus, the internal surface inside the contact line does not affect the apparent contact angle. There is a significant difference with the previous Wenzel model where the internal surface is considered as an important factor. From *dE=0* at equilibrium and Eq. (8), the revised Wenzel equation is derived:

$$K\cos\theta_Y = \cos\theta \tag{14}$$

where θ_Y is the equilibrium contact angle on a smooth surface and θ is the apparent contact angle.

In the revised Wenzel equation, the definition on the surface roughness factor is different with the previous one. The roughness factor *r* in the previous Wenzel model came from the total area under the drop. However, the roughness factor *K* in Eq. (14) should be obtained from local region near contact line.

2.3. Derivation of the Cassie-Baxter equation

Fig. 5 shows a drop on a composite substrate that consists of two kinds of ideal surfaces. The area fraction of the red region is f_1 and the yellow region is f_2.

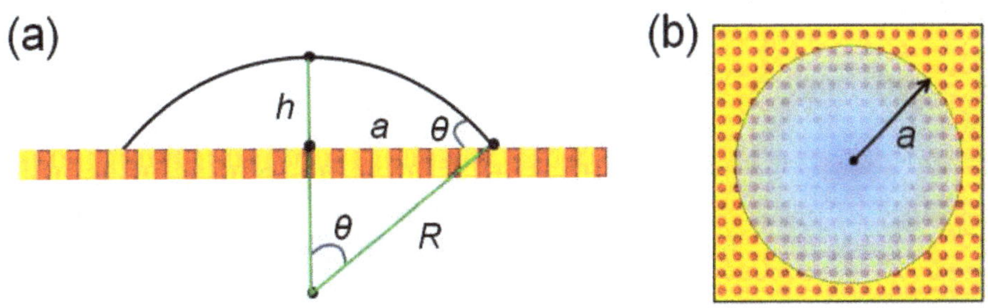

Figure 5. Schematic of the drop in the Cassie-Baxter state: (a) side-view, (b) top-view. Reprinted with permission from reference [17].

From the figure, the total energy of the system can be written as

$$E = f_1\pi a^2\left(\gamma_{sl,f_1} - \gamma_{so,f_1}\right) + f_2\pi a^2\left(\gamma_{sl,f_2} - \gamma_{so,f_2}\right) + A_s\gamma \tag{15}$$

Here, A_s is the gas/liquid interfacial area. The variation of the energy is written as

$$dE = C_1 d(R^2 \sin^2\theta) + C_2 d(R^2(1-\cos\theta)) \tag{16}$$

Here, $C_1 = f_1\pi(\gamma_{sl,f_1} - \gamma_{so,f_1}) + f_2\pi(\gamma_{sl,f_2} - \gamma_{so,f_2})$, $C_2 = 2\pi\gamma$

From $dE=0$ at equilibrium and Eq. (8), the revised Cassie-Baxter equation is derived

$$f_1\cos\theta_1 + f_2\cos\theta_2 = \cos\theta \tag{17}$$

Assuming f_2 to be a fraction for contacting with air and $\theta_2=\pi$, the revised Cassie-Baxter equation can be derived from Eq. (17). There is an important difference in the previous Cassie-Baxter equation and the revised one. The definition on the surface fraction is different. While it came from the total area under the drop in the previous model, the surface fraction in Eq. (17) should be obtained from the local region near the contact line. From the derivation, it can be also deduced that the property or the state of internal surface inside the triple line does not affect the apparent contact angle.

3. Derivation with calculus of variations

In the previous chapter, an external field, such as gravity, was not considered for simple derivation and it was possible to deal with the shape of the drop as a part of sphere. In this chapter, variational approach is employed and the shape of the drop can be distorted by the external field. Bormashenko used this approach for the first time to derive and develop the contact angle models [19]. To understand the variational approach, fundamental formulas in calculus of variations will be introduced briefly [21].

3.1. Calculus of variations

The basic concept of variational method is searching a function that has an extreme value (maximum or minimum) of a physical quantity, such as energy, length, area, time, and so on. In mathematical expression, objective function $J[y]$ can be defined as

$$J[y] = \int_{x_0}^{x_1} F[x, y(x), y'(x), y''(x), \ldots]\, dx \tag{18}$$

Thus, the goal is to find function 'y' that makes $J[y]$ have an extreme value among countless number of y within the scope of the definite integral of a function [21].

With the variational method, we can solve many problems involving the determination of maxima or minima of functionals, such as the shortest smooth curve joining two distinct points in the plane [21], the shape of solid of revolution moving in a flow of gas with least resistance

[22], the plane curve down that a particle will slide without friction from one point to the other point in the shortest time [23], the curve passing through two given points to have minimum surface area by the rotation of the curve [21], and the shape of the flexible cable of given length suspended between two poles [24].

All of the above examples involve funtionals that can be written in the form,

$$J[y] = \int_a^b F[x,\ y(x),\ y'(x)]\ dx \tag{19}$$

There is a fundamental formula for solving the simple variational problems. This is the so-called Euler equation,

$$F_y - \frac{d}{dx} F_{y'} = 0 \tag{20}$$

When a curve passes through two fixed end points, *y(a)=A* and *y(b)=B*, the values of the arbitrary constants are simply determined by the end points.

However, as shown in Fig. 6, when both end points of the curve are always placed on *y=m(x)* and *y=n(x)*, the transversality condition is an additional constraint that must be satisfied at the endpoints of the curve in order for *J[y]* to have an extreme value. When the equation has following form,

$$J[y] = \int_{x_0}^{x_1} F[x,\ y(x),\ y'(x)]\ dx \tag{21}$$

the transversality condition is as follows:

$$\left[F + (m' - y')F_{y'}\right]\Big|_{x=x_0} = 0, \left[F + (n' - y')F_{y'}\right]\Big|_{x=x_1} = 0 \tag{22}$$

It is possible to obtain the curve that must simultaneously satisfy certain constraints as a subsidiary condition. When the form of a functional is the same as $J[y] = \int_a^b F[x,\ y(x),\ y'(x)]\ dx$, and the curve '*y*' should also satisfy the following constraint, $C = \int_a^b G[x,\ y(x),\ y'(x)]\ dx$, where C is constant, the curve '*y*' that makes *J[y]* a extreme value, satisfying the constraints, is a curve that makes the following *Q[y]* an extreme value

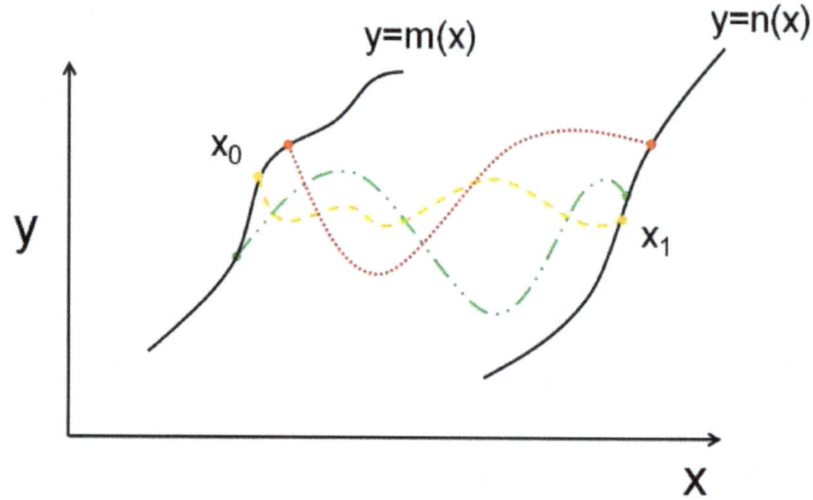

Figure 6. a) Transversality condition as additional constraints at the end points of the curve to have an extreme value required in the system.

$$Q[y] = \int_a^b (F + \lambda G)\, dx, \tag{23}$$

where λ is the Lagrange multiplier to be deduced from the subsidiary condition.

Therefore, the curve 'y' should satisfy the following equation:

$$F_y - \frac{d}{dx}F_{y'} + \lambda\left(G_y - \frac{d}{dx}G_{y'}\right) = 0. \tag{24}$$

These fundamental equations will be used to derive the contact angle models. How a drop takes its contact angle can be understood more clearly from the variational approach.

3.2. Derivation of Young's equation

Fig. 7 shows a drop on an ideal surface in a three-dimensional (3D) system. The symmetrical 3D drop sitting on the surface subject to energy density $U(r,\ h(r))$, due to an external field such as gravity, is considered. The free energy for the drop, $G(h,\ h',\ r)$, is given by the integral.

$$G(h,\ h',\ r) = \int_0^a \left[2\pi r\gamma \cdot \sqrt{1+\left(\frac{dh}{dr}\right)^2} + U(r,h) + \left[2\pi r(\gamma_{SL} - \gamma_{SO})\right] \right] dr \tag{25}$$

The linear density $U(r,\ h(r))$ of the additional energy with the dimension of (J m^{-1}) is defined by the expression, $U(r,\ h(r))=\int_0^{h(r)} 2\pi r w(r,\ y)dy$, where $w(r,\ y)$ is the volume energy density of the drop in the external field (J m^{-3}), $h(r)$ is the local height of the liquid surface, and h' is $\frac{dh}{dr}$.

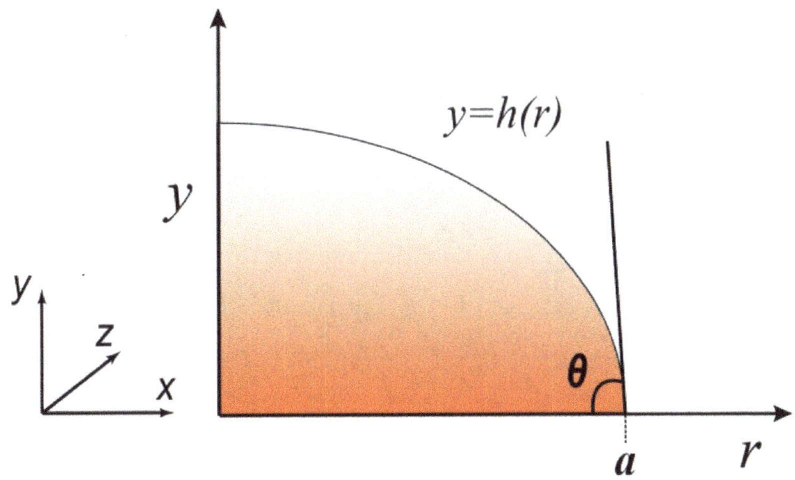

Figure 7. The cross-section of the drop deposited on an ideal surface.

The constant volume V_0 of a drop is assumed as a subsidiary condition.

$$V_0 = \int_0^a 2\pi r \cdot \left[h(r)\right] dr \tag{26}$$

From Eq. (25) and Eq. (26), the problem of energy minimization in the total system is reduced to the one of minimization of the following functional:

$$G = \int_0^a \left[\tilde{G}(h, h', r)\right] dr \tag{27}$$

where,

$$\tilde{G}(h,h',r) = 2\pi\gamma r\sqrt{1+h'^2} + U(r,h) + 2\pi r\left(\gamma_{SL} - \gamma_{SO}\right) + 2\pi\beta r h \tag{28}$$

and β is the Lagrange multiplier to be deduced from Eq. (26). Transversality condition is the one that should be satisfied at the end points in order to minimize the total energy of the system. The transversality condition at the end point a is written as

$$\left[\tilde{G} - h'\tilde{G}_{h'}\right]_{r=a} = 0 \tag{29}$$

where $\tilde{G}_{h'}$ denotes $\frac{\partial \tilde{G}}{\partial h'}$. Considering $h(a)=0$ and $U(a, h(a))=0$, substitution of Eq. (28) into the transversality condition gives

$$\left[\gamma \cdot \sqrt{1+h'^2} + \left(\gamma_{SL} - \gamma_{SO}\right) - h' \cdot \left[\frac{\gamma \cdot h'}{\sqrt{1+h'^2}}\right]\right]_{r=a} = 0 \tag{30}$$

Rearranging Eq. (30) gives

$$\gamma \cdot \left(\frac{1}{\sqrt{1+h'^2}}\right) + \left(\gamma_{SL} - \gamma_{SO}\right) = 0 \tag{31}$$

Taking into account $h' = \tan(\pi - \theta) = -\tan\theta$ at the triple line in Fig. 7, Eq. (31) is rewritten as

$$\gamma \cdot \left(\frac{1}{\sqrt{1+\tan^2\theta}}\right) + \left(\gamma_{SL} - \gamma_{SO}\right) = 0 \tag{32}$$

Rearrangement of Eq. (32) gives Young's equation:

$$\cos\theta = \frac{\gamma_{SO} - \gamma_{SL}}{\gamma} \tag{33}$$

where the apparent contact angle, θ, is the equilibrium contact angle (θ_Y).

Young's equation was derived from the transversality condition. It means that the equilibrium contact angle (Young's equation) must be satisfied at the contact line in order to minimize the total energy of the system. The variational approach assures that the contact angle is determined at the contact line. It should be noted that the external field, such as gravity, cannot affect the equilibrium contact angle, although it distorts the shape of the drop.

3.3. Derivation of the Wenzel equation

Fig. 8 shows a symmetrical 3D drop in the Wenzel state. The drop is placed on a rough surface with full contact with the solid surface (no air gap). K is a surface roughness factor that is defined as the ratio of the actual area to the projected area of the substrate. The radius of the drop is a. The smooth region under the drop is b. The contact line of the drop is assumed to be located on the rough region of the surface. The free energy for the drop, $J(h,\ h',\ r)$, is given by the integral

$$J(h,\ h',\ r)=\int_0^a\left[2\pi r\gamma\cdot\sqrt{1+\left(\frac{dh}{dr}\right)^2}+U(r,h)\right]dr+ \\ +\pi b^2(\gamma_{SL}-\gamma_{SO})+K\int_b^a\left[2\pi r(\gamma_{SL}-\gamma_{SO})\right]dr \tag{34}$$

The above equation can be rearranged as

$$J(h,\ h',\ r)=\int_0^a\left[2\pi r\gamma\cdot\sqrt{1+\left(\frac{dh}{dr}\right)^2}+U(r,h)\right]dr+K\int_b^a\left[2\pi r(\gamma_{SL}-\gamma_{SO})\right]dr+ \\ +K\int_0^b\left[2\pi r(\gamma_{SL}-\gamma_{SO})\right]dr-K\int_0^b\left[2\pi r(\gamma_{SL}-\gamma_{SO})\right]dr+\pi b^2(\gamma_{SL}-\gamma_{SO}) \tag{35}$$

where the last two terms are constants, giving

$$C=(1-K)\pi b^2(\gamma_{SL}-\gamma_{SO}) \tag{36}$$

Here, since the contact angle is independent of the absolute value of the total energy, the constant energy term has no effect on the contact angle [17]. Thus, the free energy can be redefined as follows:

$$G(h,\ h',\ r)=J(h,\ h',r)-C=\int_0^a\left[2\pi r\gamma\cdot\sqrt{1+\left(\frac{dh}{dr}\right)^2}+U(r,h)+2\pi rK(\gamma_{SL}-\gamma_{SO})\right]dr \tag{37}$$

Now, considering the constant volume of a drop as a subsidiary condition and the transversality condition, the revised Wenzel equation is derived:

$$\cos\theta=K\frac{\gamma_{SO}-\gamma_{SL}}{\gamma}=K\cos\theta_Y \tag{38}$$

where θ_Y is the equilibrium contact angle on a smooth solid surface and θ is an apparent contact angle.

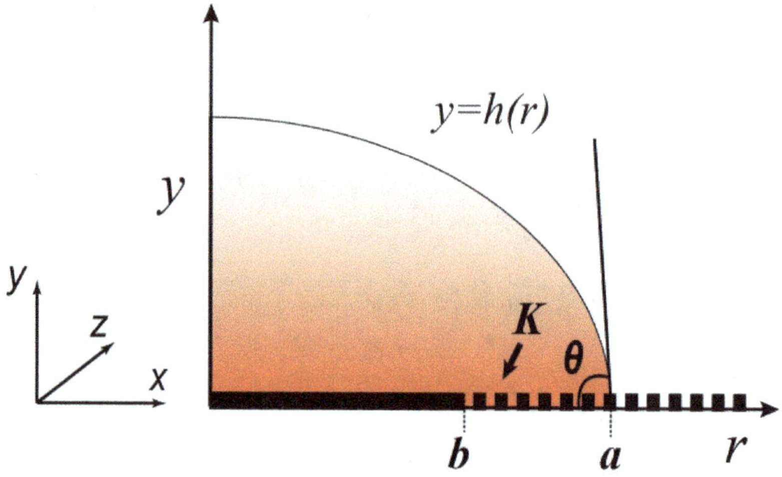

Figure 8. The cross-section of the drop placed on a rough surface.

3.4. Derivation of the Cassie-Baxter equation

Fig. 9 shows a symmetrical 3D drop in the Cassie-Baxter state. The drop is placed on a composite surface consisting of two surfaces. The radius of the drop is a. The free energy for the drop, $G(h,\ h',\ r)$, is given by the integral

$$G(h,\ h',\ r)=\int_0^a\left[2\pi r\gamma\cdot\sqrt{1+\left(\frac{dh}{dr}\right)^2}+U(r,h)\right]dr+ \\ +f_1\int_0^a\left[2\pi r\left(\gamma_{SL}^1-\gamma_{SO}^1\right)\right]dr+f_2\int_0^a\left[2\pi r\left(\gamma_{SL}^2-\gamma_{SO}^2\right)\right]dr \tag{39}$$

where f_1 and f_2 are area fractions of surface 1 and surface 2, respectively. The above equation can be rearranged as

$$G(h,\ h',\ r)=\int_0^a\left[2\pi r\gamma\cdot\sqrt{1+\left(\frac{dh}{dr}\right)^2}+U(r,h)+f_1\left[2\pi r\left(\gamma_{SL}^1-\gamma_{SO}^1\right)\right]+f_2\left[2\pi r\left(\gamma_{SL}^2-\gamma_{SO}^2\right)\right]\right]dr \tag{40}$$

Now, considering the constant volume of a drop as a subsidiary condition and the transversality condition, the revised Cassie-Baxter equation is derived.

$$\cos\theta=f_1\frac{\gamma_{SO}^1-\gamma_{SL}^1}{\gamma}+f_2\frac{\gamma_{SO}^2-\gamma_{SL}^2}{\gamma}=f_1\cos\theta_Y^1+f_2\cos\theta_Y^2 \tag{41}$$

where θ_Y^1 and θ_Y^2 are the equilibrium contact angles on the surface 1 and the surface 2. θ is an apparent contact angle.

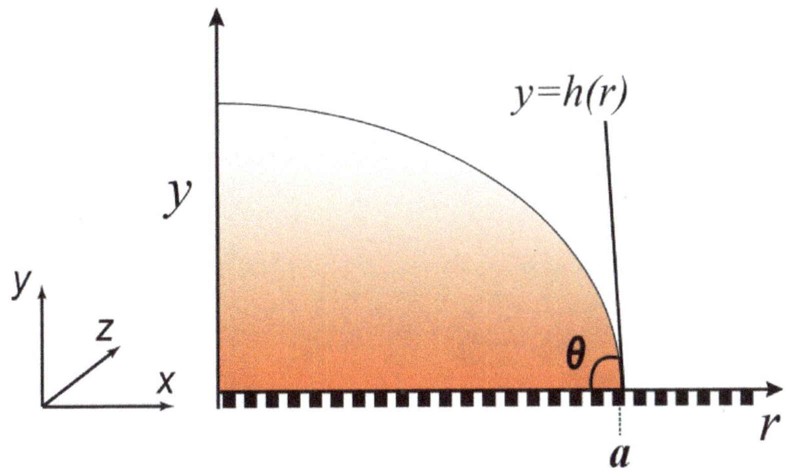

Figure 9. The cross-section of the drop deposited on a composite surface.

Likewise, a general Cassie-Baxter equation for different composite surfaces can be derived [25]. In the revised equation (Eq. (41)), the area fractions are related to only the local region at the contact line.

3.5. Contact angles under other conditions

By employing the variational approach, it is easy to understand the behavior of the contact angles under various conditions. As mentioned above, after obtaining the total energy of the system under each condition, the contact angles are obtained from the transversality condition. In this way, the contact angles of the drops on a gradient surface [19], a rotating surface [26], a curved surface [25], and a surface with an electric field [20] were studied. To conclude, it was demonstrated that if the factors or conditions to be considered do not affect the surface energy and the surface topography near the contact line, they cannot affect the contact angles.

4. Discussion

The contact angle of a drop was considered from a thermodynamic point of view. The contact angle models (Young's equation, Wenzel equation, and Cassie-Baxter equation) were re-derived by the energy minimization and the variational approach. It was clearly demonstrated from the derivations that the contact angle of a drop is a necessary condition that must be satisfied at the contact line in order to minimize the total energy of the system. In other words, a drop takes an optimal contact angle to have the lowest energy of the system. When the optimal contact angle is not satisfied, the total energy of the system is not at a local minimum. Thus, residual force exists at the contact line and changes the shape of the drop until it disappears, as shown in Fig. 2.

Two important points can be deduced from the derivations [19, 27]. Firstly, the contact angle is determined by the infinitesimal region at the contact line. The internal surface inside the contact line does not affect the contact angle. Thus, the roughness factor in the Wenzel equation or the area fraction in the Cassie-Baxter equation should be defined in the contact line region. Secondly, the contact angle is independent of the external factors that do not affect the surface energy. Fig. 10 shows the behavior of the contact angle on an ideal surface under various conditions. The contact angle is not affected by pressure, drop size, gravity, curvature of substrate, rotation of the substrate, and existence of a needle or defects.

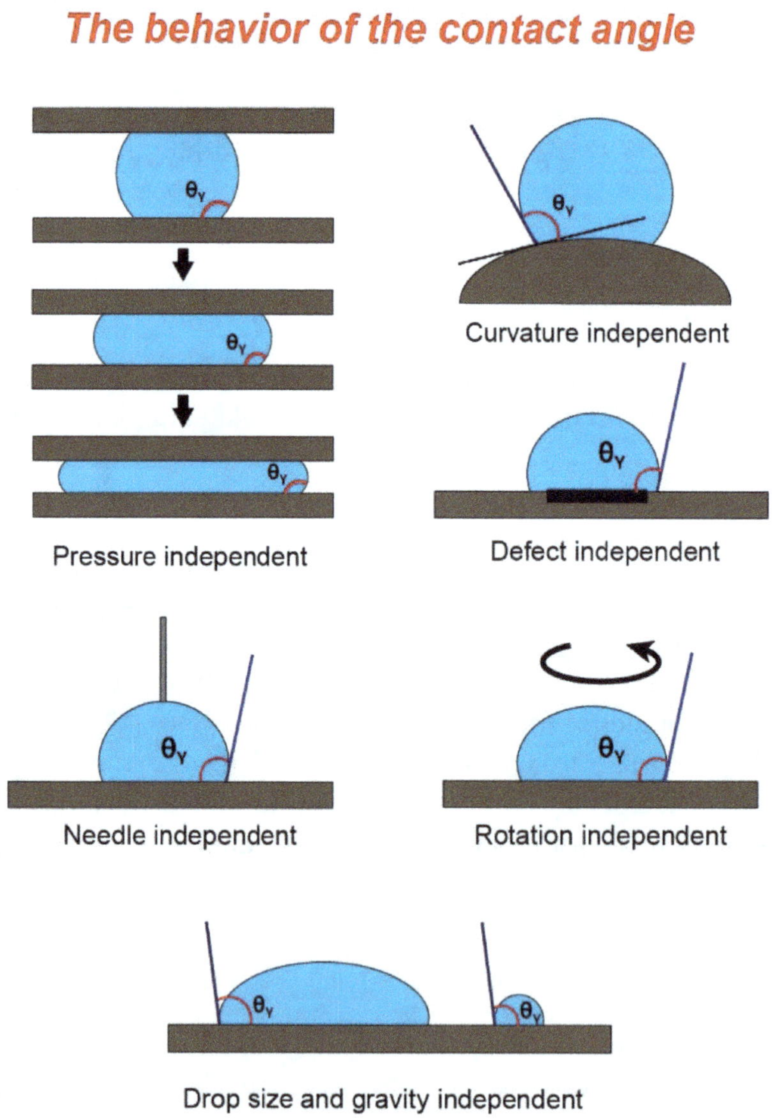

Figure 10. The behavior of the contact angle on an ideal surface under various conditions.

During the derivations, it has been assumed that the contact line moves freely on the surfaces and there is no contact angle hysteresis. However, all of real surfaces are not ideal and have contact angle hysteresis [28]. The contact line cannot move freely on them. The surfaces have a range of static contact angles between two extreme values of an advancing angle and a receding angle [29]. So, the real surfaces are hard to describe with a single equation, while Young's equation, the Wenzel equation, or the Cassie-Baxter equation yield a single contact angle. Especially, when a drop on a rough surface takes the Wenzel state, the contact angle hysteresis is very large violating the assumption of the free movement of the contact line [30, 31] and a specific static contact angle is hardly meaningful to describe the surface [32, 33]. Whereas a superhydrophobic surface has a very low contact angle hysteresis having a narrow range of static contact angles. A drop on the surface takes the Cassie-Baxter state. The contact line can easily move on it. Thus, the superhydrophobic surface can be considered as pseudo-ideal surface from the theoretical viewpoint in that the contact line can move freely. For this reason, the Cassie-Baxter equation has been widely used for the superhydrophobic surface.

In using the contact angle equations, careful attention is required. When the length scale of the pattern in the surface roughness is smaller than an order of a micrometer, additional correction factors should be considered including line tension and disjoining pressure [34-36]. However, the contact angle models do not contain these factors. In addition, they do not consider the shape and size of the pattern, which actually affect the contact angle [37]. Therefore, these equations should be used with caution and more advanced equations should be developed in order to describe the contact angle on the real surface.

5. Conclusion

Young's equation, the Wenzel equation, and the Cassie-Baxter equations were re-derived from a thermodynamic point of view. From the derivations, the behavior of the contact angle could be deduced. In an ideal situation, the contact angle is determined by the infinitesimal region in the vicinity of contact line, not by the internal surface inside the contact line. The contact angle is also independent of the external factors that do not affect the surface energy. Thus, it is not affected by pressure, drop size, gravity, curvature of the substrate surface, rotation of the substrate, and existence of a needle or defects. It was explained from the view point of the contact angle hysteresis why these equations are not proper to describe the real common surfaces although the Cassie-Baxter equation has been widely used for a superhydrophobic surface. Also, the limitations of the equations were discussed. It is expected that this study will provide a deeper understanding of the validity of the contact angle models and the nature of the contact angle.

Author details

Kwangseok Seo, Minyoung Kim and Do Hyun Kim*

*Address all correspondence to: dohyun.kim@kaist.ac.kr

Department of Chemical and Biomolecular Engineering, Korea Advanced Institute of Science and Technology, Daejeon, South Korea

References

[1] Wenzel, R.N. (1936). Resistance of Solid Surfaces to Wetting by Water. *Industrial & Engineering Chemistry,* vol. 28, pp. 988-994, ISSN 0888-5885.

[2] Cassie, A.B.D. & Baxter, S. (1944). Wettability of Porous Surfaces. *Transactions of the Faraday Society,* vol. 40, pp. 546-551, ISSN 0014-7672.

[3] Xu, W., Shi X. & Lu S. (2011). Controlled Growth of Superhydrophobic Films without any Low-Surface-Energy Modification by Chemical Displacement on Zinc Substrates. *Materials Chemistry and Physics,* vol. 129, pp. 1042-1046, ISSN 0254 0584.

[4] Tuteja, A., Choi, W., Ma, M., Mabry, J.M., Mazzella, S.A., Rutledge, G.C., McKinley, G.H. & Cohen, R.E. (2007). Designing Superoleophobic Surfaces. *Science,* vol. 318, pp. 1618-1622, ISSN 1095-9203.

[5] Noh, J., Lee, J.-H., Na, S., Lim, H. & Jung, D.-H. (2010). Fabrication of Hierarchically Micro- and Nano-structured Mold Surfaces Using Laser Ablation for Mass Production of Superhydrophobic Surfaces. *Japanese Journal of Applied Physics,* vol. 49, 106502, ISSN 1347-4065.

[6] Wang, C., Yao, T., Wu, J., Ma, C., Fan, Z., Wang, Z., Cheng, Y., Lin, Q. & Yang, B. (2009). Facile Approach in Fabricating Superhydrophobic and Superoleophilic Surface for Water and Oil Mixture Separation, *ACS Applied Materials & Interfaces,* vol. 1, pp. 2613-2617, ISSN 1944-8252.

[7] Gao, L. & McCarthy, T.J. (2007). How Wenzel and Cassie were Wrong. *Langmuir,* vol. 23, pp. 3762-3765, ISSN 1520-5827.

[8] McHale, G. (2007). Cassie and Wenzel: Were They Really So Wrong? *Langmuir,* vol. 23, pp. 8200-8205, ISSN 1520-5827.

[9] Panchagnula, M.V. & Vedantam, S. (2007). Comment on How Wenzel and Cassie Were Wrong by Gao and McCarthy. *Langmuir,* vol. 23, pp. 13242, ISSN 1520-5827.

[10] Nosonovsky, M. (2007). On the Range of Applicability of the Wenzel and Cassie Equations. *Langmuir,* vol. 23, pp. 9919-9920, ISSN 1520-5827.

[11] Marmur, A. & Bittoun, E. (2009). When Wenzel and Cassie Are Right: Reconciling Local and Global Considerations. *Langmuir,* vol. 25, pp. 1277-1281, ISSN 1520-5827.

[12] Erbil, H.Y. (2014). The Debate on the Dependence of Apparent Contact Angles on Drop Contact Area or Three-Phase Contact Line: A Review. *Surface Science Reports,* vol. 69, pp. 325-365, ISSN 0167-5729.

[13] Choi, W., Tuteja, A., Mabry, J.M., Cohen, R.E. & McKinley, G.H. (2009). A Modified Cassie-Baxter Relationship to Explain Contact Angle Hysteresis and Anisotropy on non-Wetting Textured Surfaces. *Journal of Colloid and Interface Science,* vol. 339, pp. 208-216, ISSN 0021-9797.

[14] Seo, K., Kim, M., & Kim, D.H. (2014). Candle-Based Process for Creating a Stable superhydrophobic surface. *Carbon*, vol. 68, pp. 583-596, ISSN 0008-6223.

[15] ChangWei, Y., Feng, H. & PengFei, H. (2010). The Apparent Contact Angle of Water Droplet on the Micro-Structured Hydrophobic Surface. *Science China Chemistry*, vol. 53, pp. 912-916, ISSN 1869-1870.

[16] Xu, X. & Wang, X. (2010). Derivation of the Wenzel and Cassie Equations from a Phase Field Model for Two Phase Flow on Rough Surface. *SIAM Journal on Applied Mathematics*, vol. 70, pp. 2929-2941, ISSN 0036-1399.

[17] Seo, K., Kim, M., & Kim, D.H. (2013). Validity of the Equations for the Contact Angle on Real Surfaces. *Korea-Australia Rheology Journal*, vol. 25, pp.175-180, ISSN 2093-7660.

[18] Whyman, G., Bormashenko, E. & Stein, T. (2008). The Rigorous Derivation of Young, Cassie-Baxter and Wenzel Equations and the Analysis of the Contact Angle Hysteresis Phenomenon. *Chemical Physics Letters*, vol. 450, pp. 355-359, ISSN 0009-2614.

[19] Bormashenko, E. (2009). A Variational Approach to Wetting of Composite Surfaces: Is Wetting of Composite Surfaces a One-Dimensional or Two-Dimensional Phenomenon. *Langmuir*, vol. 25, pp. 10451-10454, ISSN 1520-5827.

[20] Bormashenko, E. (2012). Contact Angles of Sessile Droplets Deposited on Rough and Flat Surfaces in the Presence of External Fields. *Mathematical Modelling of Natural Phenomena*, vol. 7, pp. 1-5, ISSN 0973-5348.

[21] Gelfand, I.N. & Fomin, S.V. (2000). *Calculus of Variations*, Dover, ISBN-13: 978-0-486-41448-5; ISBN-10: 0-486-41448-5, New York.

[22] Eggers, A.J., Resnikoff, M.M., & Dennis, D.H. (1957). Bodies of Revolutions Having Minimum Drag at High Supersonic Air Speeds, NACA Report No. 1306.

[23] Filobello-Nino, U., Vazquaz-Leal, H., Pereyra-Diaz, D., Yildirim, A., Peres-Sesma, A., Castaneda-Sheissa, R., Sanchez-Orea, J. & Hoyos-Reyes, C. (2013) A Generalization of the Bernoulli's Method Applied to Brachistochrone-like Problems. *Applied Mathematics and Computation*, vol. 219, pp. 6707-6718, ISSN 0096-3003.

[24] Mareno, A. & English, L.Q. (2009). The Stability of the Catenary Shapes for a Hanging Cable of Unspecified Length. *European Journal of Physics*, vol. 30, pp. 97-108, ISSN 1361-6404.

[25] Bormashenko, E. (2009). Wetting of Flat and Rough Curved Surfaces. *The Journal of Physical Chemistry C*, vol. 113, pp. 17275-17277, ISSN 1932-7455.

[26] Bormashenko E. (2013). Contact Angles of Rotating Sessile Droplets. *Colloids and Surfaces A: Physicochemical and Engineering Aspects*, vol. 432 pp. 38-41, ISSN 0927-7757.

[27] Seo, K., Kim, M., & Kim, D.H. (2015). Effects of Drop Size and Measuring Condition on Static Contact Angle Measurement on a Superhydrophobic Surface with Goniometric Technique. *Korean Journal of Chemical Engineering*, in print, ISSN 1975-7220.

[28] Gleiche, M., Chi, L., Geding, E. & Fuchs, H., (2001). Anisotropic Contact-Angle Hysteresis of Chemically Nanostructured Surfaces. *Chem Phys Chem*, vol. 2, pp. 187-191, ISSN 1439-7641.

[29] Marmur, A. (2006). Soft Contact: Measurement and Interpretation of Contact Angles. *Soft Matter*, vol. 2. pp. 12-17, ISSN 1744-6848.

[30] Morra, M., Occhiello, E., & Garbassi, F. (1989). Contact Angle Hysteresis in Oxygen Plasma Treated Poly(tetrafluoroethylene). *Langmuir*, vol. 5, pp.872-876, ISSN 1520-5827.

[31] Cao, L. & Gao, D. (2010). Transparent Superhydrophobic and Highly Oleophobic Coatings. *Faraday Discussions*, vol. 146, pp. 57-65, ISSN 1359-6640.

[32] Gao, L. & McCarthy, T.J. (2006) Contact Angle Hysteresis Explained. *Langmuir*, vol. 22, pp. 6234-6237, ISSN 1520-5827.

[33] Gao, L. & McCarthy, T.J. (2009). Wetting 101°. *Langmuir*, vol. 25, pp. 14105-14115, ISSN 1520-5827.

[34] de Gennes, P.G., Brochard-Wyart, F., & Quéré, D. (2003). *Capillarity and Wetting Phenomena*, Springer, ISBN 0-387-00592-7, Berlin.

[35] Wong, T.-S. & Ho, C.-M. (2009). Dependence of Macroscopic Wetting on Nanoscopic Surface Textures. *Langmuir*, vol. 25, pp. 12851-12854, ISSN 1520-5827.

[36] Marmur, A. & Krasovitski, B. (2002). Line Tension on Curved Surfaces: Liquid Drops on Solid Micro- and Nanospheres. *Langmuir*, vol. 18, pp. 8919-8923, ISSN 1520-5827.

[37] Cansoy, C.F., Erbil, H.Y., Akar, O., & Akin, T. (2011). Effect of Pattern Size and Geometry on the Use of Cassie-Baxter Equation for Superhydrophobic Surfaces. *Colloids and Surfaces A: Physicochemic*al and Engineering Aspects, vol. 386, pp. 116-124, ISSN 0927-7757.

5

Micro/Nano Hierarchical Super-Lyophobic Surfaces Against Gallium-Based Liquid Metal Alloy

Daeyoung Kim and Jeong-Bong Lee

Abstract

Against a gallium-based liquid metal alloy, we report super-lyophobic surfaces that have micro/nano hierarchical structures made of polydimethylsiloxane (PDMS) and carbon nanotube (CNT) materials. The surface oxidation of the liquid metal is a challenging problem to be utilized for various applications, as it wets most surfaces. However, the micro/nano hierarchical structures of a fabricated-PDMS micro pillar array with dual-scale surface texturing and a grown-CNT surface enable one to minimize a contact area between the liquid metal droplet and the surface. Due to the low contact area, a large static contact angle and a low contact angle hysteresis are achieved, indicating super-lyophobic surfaces. Based on these super-lyophobic surfaces, the gallium-based liquid metal alloy can be more widely utilized for undeveloped applications that rely on the liquid metal's mobility.

Keywords: Gallium-based liquid metal alloy, super-lyophobic, micro/nano hierarchical, contact angle, CNT, PDMS

1. Introduction

Gallium-based liquid metal alloy is a non-toxic liquid metal at room temperature. Recently, it has been of interest as the gallium-based liquid metal shows higher thermal and electrical conductivities, and lower vacuum pressure [1] compared to a toxic liquid metal, such as mercury. Based on these favorable material properties, it has been utilized for various applications including micro-cooling [2], electrical interconnector [3], 3D printing [4], tunable frequency antennas [5-11], metamaterials [12, 13], and electrodes [14-18]. However, most of these applications operate on the deformability of the liquid metal, not on the mobility. This is because it has a very challenging problem that the liquid metal alloy is easily oxidized when it is exposed to ambient air environment. Once it is oxidized, the surface becomes viscous and it makes the liquid metal wet on the surfaces [19]. Thus, there are two ways to utilize the oxidized gallium-based liquid metal alloy: 1) modifying or removing the oxidized surface of the liquid metal and 2) creating the non-wettable channel surface.

There have been a few studies to modify or remove the oxidized surface of the liquid metal. Zrnick and Swatik found that it was possible to remove the oxidized surface of gallium-based liquid metal by treating the surface with diluted hydrochloric acid (HCl) [20]. We also demonstrated that the HCl vapor even can modify the wettable Ga_2O_3/Ga_2O to non-wettable $GaCl_3/InCl_3$ [21]. Based on this method, the oxidized gallium-based liquid metal can be controllable in the microfluidic channels composed of a gas-permeable PDMS [22]. Sivan et al. reported that the liquid metal surface can be simply coated with nano-particles of insulators (including Teflon and silica) and semiconductors (including WO_3, TiO_2, MoO_3, In_2O_3, and carbon nanotubes) [23]. Then, this liquid metal marble can also be non-wettable by covering the oxidized surface. We also reported that the oxidized surface can be covered with ferro-magnetic micro-particles and it can be manipulated with an applied magnetic field [24]. Khan et al. also investigated that the oxidized surface of the liquid metal in the presence of electrolyte can be removed by applying a reductive potential [25].

Alternatively, if there is any surface where the oxidized liquid metal moves without any wetting problem, it may enable one to utilize the liquid metal for new applications that rely on constant and reliable motion of liquid metal. In order to utilize liquid metal's mobility, the surface should be non-wetting and super-lyophobic. The methods of making these surfaces can be analogous to those of creating the well-known super-hydrophobic surfaces.

Over the past decades, super-hydrophobic surfaces have been tremendously investigated due to its wide applicability for self-cleaning [26-28] and anti-icing effect [29, 30]. Conventionally, there are two distinctive ways to engineer the surface to be super-hydrophobic. One way is to make a rough surface to reduce the contact area between the liquid and the solid. The other way is to coat the surface with low surface energy materials such as a fluorocarbon polymer. From nature, one can find water-repelling surfaces such as shark scales and lotus leaves. Those surfaces have one thing in common—that they have micro/nano hierarchical structures to realize the rough surface [31]. This approach can also be applied to the oxidized liquid metal. Thus, the contact area between the liquid metal and the surface can be as small as possible, resulting in the super-lyophobic surface against the oxidized gallium-based liquid metal alloy.

In this chapter, the realization of super-lyophobic surfaces against gallium-based liquid metal alloy, micro/nano hierarchical structure-based surfaces using different materials such as polydimethylsiloxane (PDMS) and carbon nanotube (CNT) are discussed. Wetting phenomena such as lyophobic/lyophilic concepts and wetting models such as Cassie and Wenzel states are described. As demonstrations of super-lyophobic surfaces based on micro/nano hierarchical structure, the lyophobicity of PDMS micro pillar array with surface texturing is characterized by measuring static and dynamic contact angles and compared with those of the PDMS micro pillar and ridge arrays without any surface texturing. In addition, the lyophobicity of CNT on

PDMS is characterized by measuring static and dynamic contact angles, as well as bouncing and rolling experiments under mechanical deformation.

2. Theory

Lyophobic/lyophilic is defined as a physical property of a molecule that is repelled from or attracted to a mass of liquid. Therefore, the term 'lyophobic or lyophilic' needs to be used to indicate a target liquid except when they are 'hydrophobic/hydrophilic' or 'oleophobic/oleophilic', implying that the target liquid is water or oil, respectively. Therefore, 'lyophobic/lyophilic" is a superordinate concept to 'hydrophobic/hydrophilic' and 'oleophobic/oleophilic'.

In order to characterize lyophobicity, the static and dynamic contact angles are important parameters. The static contact angles can be measured by using a goniometer that can capture the profile of a liquid droplet on a solid substrate. The static contact angle depends on the surface tensions of liquid (liquid/gas interfacial energy), solid (solid/gas interfacial energy), and liquid/solid interactions. As shown in Fig. 1a, the static contact angle is defined as an angle formed by the intersection of the liquid-solid interface and the liquid-gas interface. Generally, when the contact angle is smaller than 90° the solid surface is considered to be 'lyophilic', which means wetting of the surface is favorable. When the contact angle is larger than 90° the solid surface is considered 'lyophobic', which indicates the liquid minimizes its contact on the surface and form a compact liquid droplet. In addition, when the contact angle is greater than 150°, the solid surface is considered to be 'super-lyophobic'.

For dynamic contact angle measurements, the volume of liquid was increased or decreased until the droplet had a maximum or minimum contact angle without changing the surface area between the liquid and the solid substrate while the side view of a droplet was recorded. Each of maximum and minimum contact angles is the advancing and receding contact angle, respectively; and the advancing and receding contact angles were obtained from the frame of the recorded movie just before the contact line is changed. The difference between the advancing angle and the receding angle is the contact angle hysteresis. When the advancing and receding contact angle of liquid on a solid substrate are close to each other (lower contact angle hysteresis), it indicates the substrate is lyophobic to the liquid.

Theoretically, as shown in Fig. 1a, Thomas Young described in 1805 the contact angle of a liquid droplet on an ideal solid substrate[32]. It is defined by the mechanical equilibrium of surface tensions of three interfaces under the droplet:

$$\cos\theta = \frac{\gamma_{SG} - \gamma_{SL}}{\gamma_{LG}} \tag{1}$$

where θ is the contact angle, and γ_{SG}, γ_{SL} , and γ_{LG} are surface tension of solid-gas, solid-liquid, and liquid-gas interfaces, respectively. Equation (1) is usually referred as Young's equation.

Since Young's equation only works with a flat homogeneous surface, the other expanded approaches are needed to describe contact angle of a droplet on rough and heterogeneous surfaces: Wenzel model [33] and Cassie model [34].

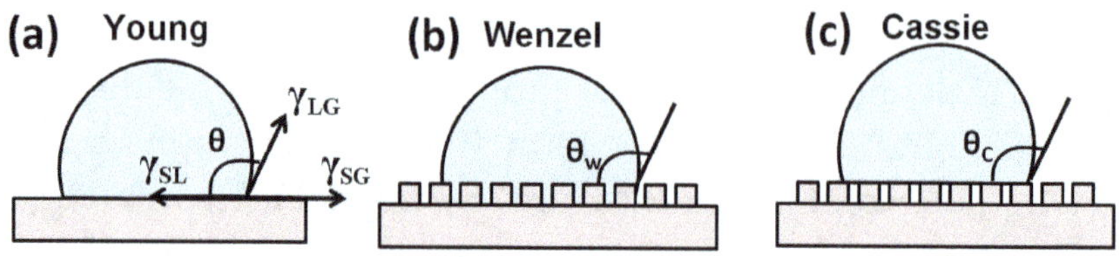

Figure 1. Schematics of different wetting regimes: (a) Young's model, (b) Wenzel model, and (c) Cassie model.

As shown in Fig. 1b, when a liquid can fully wet the surface texture, the thermodynamic equilibrium contact angle of a liquid droplet is described by the Wenzel model [33] as the following equation:

$$\cos\theta_W = r\cos\theta \tag{2}$$

where θ_w is the apparent contact angle of the Wenzel model, θ refers to Young's contact angle, and r is a roughness factor, which is defined as the ratio of an actual area of a rough surface to the flat, projected area. Droplets in this fully wetted Wenzel state typically display very high hysteresis because the contact line of the droplets becomes severely pinned on surface asperities.

On the other hand, when a liquid cannot penetrate into the surface texture, the droplet forms a highly non-wetting regime known as a Cassie state (Fig. 1c). As the surface texture entraps myriads of air pockets underneath the droplet, it can have low adhesive force. Therefore, the droplet in the Cassie state displays a very high contact angle, as well as a low hysteresis, resulting in easy rolling off of a droplet. If f_S is the fraction of the solid in contact with the liquid, the Cassie equation can be represented by Equation (3) [34].

$$\cos\theta_C = f_S(1+\cos\theta)-1 \tag{3}$$

Therefore, in order to achieve a super-lyophobic surface, the liquid droplet on the roughened structures should be in the Cassie state. It is well known that micro/nano hierarchical surface textures are critically important for this effect.

3. Micro/nano hierarchical PDMS micro pillar array with surface texturing[1]

We studied super-lyophobicity of a micro/nano-sized hierarchical PDMS pillar array against an oxidized gallium-based liquid metal alloy (Galinstan®) [35]. We first tested two types of

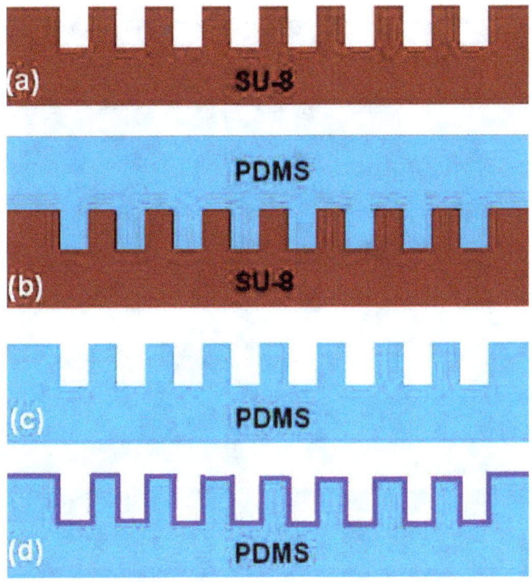

Figure 2. Fabrication sequence of the super-lyophobic micro-textured PDMS surfaces: (a) patterned SU-8 mold, (b) PDMS casting, (c) replicated PDMS, and (d) FC polymer deposition on the replicated PDMS micro arrays.

surface textured PDMS micro arrays: micro pillar arrays and micro ridge arrays. The vertical height of all PDMS micro textures was designed to be 45 μm. The fabrication of PDMS micro pillar and micro ridge arrays started with spin coating of SU-8 2025 photoresist (PR) (MicroChem Corp.) on a thermally grown oxidized Si wafer to get approximately 45 μm-thick PR. The PR was soft-baked on a hot plate at 65°C for 2 minutes, 95°C for 3 minutes, and finally 65°C for 2 minutes. Then, a UV exposure dose of 140 mJ/cm^2 was applied by standard photolithography to create an inverse pattern of the target structure, and finally a post bake was applied with the same temperature and time of the soft bake. Finally, it was developed in SU-8 developer (propylene glycol methyl ether acetate) for 3 minutes under gentle stirring condition, and was cleaned with isopropyl alcohol and deionized (DI) water (Fig. 2a). After developing, the PDMS was casted over the PR mold; then it was cured at room temperature for a day to have flexible characteristics (Fig. 2b). After curing, PDMS was peeled off from the mold and it replicated arrays (Fig. 2c). An 80 nm-thick fluorocarbon (FC) polymer was deposited on top of the PDMS arrays to enhance lyophobicity (Fig. 2d).

PDMS micro pillar arrays had various dimensions including 25 μm pillar diameter and 25 μm gap, 75 μm diameter and 25 μm gap, 75 μm diameter and 100 μm gap, 75 μm diameter and 150 μm gap, 75 μm diameter and 200 μm gap, and 75 μm diameter and 250 μm gap. PDMS micro ridge arrays had various widths and gaps (25 μm width and 25 μm gap, 75 μm width and 25 μm gap, 75 μm width and 100 μm gap, 75 μm width and 150 μm gap, 75 μm width and 200 μm gap). Fig. 3 shows scanning electron microscope (SEM) images of the micro pillar arrays (Figs. 3a and 3b) and micro ridge arrays (Figs. 3c and 3d).

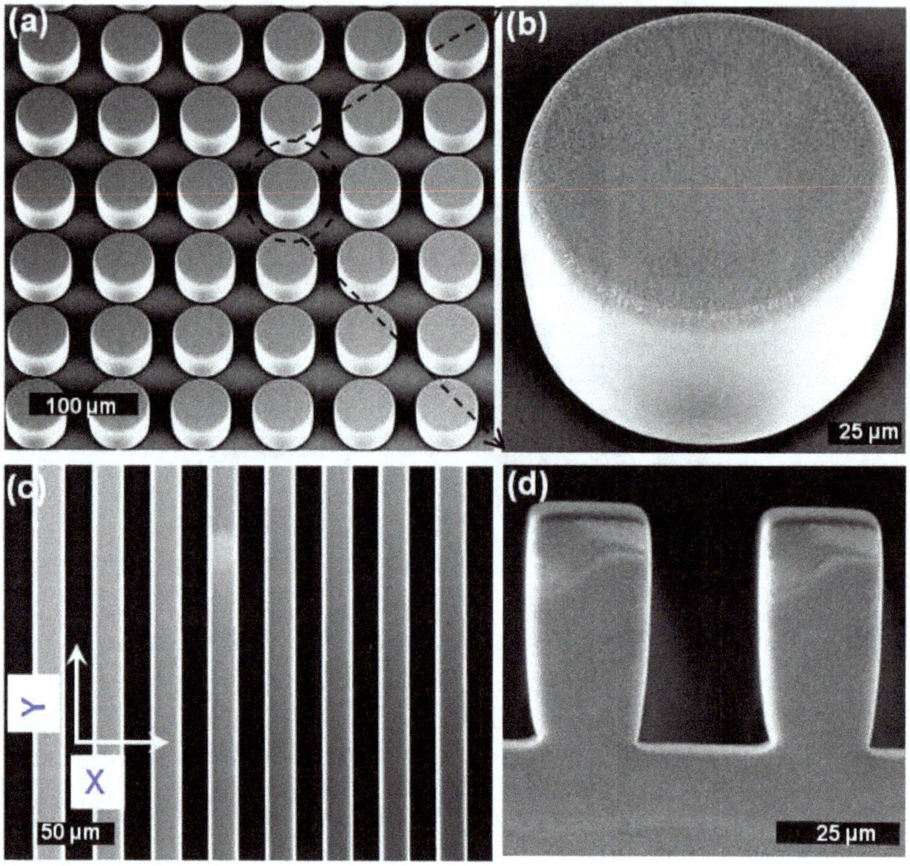

Figure 3. SEM images of PDMS micro arrays: (a) micro pillar array and (b) a close-up, 15° tilted view of a micro pillar; (c) micro ridge array and (d) a close-up side view of micro ridges.

Lyophobicity of textured surfaces was studied by measuring static and dynamic contact angles. Fig. 4 shows the images of static droplets of 7.8 μL oxidized Galinstan® on micro pillar arrays with the same pillar diameter (75 μm) but with various gap distances. Galinstan® droplet is not spherical due to the viscoelastic oxidized layer on the surface. The droplet on a 225 μm pitch distance (75 μm diameter and 150 μm gap) micro pillar array clearly formed a non-wetting Cassie state, confirmed by the presence of air pockets underneath the droplet (Fig. 4a). We observed the droplets on all micro pillar arrays with gap distances smaller than 150 μm formed a Cassie state, as it becomes more difficult for a Galinstan® to penetrate between micro pillars when the gap distance becomes smaller. On the other hand, the image of the Galinstan® droplet on a micro pillar array with 250 μm gap distance showed that the droplet penetrated between pillars, forming a fully wetted Wenzel state (Fig. 4c). We also found that the droplets on any micro pillar array with gap distance larger than 250 μm showed Wenzel state. There were intermediate transition states between the Cassie and the Wenzel states. Figure 4b shows Galinstan® droplet on a 200 μm gap distance. The droplet penetrated into a portion of the texture, but we could observe some intact air pockets as well. It was also found that micro ridge arrays with gap distances smaller than 150 μm (75 μm ridge width) showed Cassie state. Table 1 shows static contact angles for various pitch distances in the range of 50 to 225 μm.

Since the contact area of the micro pillar arrays is smaller than that of the micro ridge arrays for the same pitch distances, the static contact angle of the micro pillar arrays is relatively larger than that of the micro ridge arrays. The static contact angle of 139.1° on a pillar array with 75 μm diameter and 150 μm gap distance (solid fraction = 0.087) is not remarkably high, considering the extremely low solid fraction, due to the highly hysteretic behavior of the oxidized Galinstan®.

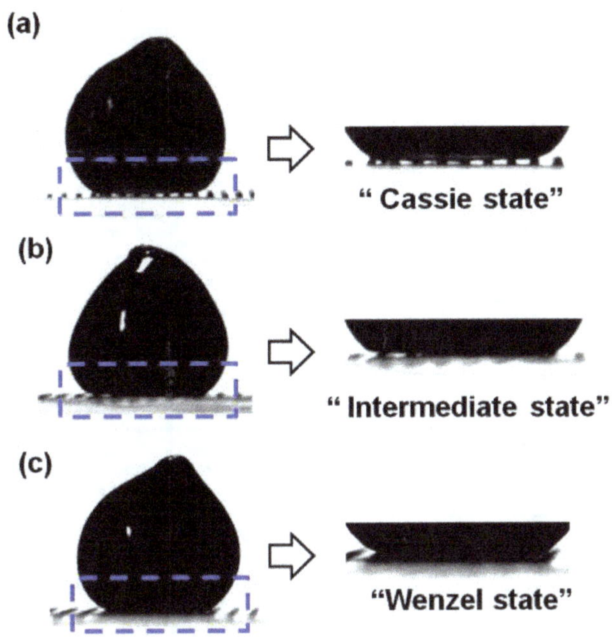

Figure 4. A 7.8 μL oxidized Galinstan® droplet on (a) 150 μm gap micro pillar array showing a non-wetting Cassie state, (b) 200 μm gap micro pillar array showing an intermediate transition state, and (c) 250 μm gap micro pillar array (no air pockets) showing a fully wetted Wenzel state. The diameter of all micro pillars is 75 μm.

<table>
<tr><th>Substrates</th><th>Diameter / width (μm)</th><th>Pitch distance (μm)</th><th>Solid fraction (f_s)</th><th>Static contact angle (°)</th></tr>
<tr><td rowspan="4">Micro pillar array</td><td>25</td><td>50</td><td>0.196</td><td>141.6 ± 0.2</td></tr>
<tr><td rowspan="3">75</td><td>100</td><td>0.442</td><td>135.8 ± 1.2</td></tr>
<tr><td>175</td><td>0.144</td><td>140.6 ± 3.5</td></tr>
<tr><td>225</td><td>0.087</td><td>139.1 ± 4.4</td></tr>
<tr><td rowspan="4">Micro ridge array</td><td>25</td><td>50</td><td>0.500</td><td>139.5 ± 1.1</td></tr>
<tr><td rowspan="3">75</td><td>100</td><td>0.750</td><td>131.4 ± 1.8</td></tr>
<tr><td>175</td><td>0.429</td><td>133.5 ± 3.7</td></tr>
<tr><td>225</td><td>0.334</td><td>137.9 ± 1.2</td></tr>
</table>

Table 1. Static contact angles for a 7.8 μL oxidized Galinstan® droplet on PDMS micro pillar and micro ridge arrays.

The dynamic contact angles were measured in the same way as described in the previous section. Table 2 shows dynamic contact angles of the oxidized Galinstan® droplet on micro pillar arrays and micro ridge arrays with various dimensions. For the micro ridge arrays, contact angles were measured from two different viewing directions when the droplet was advancing/receding: across the ridges (X direction; see Fig. 3c) and along the ridges (Y direction).

Substrates	Pitch distance (μm)	Solid fraction (f_s)	Dynamic contact angles (°)		Contact angle hysteresis (°)
			Adv.	Rec.	
Micro pillar array	50	0.196	179.0	75.3	103.7
	100	0.442	163.5	51.3	112.2
	175	0.144	160.5	77.9	82.6
	225	0.087	153.5	51.0	102.5
Micro ridge array (X direction, across the ridges)	50	0.500	161.0	50.0	111.0
	100	0.750	161.8	20.5	141.3
	175	0.429	161.5	43.3	118.2
	225	0.334	163.5	40.0	123.5
Micro ridge array (Y direction, along the ridges)	50	0.500	155.3	38.3	117.0
	100	0.750	156.8	16.8	140.0
	175	0.429	156.8	32.5	124.3
	225	0.334	158.3	37.0	121.3

Table 2. Dynamic contact angles of the oxidized Galinstan® droplet.

For simple fluids, such as water or hexadecane, the contact angle hysteresis when a droplet advances/recedes along the ridges (Y direction) is less than that on micro pillar arrays with the same solid fraction as, for micro pillar array, the solid fraction near the contact line significantly changes between advancing and receding droplets [36]. However, for Galinstan®, we found that the contact angle hystereses on the micro ridge arrays in both directions (X and Y) are larger than those on the micro pillar arrays. We suspect that the viscoelasticity of Galinstan®, combined with its extreme wettability, causes the receding of the droplet from the top of pillars and ridges virtually impossible, leading to the discrepancy between the conventional model of hysteresis and the observed values. The pillar array with 75 μm diameter and 100 μm gap distance showed the smallest contact angle hysteresis (82.6°) among all textured surfaces. Nonetheless, the residue of the oxidized Galinstan® was still left on the surface after receding even though the droplet formed the Cassie state (Fig. 5).

Although there was decrease in contact angle hysteresis from those of the various flat surfaces to the surfaces textured with PDMS micro arrays, it turned out that the naturally oxidized Galinstan® droplet did not completely dewet from all tested surfaces, regardless of the shape and pitch distance of the PDMS micro arrays.

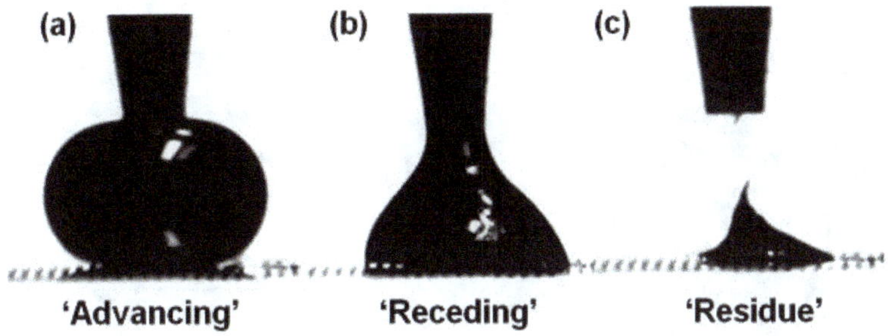

Figure 5. Advancing angle (160.5°), receding angle (77.9°), and residual state of the oxidized Galinstan® on the micro pillar array with 75 μm diameter and 100 μm gap distance.

To further improve lyophobicity of the PDMS micro arrays, we further studied lyophobicity of multi-scale surface textured PDMS micro pillar arrays. The fabrication of multi-scale surface textured PDMS micro pillar arrays started with triple-spin coating of AZP4620 PR (AZ Electronic materials) on a thermally oxidized Si wafer to get approximately a 45 μm-thick PR. The PR was intentionally under-baked in a convection oven at 88°C (typical soft-bake temperature is 110°C on a hot plate). Then, the PR was under-exposed with the exposure dose of 800 mJ/cm^2 (typical exposure dose is > 1,600 mJ/cm^2) [37]. After developing the PR with diluted AZ400K (AZ Electronic materials) (Fig. 6a), the PDMS was casted over the PR mold and it was cured at room temperature for one day (Fig. 6b). The replicated PDMS showed multi-scale surface textured micro pillar array (Fig. 6c). The 80 nm-thick FC polymer was then deposited on the replicated multi-scale surface textured PDMS micro pillar array to enhance the lyophobicity (Fig. 6d). Fig. 7 shows the SEM image of multi-scale surface textured PDMS micro pillar arrays.

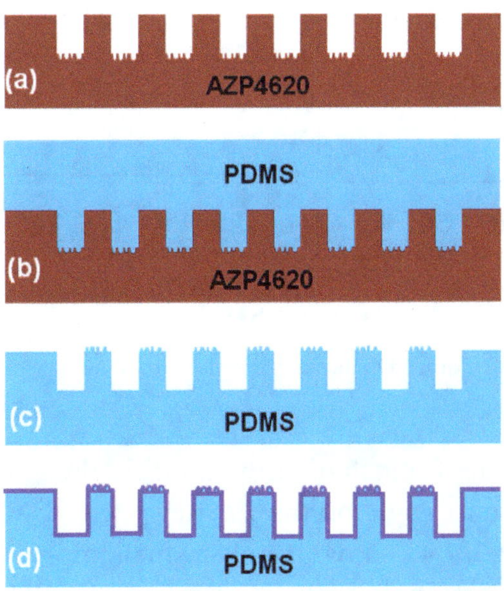

Figure 6. Fabrication sequence of the multi-scale surface textured PDMS micro pillars: (a) fabrication of AZP4620 mold with multi-scale textures (under-baked and under-exposed), (b) PDMS casting, (c) replicated PDMS, (d) FC polymer deposition on the replicated multi-scale surface textured PDMS micro pillars.

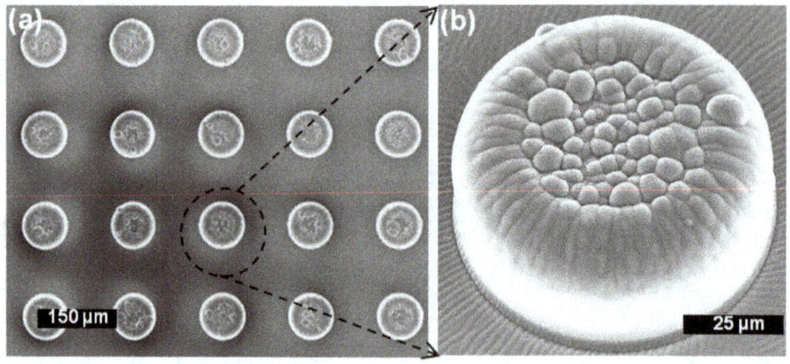

Figure 7. SEM images of (a) multi-scale surface textured PDMS micro pillar arrays and (b) close-up view of a single pillar that is textured with additional roughness.

The static and dynamic contact angles were studied on the multi-scale PDMS micro pillar arrays for the naturally oxidized Galinstan® droplet. Fig. 8 shows static contact angles as a function of pitch distance in the range of 50 ~ 225 μm for the multi-scale PDMS micro pillar arrays as well as PDMS micro pillar and ridge arrays.

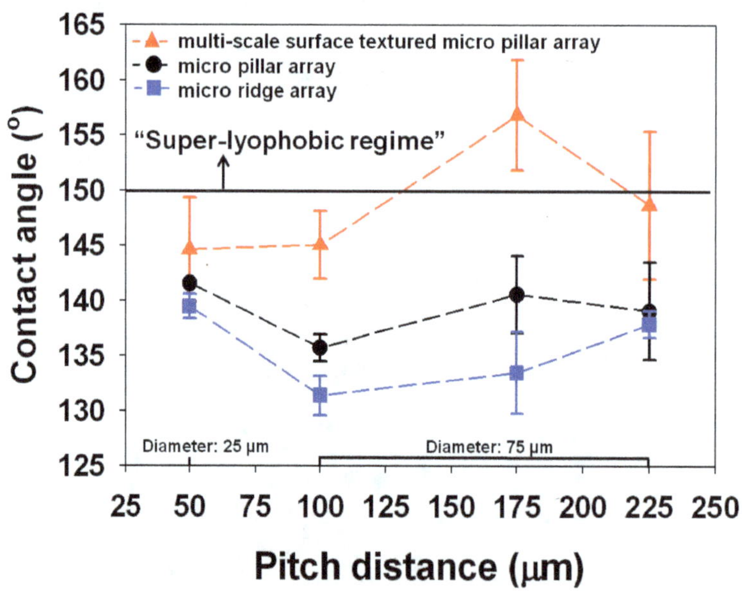

Figure 8. Contact angles of a 7.8 μL oxidized Galinstan® droplet on a multi-scale surface textured micro pillar array, micro pillar, and ridge arrays.

The static contact angle of the Galinstan® droplet on the FC polymer-deposited flat PDMS surface was 125.4°. The contact angle on the 100 μm pitch PDMS micro ridge array, on the other hand, turned out to be 131.4°, showing only a small improvement in lyophobicity of the surface. However, the contact angle of the Galinstan® droplet on the multi-scale surface patterned PDMS micro pillar array was found to be substantially higher, to be close to the super-lyophobic regime (> 150°). The 175 and 225 μm pitch distance multi-scale surface textured PDMS micro pillar arrays showed contact angles of 156.9° and 148.7°, respectively.

Regardless of the pitch distance, the static contact angle on the multi-scale surface patterned micro pillar arrays is the highest compared to other textured surfaces. We believe that this is attributed to the lower solid fraction (f_s) based on the additional roughness of the multi-scale surface pattern on the micro pillar arrays. The solid fraction of the multi-scale surface textured micro pillar array cannot be directly calculated because the surface is irregularly patterned. Therefore, the solid fraction of the multi-scale surface textured PDMS micro pillar array was estimated by comparing static contact angles of the micro pillar arrays to those of the multi-scale surface textured PDMS micro pillar arrays. The experimentally determined solid fraction of the 175 μm pitch distance multi-scale surface patterned micro pillar array is 0.105 which is much smaller than that of the micro ridge array (0.429) and of the micro pillar array (0.144).

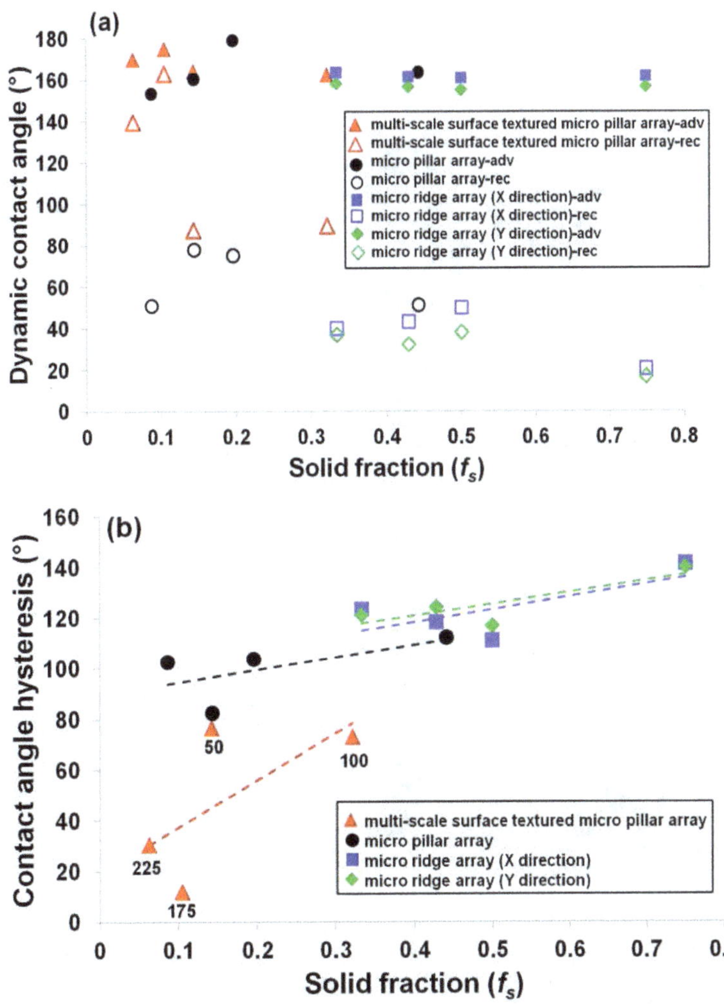

Figure 9. (a) The measured advancing and receding contact angles, and (b) contact angle hysteresis of multi-scale surface patterned micro pillar arrays, micro pillar and ridge arrays as a function of solid fraction (numbers in the graph indicates pitch distance in μm). The colored dot lines are trend lines for multi-scale surface patterned micro pillar arrays, micro pillar and ridge arrays.

Fig. 9a shows the advancing and receding contact angles as a function of the solid fraction of micro pillar arrays, micro ridge arrays and multi-scale surface patterned micro pillar arrays.

The advancing contact angles are larger than 150° for all micro arrays, but the receding contact angles vary drastically as a function of the solid fraction. Of various micro arrays, the 175 μm pitch distance (75 μm diameter pillar and 100 μm gap) multi-scale surface patterned PDMS micro pillar arrays showed the largest advancing and receding contact angles of approximately 175° and 163° (Figs. 10a and 10b), respectively. Fig. 9b shows the contact angle hysteresis as a function of the solid fraction for various micro arrays. The contact angle hysteresis increases as the solid fraction increases. The smallest contact angle hysteresis (12°) was obtained on the 175 μm pitch distance multi-scale surface patterned PDMS micro pillar array, while the 100 μm pitch distance micro ridge array (X-direction) showed the largest contact angle hysteresis of 140°.

With the smallest contact angle hysteresis, we found that there remains no residue of oxidized Galinstan® after the droplet recedes from the surface (Fig. 10c) with the 175 μm pitch distance multi-scale surface patterned PDMS micro pillar array.

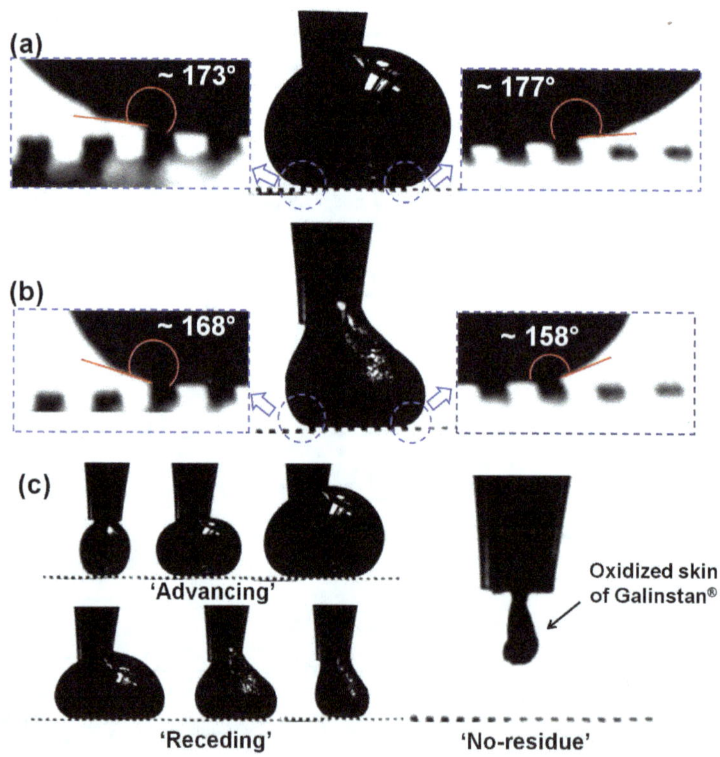

Figure 10. Optical images of an oxidized Galinstan® droplet on the multi-scale surface patterned PDMS micro pillar arrays (pitch distance 175 μm) during dynamic contact angle test: (a) advancing contact angle (θ_{adv} = 175°), (b) receding contact angle (θ_{rec} = 163°), and (c) evidence of no-residue after receding from the surface.

4. Micro/nano hierarchical carbon nanotube on PDMS[2]

CNTs show a unique material property of super-repellency against both water and oil. This is attributed to its hierarchical micro/nano structures, as well as its chemical inertness. While an individual CNT is nanometer scale, bundles of CNT forests have a micrometer scale creating a micro/nano hierarchical surface. Using this property, the super-lyophobicity of CNT on PDMS against the oxidized gallium-based liquid metal is reviewed in this section [38].

Fig. 11 shows the schematics of transferring CNT onto PDMS sheets. First, vertically aligned multi-wall carbon nanotube (MWCNT) was grown by chemical vapor deposition (CVD) process on an iron catalyst deposited by e-beam evaporation (Fig. 11a). Approximately 400 μm tall vertically aligned MWCNT was grown on a piece of Si wafer (9 mm x 13.5 mm) for 5 minutes at 780°C with C_2H_2, He, and H_2 gas [39]. RTV 615 A and B (HISCO, Inc.) were mixed in 10:1 volume ratio and the calculated volume (4.5 mL) of PDMS was poured to make approximately 500 μm-thick PDMS on a Teflon substrate (9.5 cm x 9.5 cm) placed on a petri dish. The poured PDMS was cured at room temperature for 24 hours in air environment. It is known that the higher mechanical flexibility of PDMS can be obtained with lower curing temperature [40]. In order to transfer the MWCNT onto the PDMS sheet, we utilized a nano-imprinter (Obducat Nano-Imprinter NIL-2.5" Imprint). Even though the transferring of MWCNT onto the PDMS can be simply achieved by stamping [41, 42], we applied force to have more uniform contact so that the CNT can be well transferred. We flipped the CNT-grown Si wafer upside down and placed it on top of 500 μm-thick flexible PDMS. Then, in order to imprint, we applied a force of 400 N for 20 seconds. (Fig. 11c). After the Si wafer was removed, we found that the CNT on Si was well transferred onto the PDMS sheet (Fig. 11d).

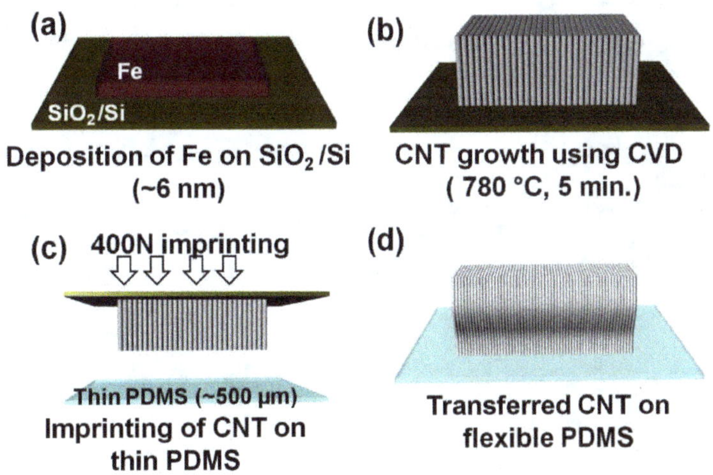

Figure 11. Fabrication sequence of the stretchable and bendable super-lyophobic CNT on PDMS sheet: (a) iron catalyst on Si, (b) vertically aligned CNT growth on Si, (c) imprinting CNT onto PDMS sheet, and (d) transferred CNT on PDMS sheet.

Fig. 12 shows optical and SEM images of the grown CNT on Si (Fig. 12a) and the transferred CNT on PDMS sheet (Fig. 12b). The grown CNT on Si was vertically aligned and its average height was 400 μm (Fig. 12c). The top-view SEM image of the grown CNT on Si clearly shows

---

2 © 2014 IOP. Portions reprinted, with permission from: D. Kim, D. Jung, J. H. Yoo, Y. Lee, W. Choi, G. S. Lee,and J.-B. Lee, "Stretchable and bendable carbon nanotube on PDMS super-lyophobic sheet for liquid metal manipulation," Journal of Micromechanics and Microengineering, vol. 24, p. 055018, 2014.

micron-scale bundled MWCNT, while individual CNTs show nanoscale surface morphology (Fig. 12d). After transferring, the height of the transferred MWCNT on the PDMS sheet, which was still vertically aligned, shrunk from 400 μm to 60 μm (Fig. 12e). We found that even after the transfer process, it maintained micro/nano hierarchical surface morphology (Fig. 12f) but the size of micron scale morphology was reduced.

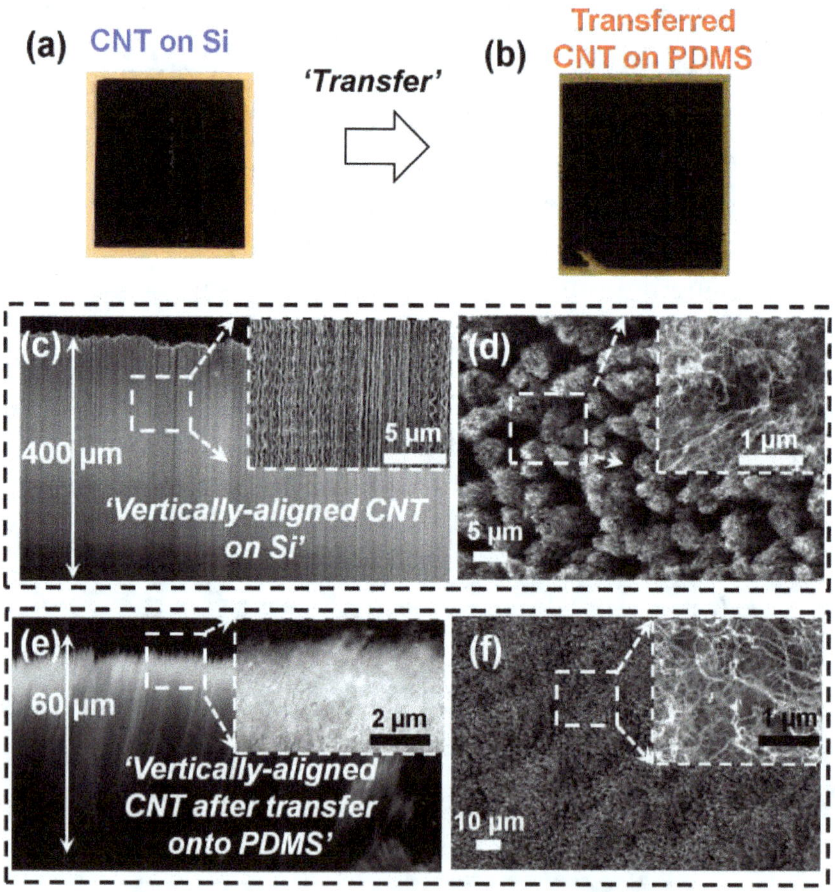

Figure 12. Optical and SEM images of CNT on Si and PDMS: optical images of (a) grown CNT on Si; (b) transferred CNT on PDMS sheet; (c) side-view SEM image of the vertically aligned CNT on Si; (d) top-view SEM image of the CNT showing micro/nano scale mixed morphology; (e) side-view SEM image of vertically aligned CNT after transfer onto the PDMS sheet; and (f) top-view SEM image of micro/nano surface morphology after the transfer.

In order to study the lyophobicity of the transferred CNT on PDMS sheet under mechanical deformation against the oxidized gallium-based liquid metal alloy (Galinstan®), we stretched and bent the CNT on the PDMS sheet. PDMS is well known for its physical flexibility due to its extremely low Young's modulus (< 4 MPa) [43]. The CNT on PDMS sheet was stretched by 50% and 100% ratio and bent with a curvature of 0.1 mm^{-1} and 0.4 mm^{-1}, respectively, as shown in Fig. 13.

Fig. 14 shows the contact angle measurement of a liquid droplet on a lyophobic flat surface and a negatively curved surface. The intrinsic contact angle (θ_{il}, θ_{ir}) on the flat surface is formed by a straight line tangential to the droplet at its left side (L_{cl}) or right side (L_{cr}) with a horizontal

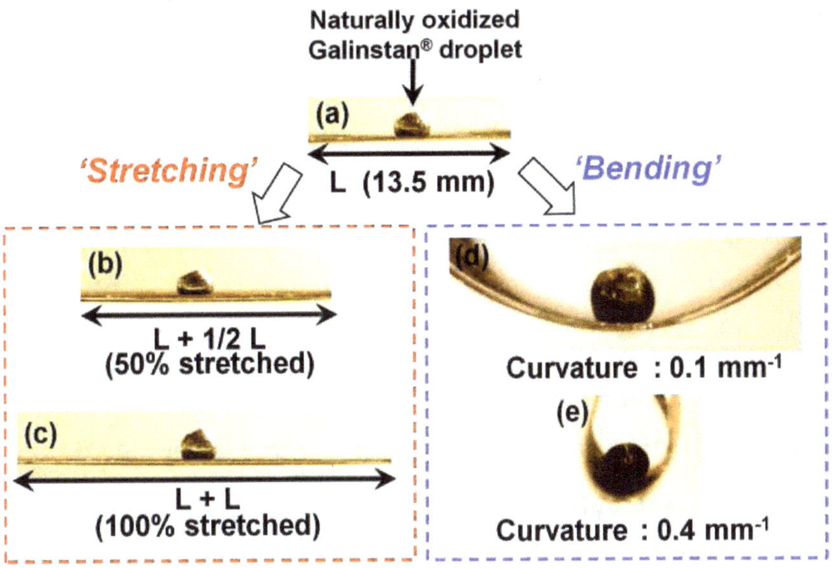

Figure 13. Stretchable and bendable CNT on PDMS sheet: (a) Galinstan® droplet on CNT on PDMS sheet, (b, c) 50% and 100% stretched CNT on PDMS sheet, and (d, e) bent CNT on PDMS sheet with a curvature of 0.1 mm^{-1} and 0.4mm^{-1}.

line (L_h). There is, however, no horizontal line for the contact angle measurement on a curved surface. Thus, we assume the shape of the curved substrate to be a part of a circle whose center is defined as '*o*'. We then identify the intersection point between the circle and the outline of the droplet. Tangential lines drawn at the left and right intersection points are used as the bottom lines to define the contact angles. The left contact angle (θ_{cl}) of a droplet on the curved surface is the exterior angle formed by a tangential line (l_{cl}) to the droplet at the left side and the tangential line to the circle at the left intersection point (l_{tl}). In the same manner, the right contact angle (θ_{cr}) of a liquid droplet on the curved surface is determined.

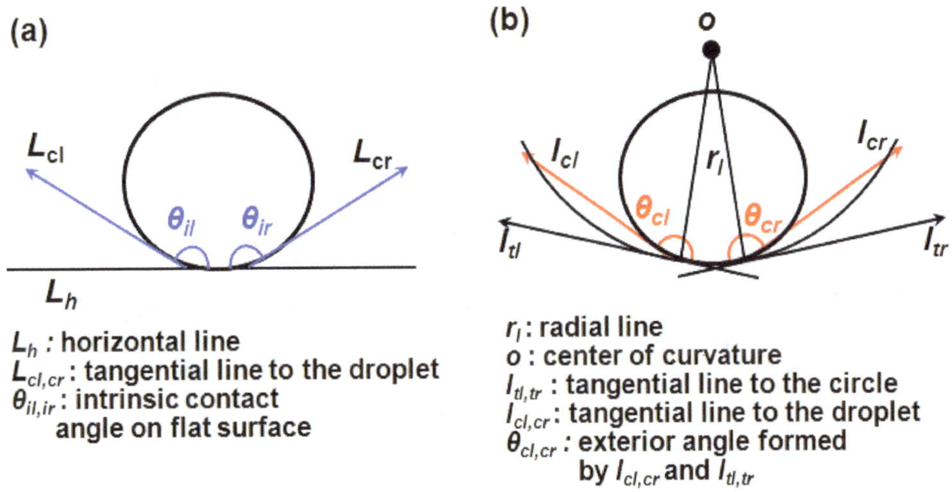

Figure 14. Schematics of contact angle measurement of a liquid droplet on (a) a lyophobic flat surface and (b) a curved surface.

We measured static and dynamic contact angles to study the lyophobicity of various surfaces against the oxidized gallium-based liquid metal alloy (Galinstan®): i) CNT grown on Si substrate; ii) transferred CNT on PDMS sheet; iii) stretched CNT on PDMS with 50% and 100% extension ratio; and iv) bent CNT on PDMS with a curvature of 0.1 mm^{-1} and 0.4 mm^{-1}.

Fig. 15 shows static contact angles of ~7.8 μL oxidized Galinstan® droplet on various surfaces. We found that all static contact angles were very close to or greater than 150°, which is a super-lyophobic regime. This is attributed to micro/nano scale combined hierarchical structure of the vertically aligned CNT. As shown in Figs. 12e and 12f, the transferred CNT on PDMS sheet maintained a vertically aligned structure as well as micro/nano scale combined surface morphology. As a result, on the transferred CNT on PDMS a very similar static contact angle of 156.2 ± 5.8° compared to that of the CNT on Si substrate was obtained.

Comparing to that of the CNT on PDMS without stretching, the contact angles on the stretched CNT on PDMS sheet with a 50% and 100% extension ratio were slightly decreased. We believe that the vertically aligned CNT may be slightly tilted or collapsed down so that the surface morphology was changed by stretching the PDMS substrate. However, the static contact angle change before and after the stretching is in the range of error bar and there is no clear trend of change in the static contact angle due to stretching. It is noted that they are still in the super-lyophobic regime even with the decreased static contact angles. The contact angle of the oxidized gallium-based liquid metal (Galinstan®) droplet placed on the negatively curved surface with a curvature of 0.1 mm^{-1} was slightly decreased to 153.8 ± 5°. The contact angle on the curved surface with a curvature of 0.4 mm^{-1}, however, is essentially the same (156.8 ± 3.3°) compared to that of the flat CNT on PDMS which is 156.2 ± 5.8°. Again, as those contact angles are in the range of error bar, we can conclude that there is no significant change in hierarchical micro/nano dual scale surface morphology of the CNT on PDMS by bending or stretching.

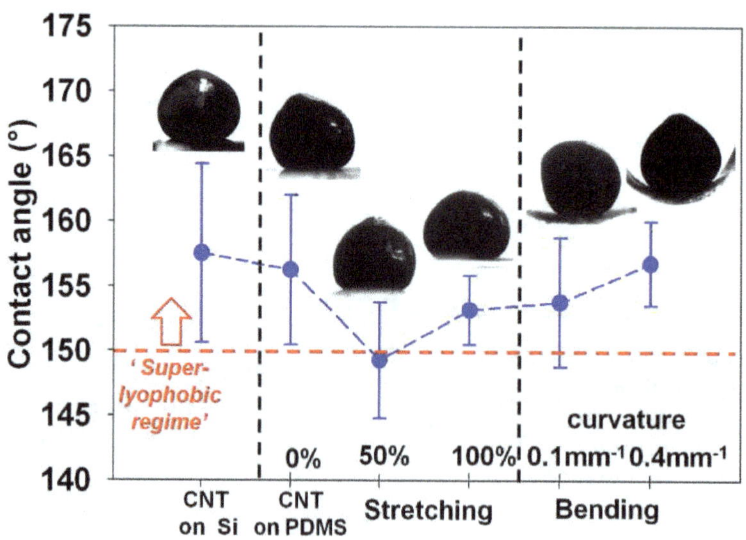

Figure 15. Static contact angle of ~7.8 μL oxidized Galinstan® droplet on various surfaces.

In conjunction with static contact angle, dynamic contact angle is another important indicator of lyophobicity, especially for the droplet manipulation. We measured dynamic contact angles of the gallium-based liquid metal alloy (Galinstan®) on the CNT on Si substrate, as well as CNT on PDMS with and without bending and stretching (Fig. 16). The red-colored circles indicate advancing contact angles and blue-colored squares indicate receding contact angles, respectively. The advancing and receding contact angles of the oxidized Galinstan® droplet on the CNT on Si substrate were 165° and 161° (optical images of the advancing and receding angle measurement are shown in the left-side inset), respectively. As a result, the contact angle hysteresis (advancing contact angle - receding contact angle) was found to be 4°. The lyophobicity seemed somewhat decreased after the CNT transfer process. The advancing contact angles on the transferred CNT on PDMS sheet, the stretched and bent CNT on PDMS sheet were all greater than 150°. The receding contact angles of the CNT on PDMS before and after bending and stretching were in the range of 133° ~ 140°. The smallest contact angle hysteresis of ~11° (optical images of the advancing and receding angle measurement are shown in the right-side inset) was found on the 100% stretched CNT on PDMS sheet and the largest contact angle hysteresis of ~22° was found on the 50% stretched CNT on PDMS sheet. The contact angle hysteresis of the transferred CNT on PDMS was ~19°.

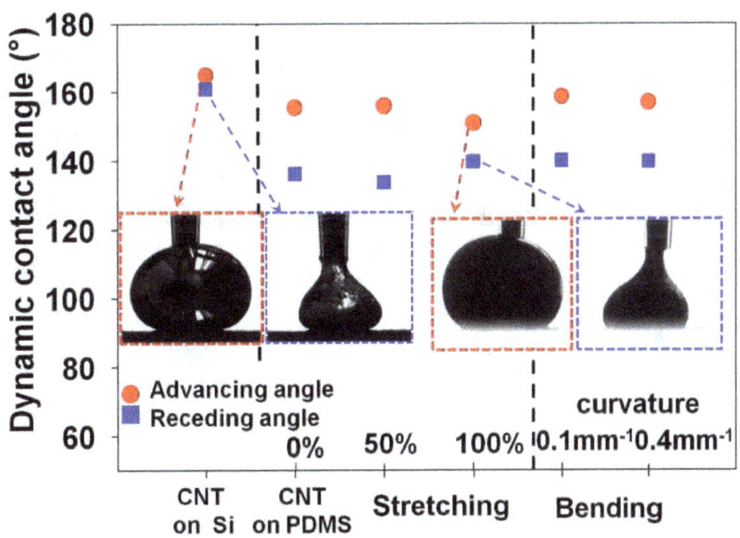

Figure 16. Dynamic contact angle of oxidized Galinstan® on various surfaces.

Based on the static and dynamic contact angle measurements of the oxidized Galinstan® droplet, we expected that the transferred CNT on PDMS sheet can be used as a stretchable and bendable super-lyophobic surface against the oxidized gallium-based liquid metal droplet. To confirm this, we carried out a series of bouncing tests for the transferred CNT on PDMS sheet with and without stretching.

The oxidized gallium-based liquid metal droplet (~7.8 μL) was dropped 3.5 cm above the surface and impacted the substrate while the movement of the droplet was recorded using a high-speed camera (Fastcam SA4, Photron Limited, Tokyo, Japan) with 1,000 frames per second. Note that the oxidation of the droplet is nearly instant in ambient air environment that the Galinstan® droplet becomes oxidized on the fly and it is already fully oxidized by the moment of impact at the surface. Fig. 17 shows a series of time-lapse images of bouncing of the oxidized Galinstan® droplet from the surface of the CNT on Si substrate, the transferred CNT on PDMS sheet with and without stretching. Before the droplet hit the surface, the shape of the droplet is not spherical due to its viscoelastic oxide layer. Nonetheless, after the oxidized liquid metal droplet hit the surfaces, the droplet spread on the surfaces and readily bounced off from all surfaces. The results from the bouncing tests indicated that these surfaces have anti-wetting characteristics against the oxidized gallium-based liquid metal droplet. In conjunction with the measured static and dynamic contact angles, this confirms that the transferred CNT on PDMS sheet is a stretchable super-lyophobic surface against the oxidized gallium-based liquid metal droplet.

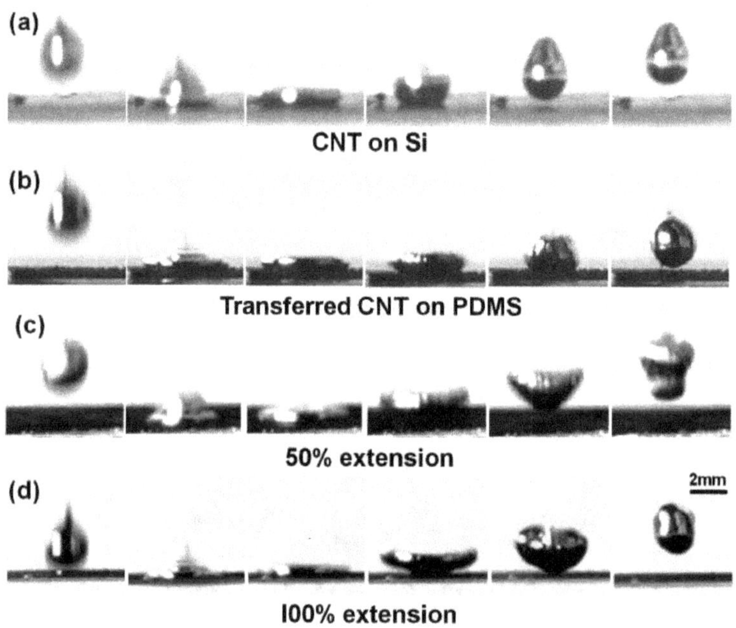

Figure 17. A series of time-lapse images of bouncing of the Galinstan® droplet from the surface of (a) CNT on Si, (b) transferred CNT on PDMS sheet without strain, (c) CNT on PDMS with 50% strain, and (d) with 100% strain.

To test anti-wetting characteristics of the bent CNT on PDMS sheet, we carried out rolling tests. The CNT on PDMS sheet were bent to have negatively curved surfaces with curvature of 0.1 mm^{-1} and 0.4 mm^{-1} as shown in Fig. 13. On the curved surface with a curvature of 0.1 mm^{-1} and 0.4 mm^{-1}, the oxidized liquid metal droplet with volume of ~7.8 μL and ~1.6 μL were utilized, respectively, for rolling tests.

Fig. 18 shows a series of time-lapse images of the rolling Galinstan® droplet on the negatively curved CNT on PDMS surfaces. The droplet was dropped from the left-end edge of the bent CNT on PDMS. On the curved surface with a curvature of 0.1 mm^{-1} and 0.4 mm^{-1}, the droplet was readily moved to the right side of the curved surface with an average speed of ~11.9 and ~26.2 cm/sec, respectively, and it was stopped at the lowest point of the curved surface. After the rolling test, the surface of the CNT on PDMS was thoroughly inspected and it was found that there was no oxidized liquid metal residue remaining after the rolling of the liquid metal droplet.

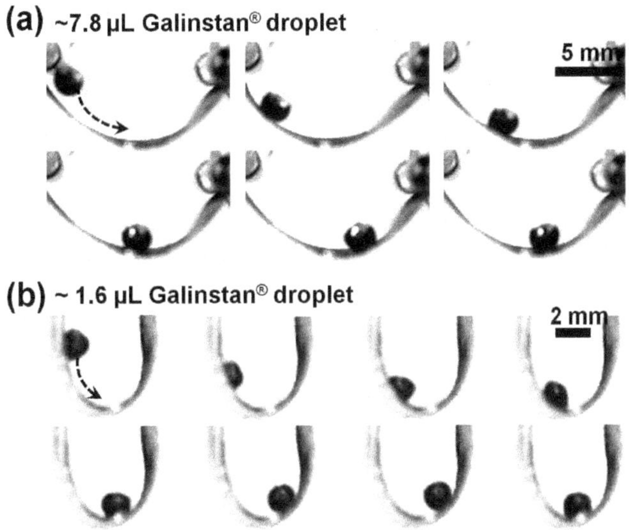

Figure 18. A series of time-lapse images of rolling of Galinstan® droplet on the bent CNT on PDMS sheet with a curvature of (a) 0.1 mm^{-1} and (b) 0.4 mm^{-1}.

5. Conclusions

We have reviewed concepts of 'lyophobic/lyophilic' and studied wetting phenomena on the planar and roughened surfaces: Young's relation and Cassie/Wenzel model. These theoretical studies emphasize that micro/nano hierarchical structure can be attributed to the super-lyophobicity against the oxidized gallium-based liquid metal alloy. As demonstrations of micro/nano hierarchical structure-based surfaces, the PDMS micro pillar array with dual scale surface texturing and CNT on PDMS are described.

Author details

Daeyoung Kim[1*] and Jeong-Bong Lee[2]

*Address all correspondence to: jokuksarang64@gmail.com

1 Department of Electrical Engineering, Korea Army Academy, Yeong-cheon, Kyungsang-buk-do, South Korea

2 Department of Electrical Engineering, The University of Texas at Dallas, Richardson, TX, USA

References

[1] T. Liu, P. Sen, and C.-J. Kim, "Characterization of Nontoxic Liquid-Metal Alloy Galinstan for Applications in Microdevices," *Journal of Microelectromechanical Systems*, vol. 21, p. 443, 2012.

[2] K.-Q. Ma and J. Liu, "Heat-driven liquid metal cooling device for the thermal management of a computer chip," *Journal of Physics D: Applied Physics*, vol. 40, p. 4722, 2007.

[3] H.-J. Kim, C. Son, and B. Ziaie, "A multiaxial stretchable interconnect using liquid-alloy-filled elastomeric microchannels," *Applied Physics Letters*, vol. 92, p. 011904, 2008.

[4] C. Ladd, J.-H. So, J. Muth, and M. D. Dickey, "3D Printing of Free Standing Liquid Metal Microstructures," *Advanced Materials*, vol. 25, p. 5081, 2013.

[5] B. Aissa, M. Nedil, M. A. Habib, E. Haddad, W. Jamroz, D. Therriault, *et al.*, "Fluidic patch antenna based on liquid metal alloy/single-wall carbon-nanotubes operating at the S-band frequency," *Applied Physics Letters*, vol. 103, p. 063101, 2013.

[6] S. Cheng, A. Rydberg, K. Hjort, and Z. Wu, "Liquid metal stretchable unbalanced loop antenna," *Applied Physics Letters*, vol. 94, p. 144103, 2009.

[7] G. J. Hayes, J.-H. So, A. Qusba, M. D. Dickey, and G. Lazzi, "Flexible Liquid Metal Alloy (EGaIn) Microstrip Patch Antenna," *Antennas and Propagation, IEEE Transactions on*, vol. 60, p. 2151, 2012.

[8] A. J. King, J. F. Patrick, N. R. Sottos, S. R. White, G. H. Huff, and J. T. Bernhard, "Microfluidically Switched Frequency-Reconfigurable Slot Antennas," *Antennas and Wireless Propagation Letters, IEEE*, vol. 12, p. 828, 2013.

[9] M. Kubo, X. Li, C. Kim, M. Hashimoto, B. J. Wiley, D. Ham, *et al.*, "Stretchable Microfluidic Radiofrequency Antennas," *Advanced Materials*, vol. 22, p. 2749, 2010.

[10] C. Shi, W. Zhigang, P. Hallbjorner, K. Hjort, and A. Rydberg, "Foldable and Stretchable Liquid Metal Planar Inverted Cone Antenna," *Antennas and Propagation, IEEE Transactions on*, vol. 57, p. 3765, 2009.

[11] D. Kim, R. G. Pierce, R. Henderson, S. J. Doo, K. Yoo, and J.-B. Lee, "Liquid metal actuation-based reversible frequency tunable monopole antenna," *Applied Physics Letters*, vol. 105, p. 234104, 2014.

[12] T. S. Kasirga, Y. N. Ertas, and M. Bayindir, "Microfluidics for reconfigurable electromagnetic metamaterials," *Applied Physics Letters*, vol. 95, p. 214102, 2009.

[13] J. Wang, S. Liu, S. Guruswamy, and A. Nahata, "Reconfigurable liquid metal based terahertz metamaterials via selective erasure and refilling to the unit cell level," *Applied Physics Letters*, vol. 103, p. 221116, 2013.

[14] N. Hallfors, A. Khan, M. D. Dickey, and A. M. Taylor, "Integration of pre-aligned liquid metal electrodes for neural stimulation within a user-friendly microfluidic platform," *Lab on a Chip,* vol. 13, p. 522, 2013.

[15] H. Huan and L. Chang, "Characterizations and optimization of electrical contact between nanocomposite elastomer and metal," in *Solid-State Sensors, Actuators and Microsystems Conference, 2009. TRANSDUCERS 2009. International,* 2009, p. 1103.

[16] B. L. Mellor, N. A. Kellis, and B. A. Mazzeo, "Note: Electrode polarization of Galinstan electrodes for liquid impedance spectroscopy," *Review of Scientific Instruments,* vol. 82, p. 046110, 2011.

[17] J.-H. So and M. D. Dickey, "Inherently aligned microfluidic electrodes composed of liquid metal," *Lab on a Chip,* vol. 11, p. 905, 2011.

[18] J.-H. So, H.-J. Koo, M. D. Dickey, and O. D. Velev, "Ionic Current Rectification in Soft-Matter Diodes with Liquid-Metal Electrodes," *Advanced Functional Materials,* vol. 22, p. 625, 2012.

[19] F. Scharmann, G. Cherkashinin, V. Breternitz, C. Knedlik, G. Hartung, T. Weber, *et al.,* "Viscosity effect on GaInSn studied by XPS," *Surface and Interface Analysis,* vol. 36, p. 981, 2004.

[20] D. Zrnic and D. S. Swatik, "On the resistivity and surface tension of the eutectic alloy of gallium and indium," *Journal of the Less Common Metals,* vol. 18, p. 67, 1969.

[21] D. Kim, P. Thissen, G. Viner, D.-W. Lee, W. Choi, Y. J. Chabal, *et al.,* "Recovery of Nonwetting Characteristics by Surface Modification of Gallium-Based Liquid Metal Droplets Using Hydrochloric Acid Vapor," *ACS Applied Materials & Interfaces,* vol. 5, p. 179, 2013.

[22] G. Li, M. Parmar, D. Kim, J.-B. Lee, and D.-W. Lee, "PDMS based coplanar microfluidic channels for the surface reduction of oxidized Galinstan," *Lab on a Chip,* vol. 14, p. 200, 2014.

[23] V. Sivan, S.-Y. Tang, A. P. O'Mullane, P. Petersen, N. Eshtiaghi, K. Kalantar-zadeh, *et al.,* "Liquid Metal Marbles," *Advanced Functional Materials,* vol. 23, p. 144, 2013.

[24] D. Kim and J.-B. Lee, "Magnetic-field-induced liquid metal droplet manipulation," *Journal of the Korean Physical Society,* vol. 66, p. 282, 2015.

[25] M. R. Khan, C. Trlica, and M. D. Dickey, "Recapillarity: Electrochemically Controlled Capillary Withdrawal of a Liquid Metal Alloy from Microchannels," *Advanced Functional Materials,* vol. 25, p. 671, 2015.

[26] R. Blossey, "Self-cleaning surfaces [mdash] virtual realities," *Nature Materials,* vol. 2, p. 301, 2003.

[27] R. Fürstner, W. Barthlott, C. Neinhuis, and P. Walzel, "Wetting and Self-Cleaning Properties of Artificial Superhydrophobic Surfaces," *Langmuir,* vol. 21, p. 956, 2005.

[28] A. Nakajima, K. Hashimoto, T. Watanabe, K. Takai, G. Yamauchi, and A. Fujishima, "Transparent Superhydrophobic Thin Films with Self-Cleaning Properties," *Langmuir,* vol. 16, p. 7044, 2000.

[29] S. Farhadi, M. Farzaneh, and S. A. Kulinich, "Anti-icing performance of superhydrophobic surfaces," *Applied Surface Science,* vol. 257, p. 6264, 2011.

[30] L. Cao, A. K. Jones, V. K. Sikka, J. Wu, and D. Gao, "Anti-Icing Superhydrophobic Coatings," *Langmuir,* vol. 25, p. 12444, 2009.

[31] B. Bhushan, Y. C. Jung, and K. Koch, "Micro-, nano- and hierarchical structures for superhydrophobicity, self-cleaning and low adhesion," *Philosophical Transactions of the Royal Society A: Mathematical, Physical and Engineering Sciences,* vol. 367, p. 1631, 2009.

[32] T. Young, "An Essay on the Cohesion of Fluids," *Philosophical Transactions of the Royal Society of London,* vol. 95, p. 65, 1805.

[33] R. N. Wenzel, "Resistance of Solid Surfaces to Wetting by Water," *Industrial & Engineering Chemistry,* vol. 28, p. 988, 1936.

[34] S. Baxter and A. B. D. Cassie, "8—The Water Repellency of Fabrics and a New Water Repellency Test," *Journal of the Textile Institute Transactions,* vol. 36, p. T67, 1945.

[35] D. Kim, D.-W. Lee, W. Choi, and J.-B. Lee, "A Super-Lyophobic 3-D PDMS Channel as a Novel Microfluidic Platform to Manipulate Oxidized Galinstan," *Journal of Microelectromechanical Systems,* vol. 22, p. 1267, 2013.

[36] W. Choi, A. Tuteja, J. M. Mabry, R. E. Cohen, and G. H. McKinley, "A modified Cassie–Baxter relationship to explain contact angle hysteresis and anisotropy on non-wetting textured surfaces," *Journal of Colloid and Interface Science,* vol. 339, p. 208, 2009.

[37] Y. Yoon, D.-W. Lee, J.-H. Ahn, J. Sohn, and J.-B. Lee, "One-step fabrication of optically transparent polydimethylsiloxane artificial lotus leaf film using under-exposed under-baked photoresist mold," in *Micro Electro Mechanical Systems (MEMS), 2012 IEEE 25th International Conference on,* 2012, p. 301.

[38] D. Kim, D. Jung, J. H. Yoo, Y. Lee, W. Choi, G. S. Lee, *et al.,* "Stretchable and bendable carbon nanotube on PDMS super-lyophobic sheet for liquid metal manipulation," *Journal of Micromechanics and Microengineering,* vol. 24, p. 055018, 2014.

[39] D. Jung, K. H. Lee, D. Kim, D. Burk, L. J. Overzet, and G. S. Lee, "Highly Conductive Flexible Multi-Walled Carbon Nanotube Sheet Films for Transparent Touch Screen," *Japanese Journal of Applied Physics,* vol. 52, p. 03BC03, 2013.

[40] J. C. Lötters, W. Olthuis, P. H. Veltink, and P. Bergveld, "The mechanical properties of the rubber elastic polymer polydimethylsiloxane for sensor applications," *Journal of Micromechanics and Microengineering,* vol. 7, p. 145, 1997.

[41] Y. Zhou, L. Hu, and G. Grüner, "A method of printing carbon nanotube thin films," *Applied Physics Letters,* vol. 88, p. 123109, 2006.

[42] D. Kim and K.-S. Yun, "Patterning of carbon nanotube films on PDMS using SU-8 microstructures," *Microsystem Technologies,* vol. 19, p. 743, 2013.

[43] D. Fuard, T. Tzvetkova-Chevolleau, S. Decossas, P. Tracqui, and P. Schiavone, "Optimization of poly-di-methyl-siloxane (PDMS) substrates for studying cellular adhesion and motility," Microelectron. Eng., vol. 85, p. 1289, 2008.

6

Recent Advances in the Methods for Designing Superhydrophobic Surfaces

Harinarayanan Puliyalil, Gregor Filipič and Uroš Cvelbar

Abstract

The investigations of superhydrophobicity and self-cleaning surfaces have been given a lot of attention in the last few decades. The surfaces having water contact angle larger than 90° are termed as hydrophobic surfaces and those which exhibit contact angle higher than 150° are said to be superhydrophobic. Such surfaces were first observed in nature in various plants and animals, for example, lotus leaf-like structures. Water repellence of various materials have shown great influences on various applications such as self-cleaning, anti-ageing, water-oil separation, water corrosion in electrical industry, water proof textiles, controlled transportation of fluids, etc. Generally, surface micro/nanostructuring combined with low surface energy of materials leads to extreme anti-wetting properties. The hundreds of research articles and more than 450 patents on the subject of nature mimicking self-cleaning surfaces prove the potential of this topic.

Self-cleaning property depends on both surface morphology and surface chemistry. For achieving superhydrophobic surfaces, we can, typically, either increase the roughness of the intrinsically hydrophobic material or tune desired roughness and morphology on hydrophilic surfaces. Depending on such parameters a water droplet on the surface attains either Wenzel or Cassie-Baxter state.

For the preparation of superhydrophobic surfaces, various physical and chemical methods have been successfully used. Methods such as hydrothermal process, using various templates, plasma surface modifications, physical and chemical vapour deposition, layer by layer deposition, electrospinning and sol-gel processing have been used for achieving desired roughness and surface chemistry on various substrates. In this field of research, various fabrication methods have significant role in deciding the extent of superhydrophobic behaviour of the surface, depending on the material which is used for the processing. In this chapter, the most recent developments in various methods used for fabricating superhydrophobic surfaces are presented along with their advantages and limitations.

Keywords: Superhydrophobicity, natural mimics, self-cleaning, hierarchical structure

---

1. Introduction

Wetting properties of various solid surfaces with various polar and non-polar solvents are getting attention in recent decades because of the increased demand for them in various fields of applications. Superhydrophilicity and superhydrophobicity are two important terms used to explain the wetting behaviour of solid surfaces. Any surface which allows the spreading of a water droplet to provide a contact angle value below 5° is called superhydrophilic surface, whereas the criteria for the surface to be superhydrophobic is to get the water contact angle to a value above 150° [1]. On the other hand, observed values of static contact angle measurements depend on the used fitting mode [2]. For a proper comparison, one should repeatedly use the same volume of water droplet since the volume can also affect the measurement results. In addition, contact angle hysteresis needs to be taken into consideration because hydrophobic surfaces are widely used for self-cleaning applications. Contact angle hysteresis is defined as the difference between the advancing and receding contact angles. The lower the value, the easier the drop will roll off from the surface.

The idea of hydrophobic surfaces came from nature itself —observing the self-cleaning ability of plant leaves and animal wings to roll off water droplet from the surface, termed as lotus effect. The naturally occurring hydrophobic surfaces achieve the high contact angle values because of the rough surfaces coated with the low surface energy wax, which has high repelling interactions with the polar solvents including water [3]. Many such surfaces have a coating composed of particles with size ranging from 20-40 μm, with or without hierarchical structures to provide additional surface area [4]. To explain the stability of the water drop on the surfaces there are two major theories put forward in which the water exists in either Wenzel or Cassie state. In Wenzel model, the water drop fit inside the grooves of the rough surface [Figure 1A] and is described by equation 1,

$$\gamma_{lv} Cos\theta_w = r\left(\gamma_{sv} - \gamma_{sl}\right) \tag{1}$$

where $_w$ is the water contact angle, r is the roughness factor and γ_{lv}, γ_{sv} and γ_{sl} are the different surface tensions involved in the system (liquid/vapour, solid/vapour and solid/liquid, respectively) [5].

In Cassie state, the droplet is stabilized on the top of a surface by the air pockets in between the grooves (Figure 1B) following the equation 2,

$$Cos\theta_c = -1 + f_s\left(\frac{L}{l}\right)^{D-2}(Cos\theta + 1) \tag{2}$$

where θ_c is the contact angle correlated to the chemical heterogeneity, f_s is the area fractions of solid material, L and l are the upper and lower limits of fractal behaviour and D is the fractal dimension [6].

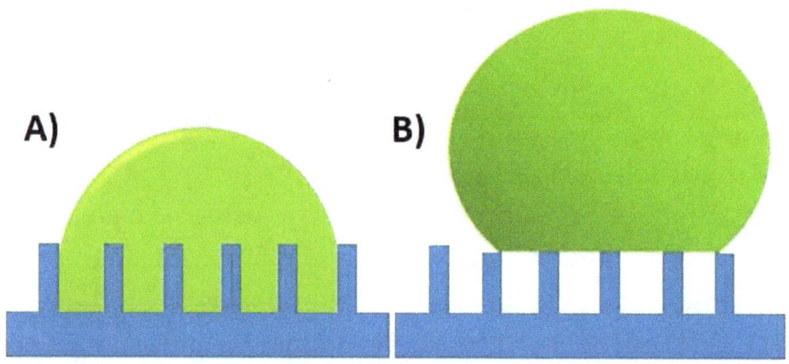

Figure 1. Existence of water droplet on a rough surface: A) Wenzel state, B) Cassie state.

There are some reports which questions the co-existence of both Wenzel and Cassie states. However, these two models are well accepted to explain the existence of solvent on a solid surface [7].

2. Methods for generating superhydrophobic surfaces

The water repelling ability of various materials have shown great influence on various applications such as self-cleaning, anti-ageing, water-oil separation, water corrosion in electrical industry, water proof textiles, controlled transportation of fluids, etc. [2, 4, 8, 9]. Generally, surface micro/nanostructuring combined with low surface energy of materials leads to extreme anti-wetting properties. Hundreds of research articles and more than 450 patents on the subject of nature mimicking self-cleaning surfaces show the potential of this topic. For the fabrication of superhydrophobic surfaces, various physical and chemical methods have been used. Methods such as templating, plasma surface modifications, physical and chemical deposition, layer by layer deposition, electrospinning, hydrothermal process and sol-gel processing have been used for achieving desired roughness and surface chemistry on various substrates. The texturing and coating are explained in short in the following text on the basis of very recent articles in terms of application point of view to give the reader a general idea of the field.

Method	Reference
Templating	[3, 10-23]
Electrospinning	[24-35]
Hydrothermal synthesis	[38-44]
Sol-Gel	[45-52]
Layer by layer method	[45, 54-61]
Physical and chemical vapour deposition methods	[4, 23, 62-70, 85-88]
Plasma etching	[3, 45, 75-79]

Table 1. List of recent publications on various methods for creating anti-wetting surfaces

2.1. Templating

Templating method refers to the copying of a rough surface with superhydrophobic properties. With this technique, a suitable 2D or 3D surface pattern is filled with certain soft material as the first step, then the material is hardened and the template removed by a suitable method keeping the replica intact [10]. The main advantage of this method is that it is very easy to copy water repelling natural surfaces in a large-scale level. This method is comparatively cheap and very suitable for patterning soft materials such as polymers where, in most of the cases, the template can be recycled [10]. Feng et al. created fine superhydrophobic nano fibre structures of poly vinyl alcohol (PVA) with aluminium oxide membrane as a template [11]. When the PVA precursors were extruded in solution through the nano porous template membrane, the polymer molecules underwent an alignment with the hydrophobic methylene groups pointing towards the external environment. A number of factors, including intermolecular hydrogen bonding, made it possible to attain a superhydrophobic surface of an amphilic material. Superhydrophobic polymer materials, especially those that have large surface area, are interesting for large-scale applications for separating hydrocarbon solvents from water, oil from water sources, etc. [12]. By applying a suitable soft material we can directly replicate the topography of various naturally occupying superhydrophobic surfaces. For example, *Xanthosoma Sagittifolium* leaf template was used to create electroactive epoxy surface to increase the water contact angle from 81° to 155° [13]. The comparison of the SEM images of the template and the replica is presented in Figure 2.

On the other hand, heat and pressure-driven nanostructuring on the polystyrene substrate with alumina template gave raise to specifically aligned nanoemboss, nanopost array with embossed base and aligned polymer nanofibres. Such nanofibres with small aspect ratio resembled the wings of *Cicada orni*, which has the self-dust cleaning capability achieved by the sculptured nanostructures on the surface, whereas the nanofibres with reasonably high aspect ratio showed self-cleaning ability towards water droplets. By using the same methodology, with additionally applied pressure, polymer materials have been nanostructured to achieve contact angle above 150^0 [14]. Template-assisted methods are also known for designing 3D ordered superhydrophobic microstructures from water insoluble materials (including

Figure 2. Schematic of the processing method and (a) Photograph of the *Xanthosoma Sagittifolium* leaves. (b) SEM image of the leaf. (c) SEM images of the imprinted layers of composite showing a top view of the surface. (d) SEM images of the imprinted layers of composite showing a cross-section view of the surface. [Reproduced with the permission of American Chemical Society, *Chem. Mater.* 2011, 23, 2075-2083]

polytetrafluoroethylene), which are promising as a host cage for various organic and inorganic materials for nanostructuring [15].

However, the template removing step — especially for delicate materials including polymers — can be really challenging since it is associated with high temperature processes or strong chemical etch [15, 16]. Exposure to non-friendly chemicals or to very high temperature can easily disorder the mesostructures or make unfavourable chemical changes to the polymer chains. To avoid these complications, use of easily removable templates is preferred. The new developments in the field focus on templates which undergo chemical changes and completely disappear or are naturally removed from the replica into the reaction medium. For instance, various metal oxide templates are utilized as a catalyst for the polymerisation process, where the metal undergoes change in oxidation state with the progress of reaction and completely converts into a water soluble form [17, 18]. These by-products are easily washed away. In some other cases, the template is formed in situ as a complex which decomposes with the ion consumption during the progress of the reaction [19]. Template-assisted micro and nanostructuring is also used for fabrication of superhydrophobic inorganic materials [20, 21]. Introducing thermal annealing as another step into the assisted sol-gel process resulted in much higher water contact angle of the 2D hierarchical structure replica due to heat deformation of the template, prior to the process [22]. Using inexpensive template methods, both doubly and triply scaled hydrophobic metal oxide as well as metal interfaces could be generated [21]. In contrast, embedding of desired material into the template can be achieved also by non-wetting methods. In a recent study, Ganesh et al. used sputtering at glancing angle to directly deposit a metal into the aluminium template pores [23]. This method has significant control over the morphology and distribution for the generation of nanorods of aluminium and tungsten through the template and the length of the wires depend solely on the deposition time.

2.2. Electrospinning

The process of electrospinning makes the practical use of electrostatic force to generate continuous and fine filaments from melt, sol-gel or solutions of various polymeric materials. The schematic of the process is shown in Figure 3. The main advantage of electrospinning is that it allows the addition of suitable fillers, including clay or other inorganic particles, directly into the solution to get a composite filament, while the diameter can be kept at around 10 nm [24]. Moreover, electrospinning is a low cost process and can be easily scaled up for industrial production. The surface energy and topological characteristics can be controlled by choosing the right material and suitable process parameters [24]. The material which is released from the needle tip gets deposited directly on the collector and there is no material loss during the production. Electrospinning forms nanofibres, which provide a high degree of roughness on the surface. In addition to roughness, surface tension can also be tailored by choosing the appropriate material or adding sufficient amount of additives in to the spinning solution. Appropriate combination of both effects gives superhydrophobic surface.

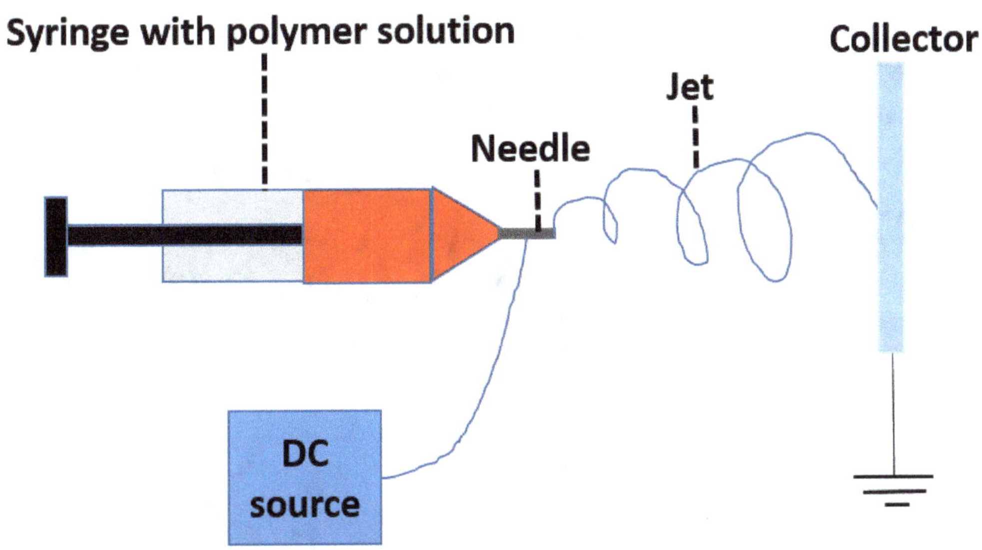

Figure 3. Schematic of the electrospinning process.

Electrospinning technique has been utilized to generate natural mimic surfaces. Lei *et al.* developed the electrospun fibres composite with porous micro particles inside, which imparted lotus leaf-like hydrophobic surface. The hierarchical structures generated by the accumulation of the fillers on the surface resembled the branch like nanostructures on the top of the micro papillae on the lotus leaf which imparted sufficient roughness to the surface to stabilise the water droplet at a contact angle above 162° [25]. On the other hand, addition of inorganic additives into the spinning solution made a nanocoating of the periphery of the fibres and increased the surface roughness. Furthermore, the time required for the trapped solvent to diffuse from the bulk of the deposited filament to its surface (drying procedure) can influence the migration of the fillers towards the outer surface of the fibres [26]. However, hydrophobic properties can be further improved by functionalising the silica before it is put into the spinning solution. For example, epoxy-siloxane modification of silica particles causes a homogeneous self-assembly of the particles on the surface and provides better durability of the hydrophobic surface [27]. As an alternative to additives, different spinning techniques are also employed for increasing the surface energy of electro- spun materials. Changing the spinning material to polymethylsilsesquioxane, the properties were not just restricted to higher water contact angle, but the difference also improved solvent resistance and thermal stability [28]. Hydrophobisation of electrospun fibres is even applicable for block copolymers. Even with 20 mL volume of water, drops exhibit contact angle values high enough to satisfy superhydrophobic behaviour on polystyrene-block-dimethylsiloxane fibres [29]. The existence of the water droplet on the spun surface and SEM images of the spun fibres are as shown in Figure 4.

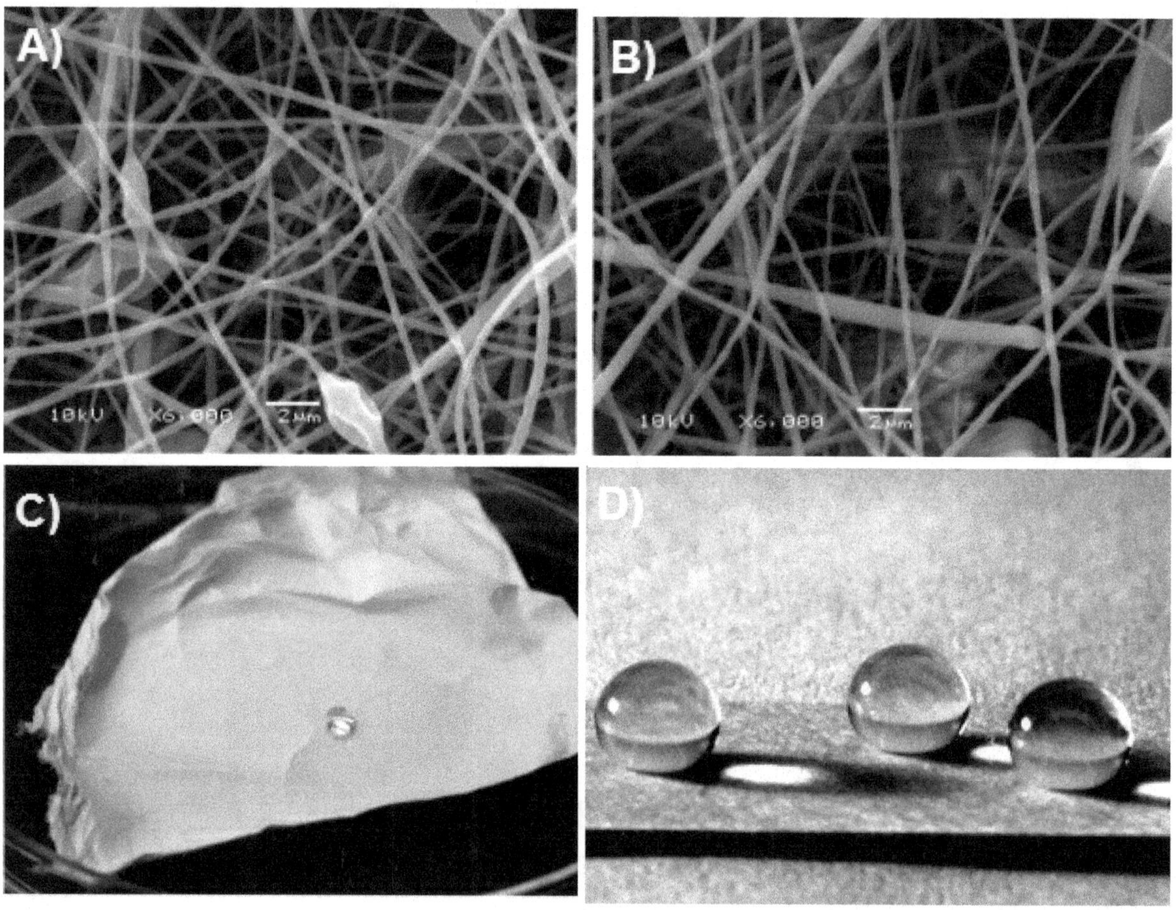

Figure 4. SEM images of electrospun fibres. (a) PS-PDMS/PS sample 6000× magnification (scale bar 2 μm), (b) PS fibres 6000× magnification (scale bar 2 μm), C) Free-standing mat composed of the PS-PDMS/PS electrospun fibres with a water droplet on it, D) several 20 mL water droplets on the mat, showing the superhydrophobicity. [Reproduced with the permission of American Chemical Society, *Langmuir* 2005, *21*, 5549-5554]

In addition to this, various advanced techniques such as coaxial electrospinning, electrospinning with a rotating disk, needleless electrospinning, etc. are used for improving the properties of electrospun fibres [30]. For instance, recently designed multi-nozzle electrospinning set up has an advantage to create micro sized beads from one nozzle where the other one creates fibre-like structures. The membrane maintains mechanical integrity and superhydrophobicity [31]. For further improvement in the hydrophobic character of electrospun fibres, different surface modifications are also found effective. But with many of the low energy polymeric materials, hydrophobicity can be achieved without any secondary modifications [24, 32].

Electrospinning technology is even used for fabrication of superhydrophobic inorganic surfaces. Such inorganic nanomaterials could be used in various fields including optoelectronics and sensing devices [33]. In most of the cases, precursor is created in the form of thin nanofibres of the desired oxide and hydrophobised with suitable post treatments [34]. But in certain cases, metal particles that are inserted into low energy polymers are directly used for the required applications including acid corrosion resistance [35].

2.3. Hydrothermal synthesis

Many inorganic oxides are highly insoluble in water under room temperature. The insolubility may be overcome by use of hydrothermal process which is employed with increased temperature and even pressure in certain cases. This method is very efficient in the preparation of compounds of elements in oxidation states that are difficult to obtain otherwise, synthesis of meta-stable states and varieties of crystals in high quality and good control over size and morphology [36]. Although it is widely used even in material synthesis, very high production cost and difficulty to monitor the crystal growth during the course of the process are some disadvantages.

Hydrothermal synthesis can provide hierarchical structures which can provide both hydrophilic and hydrophobic behaviour after suitable modifications. The control over the morphology is governed by the reaction time and temperature [37]. Generation of highly rough flake, flower-like or rod structures of metallic substrates are sufficient to provide enough roughness for constructing superhydrophobicity [38]. But as in the case of sol-gel synthesis, suitable surface modification is needed. Various polymeric supports, such as polyethylene glycol, are typically used for the improvement of topological properties. These supports, when adsorbed on the metal substrates, inhibit the crystal growth on certain spots and provide micro/nanoroughness to the material [39]. Hydrothermal process with only ultra-pure water as the reaction medium is a green technique and efficient in the production of complex biomimics with rough textures directly on metal surfaces. With additional hydrophobisation with suitable methods, hydrothermal process has been proven to be a great manner to protect various metals from corrosion [40, 41]. Very recent reports showed some interesting hierarchical mixed metal oxide nanoarrays, in which the hairy-like structures out of ZnO was grown on copper oxide nanowires by hydrothermal process followed by thermal oxidation as shown in Figure 5. After silynization of the surface, the contact angle went to a value of about 170°. These kinds of materials show a shift between Wenzel and Cassie state and are self-cleanable [42]. Similarly, hierarchical structured topology is applied even for the inner and outer surfaces of metallic cylinders for increasing the water repellence and thereby corrosion resistance and iceophobic properties [43, 44].

Figure 5. Component analysis of ZnO/CuO heterohierarchical nanostructure. (a) SEM image. (b) TEM image (Ni grid used here). (c-e) O, Cu and Zn elemental mapping recorded from an individual heterohierarchical nanostructure, respectively. [Reproduced with the permission of American Chemical Society, *Langmuir* 2011, 27, 6193-6200]

2.4. Sol-gel

Sol-gel method is associated with the change of the state from colloidal suspension of solid particles into a liquid dispersed in solid. During this transformation, the repelling particle interaction has been changed into attractive interaction between the particles within the colloidal state. Sol-gel processing can transmit multifunctional characteristics to the surface because of the presence of more than single functionalities in the sol state. This is typically carried out by hydrolysis and polycondensation in the presence of a suitable solvent. The solvent must have desired polarity and gets trapped inside the solid particles and forms a gel during the procedure [45]. Such gels can be easily applied on to the substrate, for example dip or spray coating with or without additives, depending on the nature of the material and the synthesised sol. The topological and chemical properties of so created surfaces largely depend on the physical properties and chemical composition of the reaction mixture. The main advantage of this method is that it is very easy to apply on to a large area of complicated structures. The hydrophobicity of composites has been improved by the incorporation or functionalization of the nanoparticles used for the processing or by various surface modifications, such as etching, to tailor the roughness [46]. Silica is one of the major nanocoatings for hydrophobisation and can be easily synthesised by sol-gel method. However, silica particles can behave as hydrophilic material because of the large amount of silanol groups on the surface. Thus, in many studies silanol groups have been functionalized with non-polar chains to induce better hydrophobicity [47]. But one disadvantage of such functionalization is that the moisture absorption can be increased with surface contamination and ageing [48]. However, metal oxides do not behave as hydrophobic surfaces without suitable functionalization in most of the cases. In case of metal oxides, functionalization could be their treatment with suitable acids and consequential surface conversion into salt, which provides improved water repellence [49].

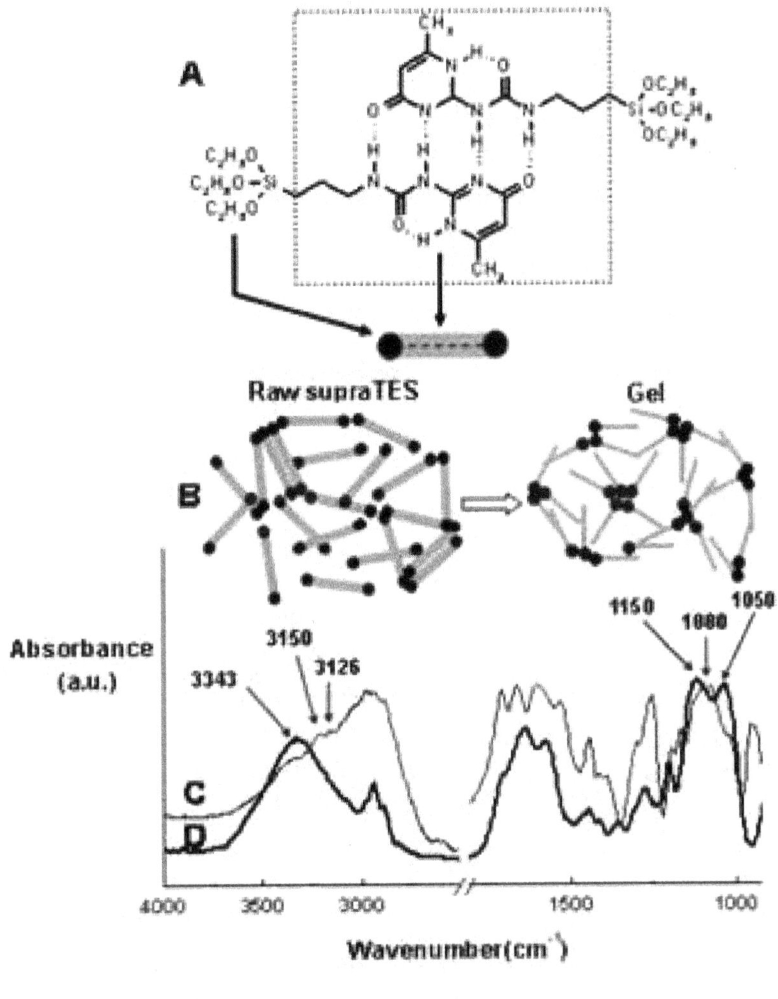

Figure 6. A) Possible structure of a synthesised supramolecular organosilane, B) Schematic of gelation with FT-IR spectrum of the corresponding formed gel. [Reproduced with the permission of American Chemical Society, *J. Am. Chem. Soc.*2004, 126, 4796-4797]

One of the important applications of the sol-gel method in the recent years is the coating of fabrics of natural fibres with hydrophobic metal oxide colloids. This provides an open way for the fabrication of water repellent textiles for mass production. Various organosilanes and mono-carboxylic acid functionalised alumina were found very efficient [50, 51]. Such organically modified nanoparticle coating gives an additional benefit of UV blocking, and thus makes this method more interesting for textile industries [51]. Sol-gel synthesis started to utilise supramolecular assembly of long chain molecules, formed as a result of intra molecular hydrogen bonding (Figure 6) [52]. Another advantage of this method is that it is easy to conduct template-assisted sol-gel synthesis, which makes the replication much easier.

2.5. Layer by layer method

In this technique, the generation of one layer of material over the other is performed to get a multilayer composite material surface. Techniques, typically used to achieve this, include dip coating, spray coating or spin coating. The prominent application of this method is to combine the properties of various nanomaterials with polymers, which are efficiently used in biological applications [53]. Besides coating, supramolecular self-assembly can be used to generate hydrophobic surfaces. This is mostly applicable to polyelectrolyte macromolecules because of the ability of opposite charges to accumulate as layer. However, the other neutral type nanoparticles can be deposited as different layers [54]. On the other hand, charged inorganic materials along with polyelectrolytes are also employed in many cases which provide a simple route for self-assembly on various substrates and avoid the requirements for using deposition techniques [55]. Cheap scalable antifouling superhydrophobic coatings can be fabricated by introducing a lubricant into the pores of the nanostructured surface.

Layer by layer techniques are often combined with other deposition techniques for achieving higher contact angles. These deposition techniques are used for easier deposition of inorganic particles on the self-assembled macroscopic layers. Combinations of such techniques have been utilized for creating natural mimics [56]. One of the limitations for using highly charged macromolecules is that it is highly influenced by the strength of the ions present in the solution, which may result in the disassembly of the layers [57]. Such influences of the ionic concentrations can be excluded via covalent bonding of one layer over the other. Covalently bonded molecules with very long non-polar tails can thus provide water contact angle above 150° and in addition the extra stability. Other chemical interactions, such as hydrogen bonding, can also be utilized for self-assembly of layers but these are not as strong as covalent bonds. To stabilize such assemblies we need to provide additional stabilization methods such as incorporation of electrostatic copolymer with suitable multivalent ions, or suitable thermal or radiation curing of the films [58].

Recently, Huang *et al.* attempted a formation of assembled layers from colloidal dispersion by incorporating highly rough, raspberry-type particles to obtain water pinning on oleophilic surfaces [59]. Raspberry resembling films have been prepared from various micro- and nanoparticles and employed in the layer by layer assembly procedures to achieve superhydrophobicity [60]. These researchers made layer by layer method more convenient to apply irrespective of the material compared to the other commonly used techniques. This makes the method more promising for water-oil separation in industrial scale and functionalizing complex porous and membranous structures. Multilayers of strippable silver generated in lasagne-like structures took advantage of the protection from the top layer, and after modification water contact angle increased to 146° (Figures 7.1 and 7.2). Furthermore, after extreme exposure to the corrosives, the damaged layer can be removed and a fresh layer is exposed, which gives regeneration property to the material [61].

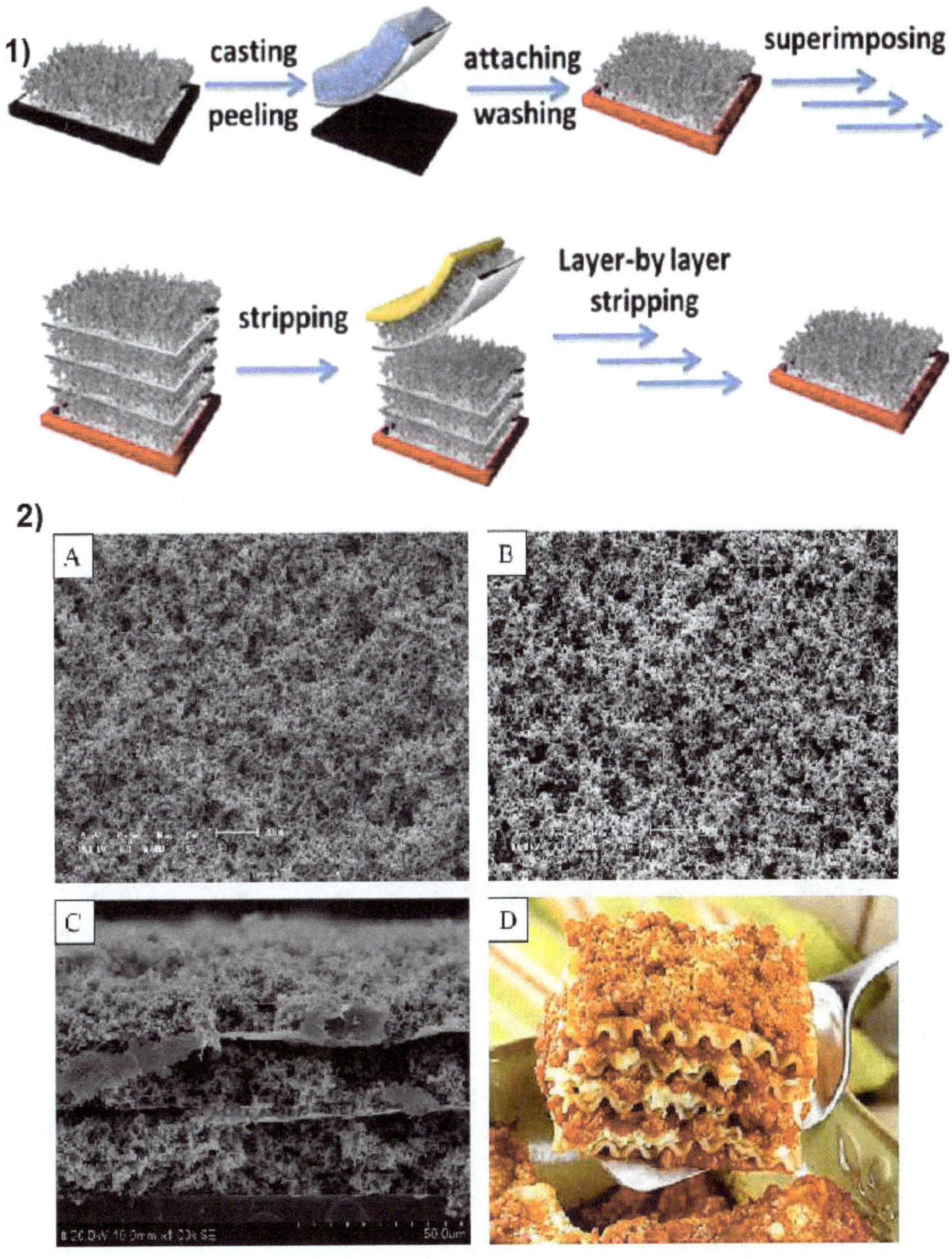

Figure 7. 1) Schematic process of fabrication and layer-by-layer stripping of the modular assembly multilayer film. 2) SEM images of the silver mirror multilayer films with three layers A), four layers (B), the cross-sectional view (C), and a photo of lasagne D). [Reproduced with the permission of American Chemical Society, *Langmuir*. 2014, 30, 548–553]

2.6. Deposition methods

2.6.1. Vapour depositions: PVD and CVD

Physical vapour deposition (PVD) and chemical vapour deposition (CVD) are the two major vapour deposition techniques used for making very thin layer of coating on various substrates. PVD utilizes the evaporated pure material whereas CVD utilizes a mixture of chemically reactive components, which are deposited on the surface, leaving the desired thin layer of coating. For the production of superhydrophobic surfaces, very low surface energy components' vapours are applied on to the surface to form a thin layer of rough and water repellent surface. For depositing such materials, various techniques including plasma, thermal evaporation or even both are used [62, 63]. Typically, plasma-assisted vaporization requires much less thermal excitation, which is a special advantage [64]. The vapour deposition coatings can be extended from inorganic metal oxides to various organic low energy materials, depending on the requirement and the substrate to be coated. Recently, surface modified silica deposition was achieved at a very low temperature of 40° C by implementing NH_3 as a catalyst into the processing chamber [65]. Developing such a low temperature processing is the best choice for coating thermally unstable surfaces. CVD has significant achievements even to control the crystallinity of the coated polymeric surfaces, tuned by controlling the deposition parameters [66]. Such control over orientation of the polymer chain deposited can significantly affect the topology of the surfaces.

Since carbon nanotubes and other graphene-related materials have excellent surface roughness and low surface energy, they are used as intrinsic hydrophobic coating, applied for water-oil separation [67]. Template-assisted vapour deposition of graphene is becoming popular for improving the porous structure with better control over the morphology, although it needs to be coated with non-polar polymer or other coatings for hydrophobisation [68]. On the other hand, carbon vapour deposited nanotubes can show superoleophilicity without any further modification [69]. Zwitter ionic deposition is another recent development for ant-fouling surfaces, where the surface energy depends on the number of exposed zwitter ions and the type of atmosphere that is exposed to the periphery, but has to be modified further to reach higher contact angle values [70].

2.7. Plasma surface treatment

Gaseous plasma is by definition partially or fully ionised gas, but it can contain also neutral gas which can be in excited state [71]. Plasma discharge can be achieved by different setups, and the reader is encouraged to read more about them [72-74]. In general, plasma particles are excited (ionised, metastabled) and even more energetic than neutral gas, and thus they can physically or chemically react with treated material. The interaction can lead to etching or deposition. Etching is achieved through physical sputtering by ions or chemical reactions with the surface and desorption (i.e. oxygen neutral atoms chemically bond to a carbon on a polymer surface and CO or CO_2 molecule is desorbed). Deposition can occur through chemical bonding of precursors from plasma to the surface or by plasma-activated chemical reactions of chemicals that are on the surface.

2.7.1. Plasma etching

Creating appropriate topography of the micro-rough surface for obtaining superhydrophobicity can be a challenge. Akinoglu *et al.* investigated plasma etching mechanism by oxygen plasma etching of polystyrene [75]. They used two setups: capacitive coupled RF low-pressure oxygen plasma (CCP) and inductively coupled RF low-pressure oxygen plasma (ICP). In both, ions and radicals and other energetically excited species contributed to the etching. Since ions are charged particles, they are directed by the electric field. The ion contribution to etching is then anisotropic, whereas the neutral species etch in isotropic manner due to diffusion governed movement. Thus, the density ratio between charged ions and neutral oxygen atoms influences the etching of the surface structure. The anisotropy in case of CPP was found to be 0.45, and in the case of ICP it was 0.21. Additionally, the etching in ICP setup was 1 order of magnitude faster than in CCP.

Despite slower etching rates, Salapare *et al.* used low-pressure CCP RF oxygen plasma for polytetrafluoroethylene (PTFE) treatment. They found a correlation between plasma power and the time of treatment with the water contact angle. With higher power (400 W) and longer time (5400 s) they managed to crate super-hydrophobic surface (151° contact angle). The FT-IR and surface roughness measurements showed that only the geometrical shape contributed to the surface energy, while the chemical composition of the surface remained the same.

PTFE was treated also with a mixture of oxygen/argon magnetron plasma [76]. With plasma etching, micro-rough surface was established, and after 4h of treatment, leaf-like protrusions were created. The contact angle changed from non-treated sample at 102° to super-hydrophobic at 158° (Figure 8). The hydrophobicity was asserted to surface roughness and not surface chemistry, which was confirmed by XPS and Raman measurements.

One has to be careful when using plasma etching, because it can also damage the surface and deteriorate the hydrophobicity as demonstrated by Zylka [77]. He prepared samples from silicone rubber with surface moulded by an acrylic paint template and left them exposed to the outside weather conditions for 45 days. The water contact angle on this surface was 149°. Then he treated the samples' surface with atmospheric air corona discharge for 100 h, and the contact angle dropped to 119°.

However, superhydrophobicity can be achieved on silicon surface too. In [78] complicated plasma etching procedure was used to create nanopillar and nanocone silicon surface. The surface was additionally passivated by octadecyltrichlorosilane monolayer. In both cases super-hydrophobic surface was achieved—contact angles were 150° for nanopillars and 165° for nanocones.

Another way of obtaining superhydrophobic properties of a surface is to deposit a thin layer on it and then modify this layers wettability. This challenge was addressed by Cai *et al.* who created polymer composites with silica spheres [79]. Then they used reactive ion etching in oxygen plasma to selectively etch the polymer and expose the silica spheres. Depending on the time of etching, different extents of the spheres were exposed. This was not sufficient for the hydrophobicity, so the liquid HF was used to etch the silica. In this manner they managed to prepare a periodic pattern of voids with the precisely selected size in the surface, and thus

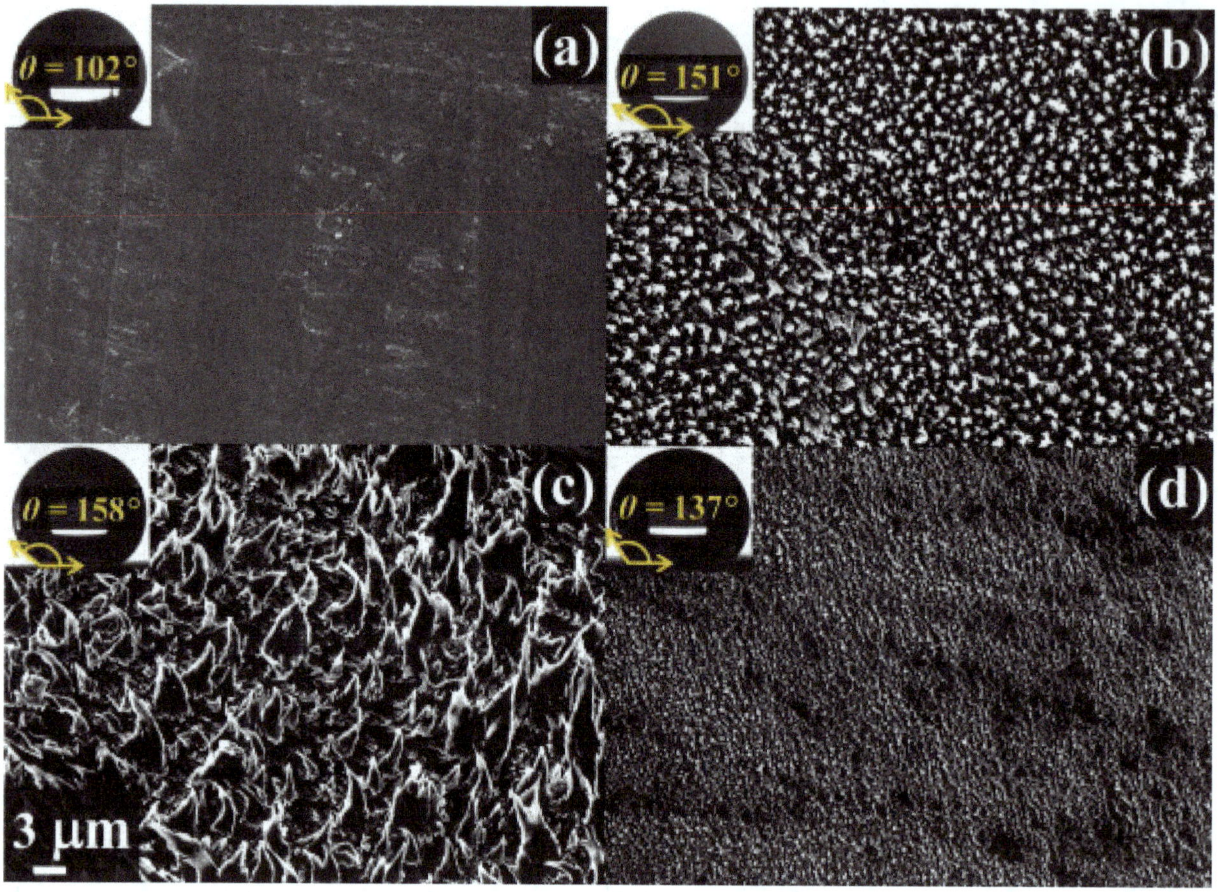

Figure 8. Insets show the FESEM surface micrographs of: (a) TF0 (untreated), (b) TF2 (2 h plasma treated), (c) TF4 (4 h plasma treated) and (d) TF5 (5 h plasma treated) PTFE samples. [Reproduced with the permission of Elsevier, *Dielectr. Electr. Insul.* 2014, 21, 1183-1188]

control the roughness necessary for the hydrophobic effect. With addition of fluorosilane the surface energy was further reduced and superhydrophobicity with contact angle of more than 150° was achieved.

2.7.2. Combination of plasma etching and deposition

To avoid many-step procedures and complicated setups, plasma etching and plasma deposition can assist in producing superhydrophobic polymer surface. Plasma etching break chemical bonds via high energy particle interaction with the surface and creates micro or nanorough surfaces by taking out the pieces of degraded polymers atom-by-atom into volatile molecules. This was nicely demonstrated for polymethyl methacrylate etched in helicon RF plasma, where etching was controlled by ion flux [80]. The ion flux was adjusted with oxygen pressure and ion energy by biasing the samples (Figure 9). The roughness was increased with lower ion flux and with higher ion energy. When bias and pressure are set, time of plasma treatment determines the RMS roughness. Additional deposition of CF_x film passivizes the surface, namely reduces the surface energy. The contact angle increases with etching time and reaches maximum value of 150° after 2 min. Similar work was performed on polystyrene, where first etching was performed with CF_4 plasma and then CF_x film was deposited by

PECVD from C_4F_8 gas [81]. In this case, CCP plasma was used. Plasma etching at 50 W was the most optimal for designing superhydrophobicity of the surface with the resulting contact angle of almost 160°. This surface property might be due to the highest fluorine-to-carbon atomic ratio on the surface after the deposition on this micro-surface structure. When the power was raised, the ion flux to the surface increased, leading to increase in the roughness and decrease in contact angle.

An interesting case is also treatment of paper (cellulose fibres) with plasma. Balu *et al.* used CCP plasma with grounded electrode for treatment as well as substrate heating to 110 °C. First, the etching was carried out with low pressure O_2 plasma, followed by deposition of fluorocarbon film from pentaflouroethane-monomer [82]. The fluorocarbon film reduced the surface energy of the surface, whereas the etching increased the roughness of the fibres' surface with nanometer-scale features. This contributed to the increase in hydrophobicity, which manifested in larger contact angle (more than 160°) compared to the samples on which only fluorocarbon film was deposited (more than 130°).

Treatment	Contact angle/°C			
	Water		Decane m m	
	Flat	Porous	Flat	Porous
Untreated	116	146	46	Wicks
O_2 plasma	131	153	55	Wicks
Plasma polymer	131	144	83	115
O_2 plasma + Plasma polymer	148	152	92	133

Table 2. Probe liquid contact angles for flat and porous PTFE substrates [Reproduced with permission of American Chemical Society, *Langmuir.* 2007, 23, 12984-12989]

With a combination of plasma etching and deposition one can tailor the surface properties in the way that, in addition to superhydrophobicity, increased repellence of non-polar liquids is achieved at the same time. Coulson *et al.* tailored surface morphology by low pressure inductively coupled oxygen plasma, and surface chemistry by deposition and polymerisation of 1H,1H,2H,2H-heptadecafluorodecyl acrylate in the same system [83]. The substrate was either flat or porous PTFE foil. The surface microstructure exhibit hydrophobicity, and polymer prevented potential capillary effect. This increased the repellence even for decane (Table 2). In this case, Cassie-Baxter relationship seemed much better to describe the system than Wenzel's theory.

Sometimes one would need hydrophobic and hydrophilic surface on same kind of sample for different use. Ruiz *et al.* demonstrated that with the appropriate selection of polymer deposition one can tune the wettability [84]. In their work they deposited photoresist on a flat surface. Then they used inductively coupled SF_6 plasma for etching and creating micro-structured surface of the photoresist leading to the increase the fluorine-to-carbon atom ratio on the surface (Figure 9a and 9b).

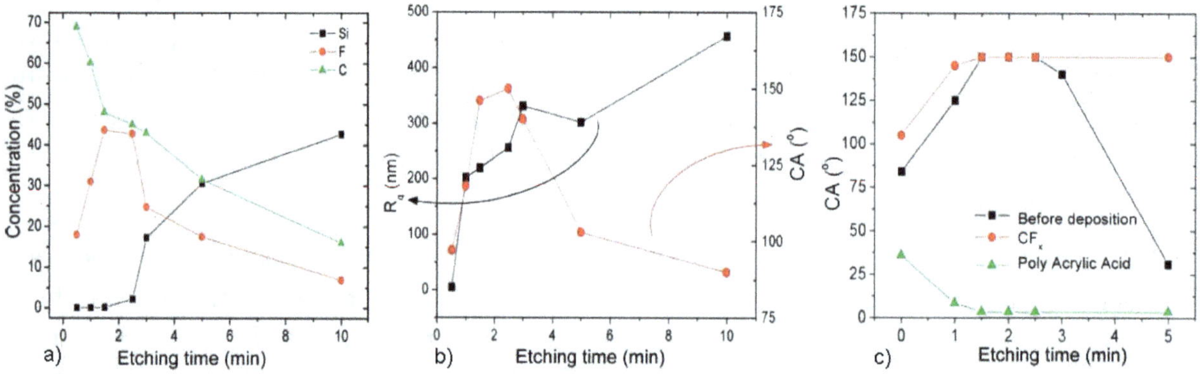

Figure 9. a) Elementary concentration obtained from XPS measurements, b) roughness and contact angles and c) contact angles with or without coatings for the SF6 etched photoresist coated silicon surfaces. [Reproduced with permission of American Chemical Society, *Langmuir*. 2007, 23, 12984-12989]

Then they coated the etched surface with CF plasma, and afterwards they deposited polyacrylic acid (PAA) by capacitively coupled plasma. The etching already caused the superhydrophobicity, which was then slightly increased by the CF coating. However, the PAA drastically changed the surface energy and caused surface superhydrophilicity (Figure 9c).

2.8. Plasma deposition

Plasma deposition is an efficient process to deposit material with low surface energy and effectively improve hydrophobic properties. The deposited films can contribute to increase in the hydrophobicity with their structure. Most of the effects that lead to surface energy modification were already described before.

Typically, highly polar material gets deposited in a way that the surface is nanorough. Examples are atmospheric pressure plasma polymerization of the toluene/HMDSO [85], expanding plasma arc deposition of Teflon [86], pulsed CCP RF plasma deposition of fluorocarbon films from C_6F_6 monomer precursor [87] and ICP plasma deposition of perfluorooctyl acrylate film [88] that all cause the superhydrophobicity.

3. Conclusions

In this chapter various methods were introduced on how to deal with the surface energy modifications. There are three major aspects for this, either by modifying the surface morphology in the way to mimic natural structures, which are effective for water repelling (lotus effect), by functionalization of the surface by non-polar group, or deposition of material with low surface energy. It was also shown that the most successful way is a combination of the techniques—deposition of the low surface energy material on the surface and creation of its microstructure, which can additionally increase the water contact angle.

Methods such as sol gel are very simple route to achieve the high water repellence and scaling up of the process is easily achievable. Herein, many techniques for the modification of surfaces or deposition of different low energy materials are presented, which easily lead to superhydrophobicity of surface. The future challenge has now shifted towards beneficial properties of the surfaces or the deposited thin layers like transparency, flexibility, electrical conductivity, catalytic properties, etc. without losing the superhydrophobic affiliation. Creation of anti-bacterial and anti-fungal surfaces for bio-medical applications are also very promising applications of such surfaces.

Author details

Harinarayanan Puliyalil[1,2], Gregor Filipič[1,2] and Uroš Cvelbar[1,2*]

*Address all correspondence to: uros.cvelbar@ijs.si

1 Jožef Stefan International Postgraduate School, Ljubljana, Slovenia

2 Jožef Stefan Institute, Ljubljana, Slovenia

References

[1] Feng. X., Jiang. L., Design and creation of superwetting/antiwetting surfaces. Advanced Materials. 2006;18:3063-3078. DOI: 10.1002/adma.200501961

[2] Zhang. X., Shi. F., Niu. J., Jiang. Y., Wang. Z., Superhydrophobic surfaces: from structural control to functional application. Journal of Materials Chemistry. 2008;18:621-633. DOI: 10.1039/b711226b

[3] Shirtcliffe. N. J., McHale. G., Atherton. S., Newton. M. I., An introduction to superhydrophobicity. Advances in Colloid and Interface Science. 2010;161:124-138. DOI: 10.1016/j.cis.2009.11.001

[4] Aminayi. P., Abidi. N., Imparting super hydro/oleophobic properties to cotton fabric by means of molecular and nanoparticles vapor deposition methods. Applied Surface Science. 2013;287:223- 231. DOI: 10.1016/j.apsusc.2013.09.132

[5] Wenzel. R. N., Resistance of solid surfaces to wetting by water. Industrial and Engineering Chemistry. 1936;28:988-994. DOI: 10.1021/ie50320a024

[6] Cassie. A. B. D., Baxter. S.,Wettability of porous surfaces. Transactions of Faraday Society. 1944;40:546-551. DOI: 10.1039/TF9444000546

[7] Gao. L., McCarthy. T. J., How Wenzel and Cassie were wrong. Langmuir. 2007;23:3762-3765. DOI: 10.1021/la062634a

[8] Yao. X., Song. Y., Jiang. L., Applications of bio-inspired special wettable surfaces. Advanced Materials. 2011;23:719-734. DOI: 10.1002/adma.201002689

[9] Nosonovsky. M., Bhushan. B. Superhydrophobic surfaces and emerging applications: non-adhesion, energy, green engineering. Current Opinion in Colloid & Interface Science. 2009;14:270-280. DOI: 10.1016/j.cocis.2009.05.004

[10] Roach. P., Shirtcliffe. N. J., Newton. M. I.,. Progess in superhydrophobic surface development. Soft Matter. 2008;4:224-240. DOI: 10.1039/B712575P

[11] Feng. L., Song. Y., Zhai. J., Liu. B., Xu. B., Jiang. L. Zhu. D.. Creation of a superhydrophobic surface from an amphiphilic polymer. Angewandte Chemie International Edition. 2003;42:800-802. DOI: 1433-7851/03/4207-0802

[12] Li. A., Sun. H. X., Tan. D. Z., Fan. W. J., Wen. S. H.,Qing. X. J., Li. G. X., Superhydrophobic conjugated microporous polymers for separation and adsorption Energy & Environmental Science. 2011;4:2062-2065. DOI: 10.1039/C1EE01092A

[13] Weng. J. C., Chang. C. H., Peng. C. W., Chen. S. W., Yeh. J. M.,. Advanced anticorrosive coatings prepared from the mimicked xanthosoma sagittifolium-leaf-like electroactive epoxy with synergistic effects of superhydrophobicity and redox catalytic capability. Chemistry of Materials. 2011;23:2075-2083. DOI: dx.doi.org/10.1021/cm1030377

[14] Lee. W., Jin. M. K, Yoo. W. C, Lee. J. K.,. Nanostructuring of a polymeric substrate with well-defined nanometer-scale topography and tailored surface wettability. Langmuir. 2004;20:7665-7669. DOI: 10.1021/la049411

[15] Wang. Y., Wang. M., Ge. X.,. Fabrication of superhydrophobic three-dimensionally ordered macroporous polytetrafluoroethylene films and its application. Langmuir. 2014;30:10804–10808. DOI: dx.doi.org/10.1021/la502866h

[16] Li. J., Fu. J., Cong. Y., Wu. Y., Xue. L., Han. Y.,. Macroporous fluoropolymeric films templated by silica colloidal assembly: a possible route to super-hydrophobic surfaces. Applied Surface Science. 2006;252:2229-2234. DOI: 10.1016/j.apsusc.2005.03.224

[17] Pan. L., Pu. L., Shi. Y., Song. S, Xu. Z., Zhang. R., Zheng. Y.,Synthesis of polyaniline nanotubes with a reactive template of manganese oxide. Advanced Materials. 2007;19:461-464. DOI: 10.1002/adma.200602073

[18] Zhang. Z., Sui. J., Zhang. L., Wan. M., Wei. Y., Yu. L.,Synthesis of polyaniline with a hollow, octahedral morphology by using a cuprous oxide template. Advanced Materials. 2005;17:2854-2857. DOI: 10.1002/adma.200501114

[19] Bormashenko. E., Stein. T., Whyman. G., Bormashenko. Y, Pogreb. R.,Wetting properties of the multiscaled nanostructured polymer and metallic superhydrophobic surfaces. Langmuir. 2006;22:9982-9985. DOI: 10.1021/la061622m

[20] Zhong. W., Li. Y., Wang. Y., Chen. X., Wang. Y., Yang. W.,Superhydrophobic polyaniline hollow bars: constructed with nanorod-arrays based on self-removing metal-

monomeric template. Journal of Colloid and Interface Science. 2012;365:28-32. DOI: 10.1016/j.jcis.2011.08.083

[21] Sun. C., Ge. L. Q., Gu. Z.,Fabrication of super-hydrophobic film with dual-size roughness by silica sphere assembly. Thin Solid Films. 2007;515:4686-4690. DOI: 10.1016/j.tsf.2006.11.027

[22] Li. Y., Cai. W., Cao. B., Duan. G., Sun. F., Li. C., Jia. L.,Two-dimensional hierarchical porous silica film and its tunable superhydrophobicity. Nanotechnology. 2006;17:238-243. DOI: 10.1088/0957-4484/17/1/040

[23] Kannarpady. G. K., Khedir. R. K., Ishihara. H., Woo. J., Oshin. O. D.,Controlled growth of self-organized hexagonal arrays of metallic nanorods using template-assisted glancing angle deposition for superhydrophobic applications. Appl. Mater. Interfaces. 2011;3:2332-2340. DOI: x.doi.org/10.1021/am200251n

[24] Sas. I., Gorga. R. E., Joines. J. A., Thoney. K. A.,Literature review on superhydrophobic self-cleaning surfaces produced by electrospinning. Journal of Polymer Science Part B: Polymer Physics. 2012;50:824-845. DOI: 10.1002/polb.23070

[25] Jiang. L., Zhao. Y., Zhai. J.,. A Lotus-leaf-like superhydrophobic surface: a porous microsphere/nanofiber composite film prepared by electrohydrodynamics. Angewandte Chemie International Edition. 2004;43:4338-4341. DOI: 10.1002/anie.200460333

[26] Lin. J., Cai. Y., Wang. X., Ding. B., Yu. J., Wang. M.,Fabrication of biomimetic superhydrophobic surfaces inspired by lotus leaf and silver ragwort leaf. Nanoscale. 2011;3:1258-1262. DOI: 10.1039/c0nr00812e

[27] Wang. S., Li. Y., Fei. X., Sun. M., Zhang. C., Li. Y., Yang. Q., Hong. X.,Preparation of a durable superhydrophobic membrane by electrospinning poly (vinylidene fluoride) (PVDF) mixed with epoxy-siloxane modified SiO_2 nanoparticles: a possible route to superhydrophobic surfaces with low water contact angle. Journal of Colloid and Interface Science. 359;2011:380-388.

[28] Xiang. H., Zhang. L., Wanga. Z., Yu. X., Long. Y., Zhang. X., et al.,Multifunctional polymethylsilsesquioxane (PMSQ) surfaces prepared by electrospinning at the sol-gel transition: superhydrophobicity, excellent solvent resistance, thermal stability and enhanced sound absorption property. Journal of Colloid and Interface Science. 2011;359:296-303. DOI: 10.1016/j.jcis.2011.03.076

[29] Ma. M., Hill. R. M., Lowery. J. L., Fridrikh. S. V., Rutledge. G. C.,Electrospun poly(styrene-block-dimethylsiloxane) block copolymer fibers exhibiting superhydrophobicity. Langmuir. 2005;21:5549-5554. DOI: 10.1021/la047064y

[30] Nuraje. N., Khan. W. S., Lei. Y., Ceylan. M., Asmatulu. R., Superhydrophobic electrospun nanofibers. Journal of Material Chemistry A. 2013;1:1929-1946. DOI: 10.1039/C2TA00189F

[31] Zhan. N., Li. Y., Zhang. C., Song. Y., Wang. H., Sun. L., A novel multinozzle electrospinning process for preparing superhydrophobic PS films with controllable bead-on-string/microfiber morphology. Journal of Colloid and Interface Science. 2010;345:491-495. DOI: 10.1016/j.jcis.2010.01.051

[32] Ma. M., Hill. R. M.,Superhydrophobic surfaces. Current Opinion in Colloid & Interface Science. 2006;11:193-202. DOI: 10.1016/j.cocis.2006.06.002

[33] Ding. B., Ogawa. T., Kim. J., Fujimoto. K., Shiratori. S.,Fabrication of a super-hydrophobic nanofibrous zinc oxide film surface by electrospinning. Thin Solid Films. 2008;516:2495-2501. DOI: 10.1016/j.tsf.2007.04.086

[34] Tang. H., Wang. H., He. J.,. Superhydrophobic titania membranes of different adhesive forces fabricated by electrospinning. Journal of Physical Chemistry C. 2009;113:14220-14224. DOI: 10.1021/jp904221f

[35] Wang. S., Liu. Q., Zhang. Y., Wang. S., Li. Y., Yang. Q., Song. Y., Preparation of a multifunctional material with superhydrophobicity, superparamagnetism, mechanical stability and acids-bases resistance by electrospinning. Applied Surface Science. 2013;279:150- 158. DOI: 10.1016/j.apsusc.2013.04.060

[36] Rabenau. A.,The role of hydrothermal synthesis in preparative chemistry. Angewandte Chemie International Edition. 24;1985:1026-1040. DOI: 10.1002/anie.198510261

[37] Liu.X., He. J.,. One-step hydrothermal creation of hierarchical microstructures toward superhydrophilic and superhydrophobic surfaces. Langmuir. 2009;25:11822-11826. DOI: 10.1021/la901426r

[38] Wang. S., Wang. C., Liu. C., Zhang. M., Ma. H., Li. J.,Fabrication of superhydrophobic spherical-like α-FeOOH films on the wood surface by a hydrothermal method. Colloids and Surfaces A: Physicochemical Engineering Aspects. 2012;403:29-34. DOI: 10.1016/j.colsurfa.2012.03.051

[39] Xiao. C., Yan. J., Li. T.,Fabrication and superhydrophobic property of ZnO micro/nanocrystals via a hydrothermal route. Journal of Nanomaterials. 2014;2014:680592. DOI: 10.1155/2014/680592

[40] Li. L., Zhang. Y., Lei. J, He. J., Lv. R., Li. R., Pan. F.,A facile approach to fabricate superhydrophobic Zn surface and its effect on corrosion resistance. Corrosion Science. 2014;85:174-182. DOI: 10.1016/j.corsci.2014.04.011

[41] Li. L, Zhang. Y, Lei. J, He. J, Lv. R., Li. N., et al.,Water-only hydrothermal method: a generalized route for environmentally-benign and cost-effective construction of superhydrophilic surfaces with biomimetic micronanostructures on metals and alloys. Chemical Communications.2014;50:7387-7554. DOI: 10.1039/c4cc00012a

[42] Guo. Z., Chen. X., Li. J., Liu. J. H., Huang. X.J.,ZnO/CuO hetero-hierarchical nanotrees array: hydrothermal preparation and self-cleaning properties. Langmuir. 2011;27:6193-6200. DOI: 10.1021/la104979x

[43] Hao. X., Wanga. L., Lv. D., Wang. Q., Li. L., He. N, et al.,Fabrication of hierarchical structures for stable superhydrophobicity on metallic planar and cylindrical inner surfaces. Applied Surface Science. 2015;325:151-159. DOI: 10.1016/j.apsusc.2014.11.014

[44] Shen. Y., Tao. H., Chen. S., Zhu. L., Wang. T., Tao. J.,Icephobic/anti-icing potential of superhydrophobic Ti6Al4V surfaces with hierarchical textures. RSC Advances. 2015;5:1666-1672. DOI: 10.1039/c4ra12150c

[45] Latthe. S. S., Gurav. A. B., Maruti. C. S., Vhatkar. R. S.,Recent progress in preparation of superhydrophobic surfaces: a review. Journal of Surface Engineered Materials and Advanced Technology. 2012;2:76-94. DOI: 10.4236/jsemat.2012.22014

[46] Zhang. X., Jin. M., Liu. Z, Tryk. D. A., Nishimoto. S.,Superhydrophobic TiO_2 surfaces: preparation, photocatalytic wettability conversion, and superhydrophobic-superhydrophilic patterning. Journal of Physical Chemistry C. 2007;111:14521-14529. DOI: 10.1021/jp0744432

[47] Rao. A. V., Haranath. D.,Effect of methyltrimethoxysilane as a synthesis component on the hydrophobicity and some physical properties of silica aerogels. Microporous and Mesoporous Materials. 1999;30:267-273. DOI: 10.1016/S1387-1811(99)00037-2

[48] Rao. A. V., Gurav. A. B., Latthe. S. S., Vhatkar. R. S., Imai. H., et al.,Water repellent porous silica films by sol-gel dip coating method. Journal of Colloid and Interface Science. 2010;352:30-35. DOI: 10.1016/j.jcis.2010.08.003

[49] Lakshmi. R. V., Basu. B. J.,Fabrication of superhydrophobic sol-gel composite films using hydrophobically modified colloidal zinc hydroxide. Journal of Colloid and Interface Science. 2009;339:454-460. DOI: 10.1016/j.jcis.2009.07.064

[50] Shang. S. M., Li. Z., Xing. Y., Xin. J. H., Tao. X. M.,Preparation of durable hydrophobic cellulose fabric from water glass and mixed organosilanes. Applied Surface Science. 2010;257:1495-1499. DOI: 10.1016/j.apsusc.2010.08.081

[51] Pan. C., Shen. L., Shang. S., Xing. Y.,Preparation of superhydrophobic and UV blocking cotton fabric via sol-gel method and self-assembly. Applied Surface Science. 2012;259:110-117.

[52] Han. J. T., Lee. D. H, Ryu. C. Y., Cho. K.,Fabrication of superhydrophobic surface from a supramolecular organosilane with quadruple hydrogen bonding. Journal of American Chemical Society. 2004;126:4796-4797. DOI: 10.1021/ja0499400

[53] Srivastava. S., Nicholas. A. K.,Composite layer-by-layer (LBL) assembly with inorganic nanoparticles and nanowires. Accounts of Chemical Research. 2008;41:831-1841. DOI: 10.1021/ar8001377

[54] Isimjan. T. T., Wang. T., Rohani. S.,. A novel method to prepare superhydrophobic, UV resistance and anti-corrosion steel surface. Chemical Engineering Journal. 2012;210:182-187. DOI: 10.1016/j.cej.2012.08.090

[55] Sunny. S., Vogel. N., Howell. C.,Vu. T. L., Aizenberg. J.,Lubricant-infused nanoparticulate coatings assembled by layer-by-layer deposition. Advanced Functional Materials. 2014;6658-6667.24. DOI: 10.1002/adfm.201401289

[56] Shi. F., Wang. Z., Zhang. X.,Combining a layer-by-layer assembling technique with electro chemical deposition of gold aggregates to mimic the legs of water striders. Advanced Materials. 2005;17:1005-1009. DOI: 10.1002/adma.200402

[57] Amigoni. S., Givenchy. E. T., Dufay. M., Guittard. F.,Covalent layer-by-layer assembled superhydrophobic organic-inorganic hybrid films. Langmuir. 2009;25:11073-11077. DOI: 10.1021/la901369f

[58] Quinn. J. F., Angus. P. R. J., Georgina. K. S., Alexander. N. Z., Caruso. F.,Next generation, sequentially assembled ultrathin films: beyond electrostatics. Chemical Society Reviews. 2007;36:707-718. DOI: 10.1039/b610778h

[59] Huang. H., Zacharia. N. S.,Layer-by-layer rose petal mimic surface with oleophilicity and underwater oleophobicity. Langmuir. 2015;31:714-720. DOI: 10.1021/la504095k

[60] Tsai. H., Lee. Y.,. Facile method to fabricate raspberry-like particulate films for superhydrophobic surfaces. Langmuir. 2007;23:12687-12692. DOI: 10.1021/la702521u

[61] Li. Y.,Chen. X., Li. Q., Song. K., Wang. S., Chen. X., et al.,Layer-by-layer strippable Ag multilayer films fabricated by modular assembly. Langmuir. 2014;30:548–553. DOI: 10.1021/la4045557

[62] Liu. H., Feng. L., Zhai. J, Jiang. L., Zhu. D., reversible wettability of a chemical vapor deposition prepared ZnO film between superhydrophobicity and superhydrophilicity. Langmuir. 2004;20:5659-5661. DOI: 10.1021/la036280o

[63] Liu. H., Feng. L., Zhai. J, Jiang. L., Zhu. D. Reversible wettability of a chemical vapor deposition prepared ZnO film between superhydrophobicity and superhydrophilicity. Langmuir. 2004;20:5659-5661.

[64] Kamal. S.A. A., Ritikos. R., Rahman. S. A.,Wetting behaviour of carbon nitride nanostructures grown by plasma enhanced chemical vapour deposition technique. Applied Surface Science. 2015;328:146-153. DOI: 10.1016/j.apsusc.2014.12.001

[65] Rezaei. S., Manoucheri. I., Moradian. R., Pourabbas. B.,One-step chemical vapor deposition and modification of silica nanoparticles at the lowest possible temperature and superhydrophobic surface fabrication. Chemical Engineering Journal. 2014;252:11-16. DOI: 10.1016/j.cej.2014.04.100

[66] Coclite. A. M., Shi. Y., Gleason. K. K.,Controlling the degree of crystallinity and preferred crystallographic orientation in poly-perfluorodecylacrylate thin films by initiated chemical vapor deposition. Advanced Functional Materials. 2012;22:2167-2176. DOI: 10.1002/adfm.201103035

[67] Dong. X., Chen. J., Ma. Y., Wang. J., Chan-Park. M. B.,Liu. X., et al.,Superhydrophobic and superoleophilic hybrid foam of graphene and carbon nanotube for selective

removal of oils or organic solvents from the surface of water. Chemical Communications. 2012;48:10660-10662. DOI: 10.1039/C2CC35844A

[68] Singh. E., Chen. Z., Houshmand. F., Ren. W., Peles. Y., Cheng. M., Koratkar. N.,Superhydrophobic graphene foams. Small. 2013;9:75-80. DOI: 10.1002/smll.201201176

[69] Lee. C. H., Johnson. N., Drelich. J., Yap. Y. K.,The performance of superhydrophobic and superoleophilic carbon nanotube meshes in water-oil filtration. Carbon. 2011;49:669-676. DOI: 10.1016/j.carbon.2010.10.016

[70] Yang. R., Gleason. K. K.,Ultrathin antifouling coatings with stable surface zwitterionic functionality by initiated chemical vapor deposition (iCVD). Langmuir. 2012;28:12266–12274. DOI: 10.1021/la302059s

[71] Bittencourt. J. A., Fundamentals of plasma physics. Springer, New York; 2014

[72] Popov. A. O., High density plasma sources. Matsushita Electric Works, Woburn; 1996

[73] Francombe. H. M., Plasma sources for thin film deposition and etching. Academic Press, London; 1994

[74] Ostrikov. K. K., Xu. S., Plasma-aided nanofabrication: from plasma sources to nanoassembly. Wiley-VCH, Weinheim; 2007

[75] Akinoglu. M. E., Morfa. J. A., Giersig. M., Understanding anisotropic plasma etching of two-dimensional polystyrene opals for advanced materials fabrication. Langmuir. 2014;30:12354-12361. DOI: 10.1021/la500003u

[76] Barshilia. C. H., Gupta. N., Superhydrophobic polytetrafluoroethylene surfaces with leaf-like micro-protrusions through Ar + O_2 plasma etching process. Vacuum. 2014;99:42-48. DOI: 10.1016/j.vacuum.2013.04.020

[77] Zylka. P., On the surface performance of superhydrophobic silicone rubber specimens fabricated by direct replica method. IEEE Transactions on Dielectrics and Electrical Insulation. 2014;21:1183-1188. DOI: 10.1109/TDEI.2014.003843

[78] Checco. A., Rahman. A., Black. T. C., Robust superhydrophobicity in large-area nanostructured surfaces defined by block-copolymer self assembly. Advanced Materials. 2014;26:886-891. DOI: 10.1002/adma.201304006

[79] Cai. C. Y., Lin. A. K. Y., Yang. H., Superhydrophobic anti-ultraviolet films by doctor blade coating. Applied Physics Letters. 2014;105:201913. DOI: 10.1063/1.4902547

[80] Vourdas. N., Tserepi. A., Gogolides. E., Nanotextured super-hydrophobic transparent poly(methyl methacrylate) surfaces using high-density plasma processing. Nanotechnology. 2007;18:125304. DOI: 10.1088/0957-4484/18/12/125304

[81] Mundoa. R., Benedictis. V., Palumbo. F., d'Agostino. R., Fluorocarbon plasmas for nanotexturing of polymers: a route to water-repellent antireflective surfaces. Applied Surface Science. 2009;255:5461-5465. DOI: 10.1016/j.apsusc.2008.09.020

[82] Balu. B., Breedveld. V., Hess. W. D., Fabrication of "roll-off" and "sticky" superhydrophobic cellulose surfaces via plasma processing. Langmuir. 2008;24:4785-4790. DOI: 10.1021/la703766c

[83] Coulson. S. R:, Woodward. I., Badyal. J. P. S., Super-repellent composite fluoropolymer surfaces. J. Phys. Chem. B 2000;104:8836-8840. DOI: 10.1021/jp0000174

[84] Ruiz. A., Valsesia. A., Ceccone. G., Gilliland. D., Colpo. P., Rossi. F., Fabrication and characterization of plasma processed surfaces with tuned wettability. Langmuir. 2007;23:12984-12989. DOI: 10.1021/la702424r

[85] Ji. Y. Y., Kim. S. S., Kwon. O. P., Lee. S. H., Easy fabrication of large-size superhydrophobic surfaces by atmospheric pressure plasma polymerization with non-polar aromatic hydrocarbon in an in-line process. Applied Surface Science. 2009;255:4575-4578. DOI: 10.1016/j.apsusc.2008.12.002

[86] Satyaprasad. A., Jain. V., Nema. S. K., Deposition of superhydrophobic nanostructured Teflon-like coating using expanding plasma arc. Applied Surface Science. 2007;253:5462-5466. DOI: 10.1016/j.apsusc.2006.12.085

[87] Yang. S. H., Liu. C. H., Hsu. W. T., Chen. H., Preparation of super-hydrophobic films using pulsed hexafluorobenzene plasma. Surface and Coatings Technology. 2009;203:1379-1383. DOI: 10.1016/j.surfcoat.2008.11.007

[88] Teare. D. O. H., Spanos. C. G., Ridley. P., Kinnmond. E. J., Roucoules. V., Pulsed plasma deposition of super-hydrophobic nanospheres. Chemistry of Materials. 2002;14:4566-4571. DOI: 10.1021/cm011600f

7

Solid-Liquid-Solid Interfaces

Jeffrey L. Streator

Abstract

Interfaces comprised of a liquid interposed between two solids in close proximity are common in small-scale devices. In many cases, the liquid induces large and undesired adhesive forces. It is of interest, therefore, to model the way in which forces are developed in such an interface. The following chapter presents several models of liquid-mediated adhesion, considering the roles of surface geometry, liquid surface tension, elastic deformation, surface roughness, and surface motion on the development of interfacial forces.

Keywords: Capillary film, liquid-mediated adhesion, liquid bridge

1. Introduction

Phenomena related to the wetting of solid–solid interfaces are of technological importance. When two surfaces are in close proximity, the presence of a liquid film may cause the surfaces to stick together. Such liquid-mediated adhesion can negatively affect the operation of micro/nanoscale systems [1–7]. The interfacial liquid film, which may be present due to condensation, contamination, or lubrication, may experience large concave curvatures at the liquid-vapor interface and large negative pressures. These negative pressures give rise to large adhesive forces, which can have a potentially deleterious effect on the performance of small-scale devices.

In this chapter, we will discuss the behavior of an interface comprised of a liquid interposed between two solids. Throughout this chapter, we are concerned with the role of liquid films in regimes where gravitational effects are negligible, which generally implies that the vertical length scale is small. As an illustration, it can be easily shown that the change in pressure due to gravity within a near-hemispherical water droplet (resting on a horizontal surface) from just within the top of the free surface to the bottom of the droplet is given by $(\Delta p)_{gravity} = \rho g R$, where ρ is the water mass density, g is the gravitational acceleration, and R is the approximate radius of the droplet. By comparison, the change in pressure across the free surface of the droplet is given by $(\Delta p)_{surface\ tension}=2\gamma / R$, where γ is the liquid surface tension. Thus, the ratio of gravitational effects to surface tension effects is equal to $\rho g R^2 / 2\gamma$. For water at room temperature (and ambient pressure), one has ρ = 1000 kg/m^3 and γ = 0.0727 N/m, so that for a radius of 1.0 mm, we have a ratio of about 0.07, meaning that the change in pressure due to gravity is only about 7% of that due to surface tension. Moreover, it is seen that the relative effects of gravity decrease in proportion to the square of the droplet radius. In general, the smaller the vertical scale, the less important are the effects of gravity in comparison to those of surface tension.

Of particular interest in this chapter is the topic of liquid-mediation adhesion, a mechanism by which the liquid film pulls inward on the solid surfaces. We consider the effects of liquid surface tension, liquid viscosity, surface geometry, surface roughness, surface elasticity, and surface motion on the development of adhesive forces in the interface. Our approach to discussing the recent literature on the topic of liquid-mediated adhesion is to organize things according to several basic characteristics: gross interface geometry (flat or curved), surface topography (smooth or rough), structural properties (rigid or deforming), meniscus type (constant-volume or constant-pressure) and separating process (quasi-static or dynamic). In this context, Table 1 categorizes recent research that is particularly relevant to the subject of this chapter. It is noted that an entry of "volume" under the "film constant" heading means that the volume of the liquid bridge is held fixed during the separation process, while an entry of "pressure" indicates that the liquid is assumed to remain in thermodynamic equilibrium with its vapor during the separation process.

Gross Interface Geometry	Surface Type	Deform. Behavior	Loading Process	Film Constant	Author(s)	Year Publ.	Ref. No.
flat on flat	smooth	elastic	quasi-static	volume	Zheng and Streator	2004	28
flat on flat	rough	elastic-plastic	quasi-static	pressure	Del Rio et al.	2008	19
flat on flat	rough	elastic	quasi-static	pressure	Wang and Regnier	2015	37
flat on flat	rough	elastic	quasi-static	pressure	Peng et al.	2009	21
flat on flat	rough	rigid	quasi-static	pressure or volume	de Boer and de Boer	2007	18
flat on flat	rough	elastically hard	quasi-static	pressure	de Boer	2007	17
flat on flat	rough	elastic	quasi-static	pressure	Persson	2008	20
flat on flat	rough	elastic	quasi-static	volume	Streator and Jackson	2009	34
flat on flat	rough	elastic	quasi-static	volume	Streator	2009	33
flat on flat	rough	elastic	quasi-static	volume	Rostami and Streator	2015	35

Gross Interface Geometry	Surface Type	Deform. Behavior	Loading Process	Film Constant	Author(s)	Year Publ.	Ref. No.
flat on flat	rough	elastic	quasi-static	volume	Rostami and Streator	2015	36
sphere on flat	smooth	elastic	quasi-static	pressure	Men et al.	2009	24
sphere on flat, and sphere on sphere	smooth	rigid	quasi-static	volume	Rabonivich et al.	2005	23
sphere on sphere	smooth	elastically soft	quasi-static	pressure	Butt et al.	2010	22
sphere on sphere	smooth	elastic	quasi-static	volume	Zheng and Streator	2003	30
sphere on sphere	smooth	elastic	quasi-static	volume	Zheng and Streator	2007	31
flat on flat	smooth	rigid	dynamic	n/a (flooded)	Roemer et al.	2015	15
flat on flat or sphere on flat	smooth	rigid	dynamic	volume	Cai and Bhushan	2007	9
sphere on flat	smooth	rigid	dynamic	n/a (flooded)	Streator	2006	25

Table 1. Recent research on the topic of the liquid-mediated adhesion

2. Models of solid surfaces bridged by a liquid

2.1. Liquid film between smooth, rigid, parallel flats

2.1.1. Static and quasi-static conditions

Consider the problem of a continuous liquid film that is at static equilibrium between two rigid, parallel flats in close proximity as shown in Figure 1. In this idealized case, the liquid forms an axisymmetric configuration, so that any horizontal cross section is circular. Because the liquid is in static equilibrium, the entire film must be at a single pressure. Per the Young-Laplace equation [8], the pressure drop Δp across the free surface is given by

$$\Delta p = p_a - p = \gamma\left(\frac{1}{R_1} + \frac{1}{R_2}\right). \tag{1}$$

Where p_a is the ambient pressure, p is the film pressure, and $R_{1,2}$ are the principal radii of normal curvature of the free surface at any given point on the free surface. Since we are dealing with small vertical spacing, it is reasonable to assume the radius of curvature (R_2) that exists in the plane of the figure at each free surface point is much smaller than the other principal radius of curvature (R_1), which lies in a plane that is perpendicular to the plane of the figure as well as perpendicular to the tangent plane to the free surface at the point in question. In Figure 1, we have chosen to illustrate the value of R_1 that exists in the plane of minimum horizontal diameter. Assuming R_1 is sufficiently larger than R_2 that $1/R_1$ may be neglected, the pressure drop in Eq. (1) becomes

$$\Delta p = \frac{\gamma}{R_2} \tag{2}$$

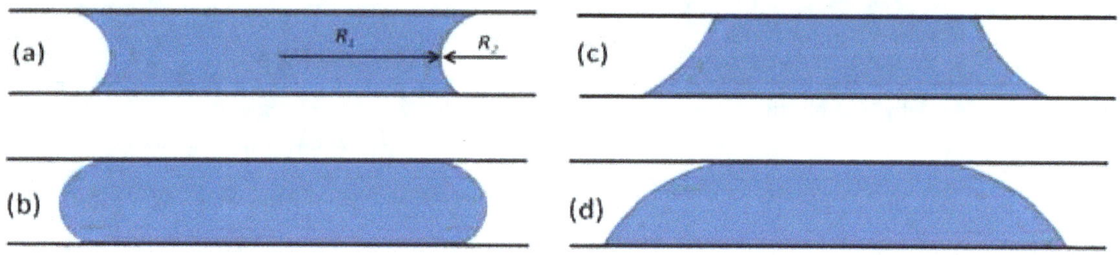

Figure 1. Profile of an axisymmetric liquid film between rigid, parallel plates. (a) The liquid wets both surfaces leading to a concave film shape. (b) The liquid wets neither surface, leading to a convex film shape. (c) The liquid wets one surface but not the other, with the wet surface closer to complete wetting than the non-wet surface is to complete non-wetting, leading to a convex film. (d) The liquid wets one surface, but not the other, with the non-wet surface closer to complete non-wetting than the wet surface is to complete wetting, leading to a convex film.

Moreover, owing to the fact that the liquid film, being continuous and in static equilibrium, must experience a uniform pressure, one may conclude that the radius of curvature R_2 is the same at every point of the free surface. Thus, the free surface profile is in the shape of a circle. Using this result leads to the geometrical relationship depicted in Figure 2, by which one concludes that

$$h = R_2(\cos\theta_1 + \cos\theta_2) \tag{3}$$

so that

$$\Delta p = \frac{\gamma(\cos\theta_1 + \cos\theta_2)}{h} \tag{4}$$

where $\theta_{1,2}$ are the liquid–solid contact angles. A liquid is considered to "wet" a given surface if its contact angle (measured from solid–liquid interface to the solid-vapor interface) is less than 90 degrees, and a liquid is considered to be non-wetting if its contact angle is greater than 90 degrees. Complete wetting is associated with a contact angle of 0 degrees and complete non-wetting corresponds to a contact angle of 180 degrees. While Figure 2 shows a case for which the liquid wets both upper and lower surfaces, Eq. (3) holds for each of the configurations depicted in Figure 1. Depending on both signs and relative magnitudes of the cosine terms in

Eq. (4), the film pressure may be greater than, equal to, or less than atmospheric pressure. For the case of Figure 1a, where both surfaces are wet by the liquid, both cosine terms are positive and the film pressure is sub-ambient. In contrast, for the case of Figure 1b, where neither surface is wet by the liquid, both cosine terms are negative and the film pressure exceeds the ambient pressure. When one of the surfaces is wet by the film while the other is not (Figures 1c and 1d), the sign of the film pressure depends on the relative magnitudes of the two cosine terms. When the contact angle associated with the surface that is wet by the liquid is closer to zero than the other contact angle is to 180 degrees (Figure 1c), then the film shape is concave and the film pressure is sub-ambient. On the other hand, if the contact angle associated with the surface that is wet by the liquid departs from zero degrees more than the opposing surface departs from 180 degrees (Figure 1d), then the film shape is convex and the film pressure is greater than ambient.

The value of the contact angle for a particular case is determined by a local thermodynamic equilibrium among the three relevant interfaces, which can be expressed in the Young-Dupree equation [8]

$$\gamma_{SV} = \gamma_{SL} + \gamma_{LV} \cos\theta \tag{5}$$

where γ_{SV}, γ_{SL}, and γ_{LV} are the surface energies per unit area of the solid–vapor, solid–liquid, and liquid–vapor interfaces, respectively. Note that γ_{LV} is the same as the surface tension of the liquid γ.

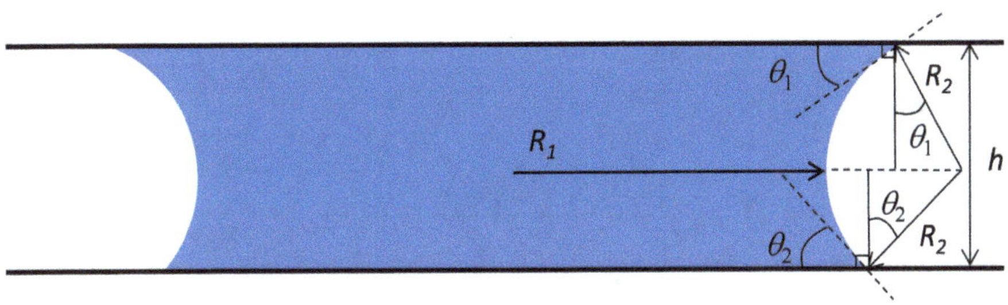

Figure 2. Geometrical relationship between plate spacing and radius of curvature of free surface for an assumed circular profile. Without loss of generality, the liquid is shown here as wetting both surfaces.

For a concave film shape (Figures 1a and 1c) the sum on the right-hand side of Eq. (4) is positive, yielding a positive pressure drop relative to atmospheric pressure. Thus, in terms of gauge pressure, the pressure within the film is negative. One important consequence is that the liquid exerts a force that pulls inward on the two plates so that the force exerted on either of the plates may be considered the force of adhesion due to the presence of the film. With reference to Figure 3, this adhesive force (F_{ad}) can be expressed as

$$F_{ad} = (-p)\pi R_1^2 + 2\pi R_1 \gamma = \frac{\gamma(\cos\theta_1 + \cos\theta_2)}{h}\pi R_1^2 + 2\pi R_1 \gamma \tag{6}$$

The first term on the right-hand side is the contribution to the adhesive force arising from the pressure drop across the free surface, while the second term is the adhesive force exerted by the free surface itself. Note that the total force exerted on the bottom of this upper section of the liquid film is simply transmitted to the upper plate, so the force given by Eq. (6) is indeed the adhesive force. Now under the assumption that R_1 is much greater than R_2, it can be shown that the force contribution to the pressure drop dominates the force contribution due to the free surface. Let the first term on the far right-hand side be denoted by $F_{\Delta p}$ and the second term be denoted by F_γ. Then $F_\gamma / F_{\Delta p}$ =2 R_2 / R_1. Thus, to the extent that $1/R_1$ can be neglected in Eq. (1), which leads to Eq. (2), the force contributed by the free surface $2\pi R_1 \gamma$ may be neglected in Eq. (6).

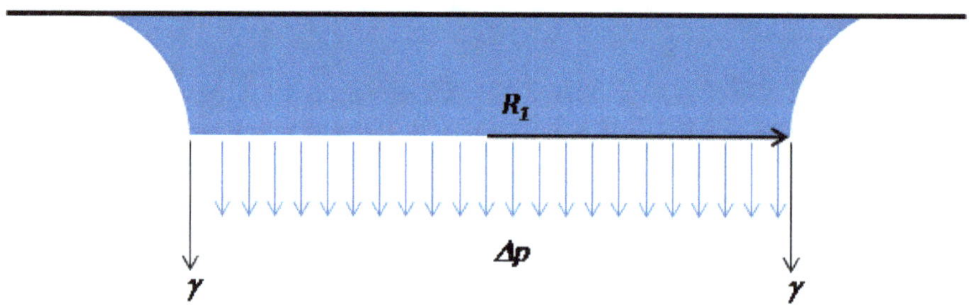

Figure 3. Sources of force exerted on the upper section of the fluid (dividing line chosen at the plane of minimum diameter).

Suppose now that the liquid film has a fixed volume V_o. To a good approximation, this volume may be expressed as $V_o = \pi R_1^2 h$, where the deviation from a cylindrical geometry has been ignored. Then the adhesive force may be written as:

$$F_{ad} = \frac{\gamma(\cos\theta_1 + \cos\theta_2)}{h^2} V_o \tag{7}$$

This equation shows that under the conditions of fixed liquid volume the adhesive force is inversely proportional to the square of the film thickness.

When a quantity of a pure liquid of given chemical species is at thermodynamic equilibrium, the partial pressure of the vapor phase of the species is equal to the vapor pressure of the liquid phase for the given temperature. For a curved free surface, there is a small deviation in the vapor pressure from that corresponding to a planar free surface. This deviation is accounted for by the well-known Kelvin equation [8]

$$ln\frac{p_s}{p_v} = \gamma\left(\frac{1}{R_1} + \frac{1}{R_2}\right)\frac{V_m}{RT} = \frac{\gamma}{R_K}\frac{V_m}{RT} \tag{8}$$

where p_s is the saturation pressure at the given temperature, p_v is the pressure of the vapor just outside of the liquid film, V_m is the molar volume, R_K is the Kelvin radius, R is the universal gas constant, and T is the absolute temperature. Assuming, as before, $R_1 \gg R_2$, and then isolating R_2, one obtains

$$R_2 = R_K = -\frac{\gamma V_m}{RTln\frac{p_v}{p_s}} \tag{9}$$

Using this result in Eq. (2) gives

$$p_a - p = -\frac{RTln\frac{p_v}{p_s}}{V_m} \tag{10}$$

Now, suppose the chemical species in question is water, so that the ratio p_v / p_s represents the relative humidity. Then Eq. (9) states that, at thermodynamic equilibrium, the radius of curvature of any free surface of the liquid film is determined by the relative humidity. For example, taking properties of water at room temperature and assuming a relative humidity of 95%, we have

$$R_2 = -\frac{\left(0.0727\frac{\text{N}}{\text{m}}\right)(0.018\ \text{kg / mol})/\left(.00100\text{m}^3\ /\ \text{kg}\right)}{8.314\frac{\text{J}}{\text{mol}-\text{K}}(293\text{K})\ln(0.95)} = 10.5\,\text{nm} \tag{11}$$

so that, from Eq. (10),

$$p_a - p = -\frac{RTln\frac{p_v}{p_s}}{V_m} = \frac{\left(0.0727\frac{\text{N}}{\text{m}}\right)}{10.5\ \text{nm}} = 6.9\ \text{MPa} \tag{12}$$

If we take the contact angles to be zero, then, from Eq. (4) and Eq. (11), $h = 2R_2 = 21$ nm. Now, for this mathematically idealized case of perfectly parallel plates, 21 nm would be the only spacing for which a liquid film could exist at thermodynamic equilibrium at 95% relative humidity. On the other hand, if the surfaces were curved, even slightly, then there would be a range of humidity values for which a film could be sustained at thermodynamic equilibrium. It should be noted that, in practical situations, the establishment of thermodynamic equilibrium may require a considerable amount of time, such as hours, or even days. In the interim, the adhesive forces will be dictated by the current amount of liquid within the interface.

2.1.2. Dynamic separation

The foregoing analysis is applicable to conditions of static (or quasi-static) equilibrium. Additional effects may arise from viscous interactions. Consider now a situation where the upper plate is pulled upward at a prescribed rate, while the lower plate is held fixed. One approach to analyzing such a situation [9] is to assume that the liquid flow is governed by the Reynolds equation of lubrication [10].

$$\frac{1}{r}\frac{\partial}{\partial r}\left(rh^3\frac{\partial p}{\partial r}\right)=12\mu\frac{\partial h}{\partial t} \tag{13}$$

where r is the radial coordinate measured from the center of the axisymmetric film cross-section. Assuming that the gap, h, is uniform (i.e., independent of r), the above equation can be integrated twice to give

$$p(r,t)=3\mu\frac{\dot{h}}{h^3}+c_1(t)\ln(r)+c_o(t) \tag{14}$$

where $\dot{h}\equiv\frac{\partial h}{\partial t}$ and $c_1(t)$ and $c_o(t)$ are constants of integration (i.e., independent of r).

To obtain the constants of integration, we assume that (1) the pressure just inside the free surface is that corresponding to the static case (see Eq. 4), and (2) the pressure is finite at $r=0$. Then letting R_1, as before, denote the inner radius of the droplet, we obtain (in terms of gauge pressure):

$$p(r,t)=-\frac{\gamma(\cos\theta_1+\cos\theta_2)}{h}+3\mu\frac{\dot{h}}{h^3}\left(r^2-R_1^2\right) \tag{15}$$

Now, the adhesive force is just given by

$$F_{ad}=\int_0^b -p(r,t)2\pi r dr=\frac{\gamma(\cos\theta_1+\cos\theta_2)}{h}\pi R_1^2+\frac{3}{2}\pi\mu\frac{\dot{h}}{h^3}R_1^4 \tag{16}$$

For a fixed liquid volume (V_o), which is approximated by $V_o=\pi R_1^2 h$, we arrive at

$$F_{ad}=\int_0^b -p(r,t)2\pi r dr=\frac{\gamma(\cos\theta_1+\cos\theta_2)}{h^2}V_o+\frac{3}{2}\mu\frac{\dot{h}}{h^5}V_o^2 \tag{17}$$

The above equation shows that adhesive force grows in proportion to the rate $\dot{h}$ at which the plates are being pulled apart and is quite sensitive to the value of separation. Small separations require a much larger separating force than what is required at larger separations for the given separating rate. One caveat, however, is that there is a practical limit as to the magnitude of negative pressure that can be sustained during the separation of surfaces. Whereas thermodynamic equilibrium suggests that the liquid will cavitate once its absolute pressure approaches zero (e.g., [11]), it has also been found that, in certain cases, the liquid film may achieve absolute pressures that are negative [7, 12–15]. For example, an analytical model was developed [15] for the dynamic vertical separation of opposing plates and, based on fitting with experimental data, a tensile strength of 35 kPa was found for a mineral oil. In any case, if one denotes p_{cav} as the cavitation pressure (relative to atmospheric pressure), then the maximum possible adhesive force can be written as:

$$F_{ad} = -p_{cav}\pi R_1^2 \tag{18}$$

2.2. Liquid film between rigid, inclined surfaces

Consider the situation depicted in Figure 4, where there is a liquid film between two flat surfaces whose planes intersect. The configuration of Figure 4a is a non-equilibrium state owing to the greater free-surface curvature on the right than on the left, and the associated lower pressure (i.e., greater reduction in pressure compared to ambient). Thus, the fluid will flow from left to right, all the way up to the edge (Figure 4b) until achieving a configuration with equal free-surface curvature at left and right ends, thereby yielding the same pressure drop. The two-dimensional depiction of Figure 4, of course, obscures the required re-configuration that happens in three dimensions. In fact, the entire free surface must attain the same curvature, which means that liquid would find its way to both the front and back edges as well as the right edge.

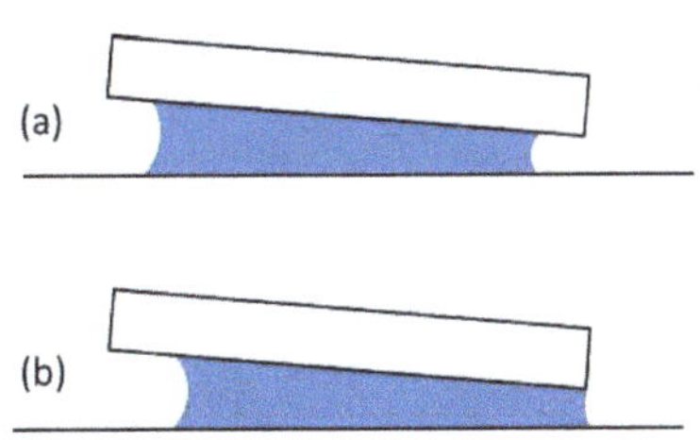

Figure 4. Liquid film between inclined surfaces: (a) non-equilibrium configuration and (b) equilibrium configuration.

2.3. Liquid film between a smooth, rigid sphere and a rigid flat

The sphere-flat configuration is of interest in its own right and as an important part of a rough surface contact model, in which contributions from various asperity-asperity liquid bridges are summed by viewing each pair as reflecting the interaction between a pair of spheres having the asperity curvatures.

2.3.1. Static and quasi-static conditions

The interaction between a sphere and flat bridged by a liquid film, as illustrated in Figure 5, has been analyzed in [16]. When the radial width of the liquid film (b) is sufficiently small compared to the radius of the sphere, the slope of the sphere at the location of the free surface may be taken as horizontal. In this case, the pressure within the film will be given by Eq. (4) with h replaced by $h(b)$, the film thickness at the radial location of the free surface. Additionally, the sphere contour can be approximated well by that of a paraboloid, so that

$$h(b) = D + \frac{b^2}{2R} \tag{19}$$

This gives

$$p_a - p = \frac{\gamma(\cos\theta_1 + \cos\theta_2)}{h(b)} = \frac{\gamma(\cos\theta_1 + \cos\theta_2)}{D + \dfrac{b^2}{2R}} \tag{20}$$

The force of adhesion is obtained by multiplying this pressure difference by the cross-section area of the liquid bridge (πb^2), giving (after re-arrangement)

$$F_{ad} = 2\pi R\gamma(\cos\theta_1 + \cos\theta_2)\left(1 - \frac{D}{h(b)}\right) = \frac{2\pi R\gamma(\cos\theta_1 + \cos\theta_2)}{1 + \dfrac{2RD}{b^2}} \tag{21}$$

Several studies have considered the role of relative humidity on the adhesion between a sphere and a flat (or sphere on sphere) [9, 17–23], where, at thermodynamic equilibrium, the radius of the curvature of the free surface of the meniscus would be equal to the Kelvin radius, per Eq. (9). Such analysis is most appropriate for volatile liquids [24]. In this case the value of $h(b)$ appearing in Eq. (21) would be determined directly by the relative humidity, via Eqs. (3) and (9).

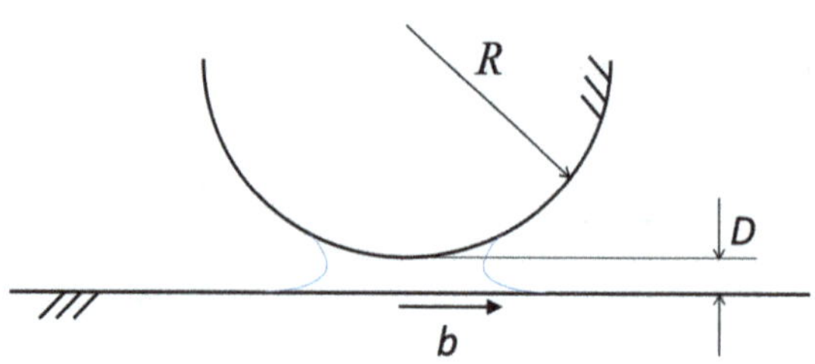

Figure 5. Liquid film between a rigid sphere and a rigid flat.

2.3.2. *Dynamic separation*

Now we consider the forces that arise when a sphere of mass m and radius R is separated from the flat in a dynamic fashion, so that the minimum spacing D is a function of time. Denoting the instantaneous vertical spacing between the sphere surface and the flat as $h(r, t)$, we have $D=h(0, t)$. When the wetted radius b is much smaller than the sphere radius, one gets

$$h(r,t)=D(t)+\frac{r^2}{2R} \tag{22}$$

When a net external force F (i.e., an applied force less the sphere weight) acts on the sphere (positive upward), the governing equation becomes [25]

$$m\frac{d^2D}{dt^2}=F-F_m-F_v \tag{23}$$

where F_m is the "meniscus" force, which accounts for the effect of the pressure drop across the curved free surface of the liquid meniscus and F_v is the "viscous" force, which arises from the deformation of the liquid bridge. It is assumed that any buoyancy forces are negligible. Following [26], the pressure field, as derived from the solution of the Reynolds equation (e.g., [10]), can be written as

$$p(r,t)=p(b,t)-3\mu R\left(\frac{1}{h(r,t)^2}-\frac{1}{h(b,t)^2}\right)\frac{1}{D}\frac{dD}{dt} \tag{24}$$

where μ is the liquid viscosity. Note that the wetted radius b is itself a function of time. Assuming that the free surface of the liquid has the same shape as when the film is quasi-static, the pressure just inside the meniscus, $p(b, t)$ is given by Eq. (4). Thus, integrating over the meniscus area gives [25]

$$F_{ad}=F_m+F_v=2\pi R\gamma(\cos\theta_1+\cos\theta_2)\left(1-\frac{D}{h(b,t)}\right)+6\pi\mu R^2\left(1-\frac{D}{h(b,t)}\right)^2\frac{1}{D}\frac{dD}{dt} \tag{25}$$

where F_{ad} is the adhesive force. Direct integration of the film thickness profile (19) provides the liquid volume:

$$V=\pi b^2D+\frac{\pi b^4}{4R}=\pi R\left[h(b,t)^2-D^2\right] \tag{26}$$

Assuming the meniscus volume is fixed, we set $V = V_o$ and express the film thickness at the free surface as

$$h(b,t) = \sqrt{D^2 + \frac{V_o}{\pi R}} \tag{27}$$

Using this result in Eq. (25) allows the force exerted by the liquid to be expressed in terms of the separation D

$$F_{ad} = F_m + F_v = 2\pi R\gamma(\cos\theta_1 + \cos\theta_2)\left(1 - \frac{D}{\sqrt{D^2 + \frac{V_o}{\pi R}}}\right) + 6\pi\mu R^2\left(1 - \frac{D}{\sqrt{D^2 + \frac{V_o}{\pi R}}}\right)^2 \frac{1}{D}\frac{dD}{dt} \tag{28}$$

In cases where the inertial term of Eq. (23) is negligible, the net applied load F is equated with the sum of the capillary and viscous forces F_{ad}. Moreover, in cases where the variation in the capillary force is small compared to the variation in the viscous force, Eq. (28) can be integrated to give [9, 27]

$$\int_0^{t_s}(F - F_m)dt = \int_0^{D_s} F_v dt = \int_0^{D_s} F_v \frac{dt}{dD} dD = 6\pi\mu R^2 ln\left[\frac{D_s\left(D_o + \sqrt{D_o^2 + \frac{V_o}{\pi R}}\right)^2 \sqrt{D_s^2 + \frac{V_o}{\pi R}}}{D_o\left(D_s + \sqrt{D_s^2 + \frac{V_o}{\pi R}}\right)^2 \sqrt{D_o^2 + \frac{V_o}{\pi R}}}\right] \tag{29}$$

where t_s is the time to completely separate the surfaces, D_s is the distance at which the sphere and flat are considered completely separated, and D_o is the sphere-flat spacing at the beginning of the separation process. It was argued in [27] that even in the case of initial solid-solid contact, the value of D_o should not be zero, but should be selected in accordance with the interface roughness. Eq. (29) suggests that the separation process for a viscous-dominated interaction is governed by the time integral of the viscous force, known as the viscous impulse, the value of which is a function only of the input parameters of the problem [27]. Now the distance of complete separation D_s is taken as infinite in [27], and finite in [9]. For $D_s \to \infty$, Eq. (29) can be simplified [27], giving

$$I_v = 6\pi\mu R^2 ln\left[\frac{\left(D_o + \sqrt{D_o^2 + \frac{V_o}{\pi R}}\right)^2}{4D_o\sqrt{D_o^2 + \frac{V_o}{\pi R}}}\right] = 6\pi\mu R^2 ln\left[\frac{\left(D_o + \sqrt{D_o^2 + \frac{V_o}{\pi R}}\right)^2}{4D_o\sqrt{D_o^2 + \frac{V_o}{\pi R}}}\right] \tag{30}$$

where I_v is the viscous impulse. Applying Eq. (30) to the case of a sphere interacting with a surface that is initially covered with a thin film of uniform thickness h_o, the above is approximated by [27]

$$I_v = 6\pi\mu R^2 \ln\left(h_o / 2D_o\right) \tag{31}$$

One important result of the above relationship is that the rate of applied loading determines the peak adhesive load developed during separation, which we label here the "pull-off force" ($F_{\text{pull-off}}$).

For example, when the externally applied force increases at a constant rate ($\dot{F}$), the pull-off force takes the form:

$$F_{\text{pull-off}} = F_m + \sqrt{2\dot{F}I_v} \tag{32}$$

A modified approach is needed to analyze the "fully-flooded" case, where the sphere interacts with a sufficiently thick lubricant film that further increases to the film thickness have negligible impact on the adhesive force. In this case, Eq. (32) still holds, but the viscous impulse becomes [25]

$$I_v = 6\pi\mu R^2 \ln\left[0.1\frac{\left(6\pi\mu R^2\right)^{\frac{3}{2}}}{mD_o\sqrt{2\dot{F}}}\right] \tag{33}$$

where m is the mass of the sphere.

It is emphasized here that Eqs. (28)-(33) presume the liquid film is not experiencing any cavitation. As discussed previously (see Eq. (18)), the potential development of a fully cavitated film would provide an upper bound for the adhesive force.

2.4. Liquid film between smooth, elastic flats

Figure 6 depicts a scenario when a liquid film interacts with two semi-infinite elastic bodies, where $E_{1,2}$ and $\nu_{1,2}$ are the elastic moduli and Poisson ratios of bodies 1 and 2, respectively and H is the uniform gap between the surfaces that exists in the absence of deformation. For this situation, the pressure within the liquid film causes elastic deformation of the half-spaces. Here we focus our attention on the case were the liquid film wets both surfaces such that they each experience a contact angle less than 90 degrees. This problem has been analyzed previously [28] and that work is summarized here. Letting Δp represent the pressure drop across the free surface of the liquid (from outside to inside), the film pressure (gauge) can be expressed as

$$p(r)=\begin{cases}-\Delta p & r\le b\\ 0 & r>b\end{cases} \tag{34}$$

This pressure field causes an associated deformation field [29]

$$u(r)=u_1(r)+u_2(r)=\begin{cases}\dfrac{4\Delta pb}{\pi E'}\displaystyle\int_0^{\pi/2}\sqrt{1-\frac{r^2}{b^2}\sin^2\psi}\,d\psi & r\le b\\ \dfrac{4\Delta pb}{\pi E'}\displaystyle\int_0^{\pi/2}\frac{b^2\cos^2\psi}{\sqrt{1-\frac{r^2}{b^2}\sin^2\psi}}\,d\psi & r>b\end{cases} \tag{35}$$

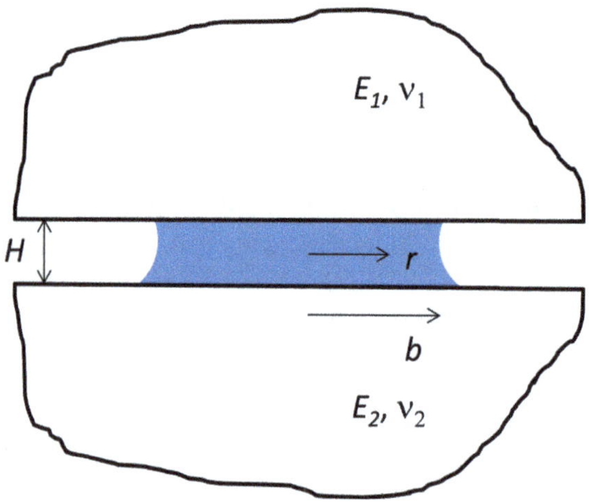

Figure 6. A liquid film bridging two elastic half-spaces.

In the above equation, $u_1(r)$ and $u_2(r)$ are the normal surface displacements of bodies 1 and 2, respectively (each positive toward the opposing body), $u(r)$ is the total displacement, and the reduced modulus E' is given by $1/E'\equiv(1-\nu_1^2)/E_1+(1-\nu_2^2)/E_2$. It is assumed here that the bulk positions of the bodies are fixed, so that opposing surface points remote to the interface are maintained at a spacing of H. At any radial position within the wetted film, the film thickness can be expressed as

$$h(r)=H-u(r) \tag{36}$$

Using Eq. (35), the volume of the liquid bridge (V_o), which is assumed to be fixed, is given by

$$V_o=\int_o^b 2\pi rh(r)dr=\pi Hb^2-\frac{16}{3E'}\Delta pb^3 \tag{37}$$

The equilibrium configuration can be determined by considering the minimization of the free energy, which is comprised of elastic strain energy (U_S) and surface energy U_E. The elastic strain energy is simply given by the work done in creating the deformation field

$$U_E = \frac{1}{2}\int_o^b \Delta p u(r) 2\pi r dr \tag{38}$$

Using Eq. (35) and carrying out the integration gives

$$U_E = \frac{8}{3E'}\Delta p^2 b^3 \tag{39}$$

Now the surface energy consists for energy contributions from the solid-vapor, solid-liquid, and liquid-vapor interfaces, so that

$$U_S = \pi\left(R_o^2 - b^2\right)\left(\gamma_{SV_1} + \gamma_{SV_2}\right) + \pi b^2\left(\gamma_{SL_1} + \gamma_{SL_2}\right) + A_{LV}\gamma \tag{40}$$

where subscripts 1 and 2 refer to the upper and lower surfaces, respectively and R_o is the radius of an arbitrary control region that encloses the interface. It is reasonable to assume here that the film thickness at the free surface is sufficiently small compared to the meniscus radius that the energy contributions from the liquid-vapor interface are negligible. Then the total free energy is given by

$$U_T = U_S + U_E = \pi\left(R_o^2 - b^2\right)\left(\gamma_{SV_1} + \gamma_{SV_2}\right) + \pi b^2\left(\gamma_{SL_1} + \gamma_{SL_2}\right) + \frac{8}{3E'}\Delta p^2 b^3 \tag{41}$$

Applying Eq. (5) to each surface and recalling that $\gamma \equiv \gamma_{LV}$ one obtains

$$U_T = \pi R_o^2\left(\gamma_{SV_1} + \gamma_{SV_2}\right) - \pi b^2 \gamma\left(\cos\theta_1 + \cos\theta_2\right) + \frac{8}{3E'}\Delta p^2 b^3 \tag{42}$$

A stable equilibrium corresponds to the minimization of the free energy U_T under the constraint of constant liquid volume V_o. Let us now introduce dimensionless quantities:

$$\eta \equiv \frac{2\Delta pb}{E'H} \tag{43}$$

$$U_T^* \equiv \frac{U_T - \pi R_o^2\left(\gamma_{SV_1} + \gamma_{SV_2}\right)}{\sqrt{\pi}E'H^{3/2}V_o^{1/2}} \tag{44}$$

$$\Gamma \equiv \frac{\gamma^2\left(\cos\theta_1 + \cos\theta_2\right)^2 V_o}{4E'^2H^5} \tag{45}$$

With these definitions, the dimensionless free energy can be expressed as

$$U_T^* \equiv \frac{2\eta^2}{3\pi\sqrt{1-(8/3\pi)\eta}} - 2\sqrt{\frac{\Gamma}{\pi}}\frac{1}{1-(8/3\pi)\eta} \tag{46}$$

Note also that from Eqs. (35), (36) and (43), the minimum film thickness is given by

$$h_{\min} = H - u(0) = H - \frac{2\Delta pb}{E'} = H(1-\eta) \tag{47}$$

So that η provides a dimensionless measure as to the degree of surface approach. When $\eta = 0$, the surfaces are at the original separation throughout, and when $\eta = 1$ the surfaces come into point contact. For a stable equilibrium, the dimensionless energy achieves a local minimum with respect to η. Thus, a necessary (but not sufficient) condition for stable equilibrium is that $dU_T^*/d\eta = 0$, which yields the following requirement:

$$\eta^2\left(1 - \frac{2}{\pi}\eta\right)^2\left(1 - \frac{8}{3\pi}\eta\right) = \frac{16\Gamma}{\pi} \tag{48}$$

The solution space of Eq. (48) is shown in Figure 7. An investigation of $d^2U_T^*/d\eta^2$ at the various equilibrium values of η reveals that values to the left of the peak ($\eta < 0.5577$) correspond to stable equilibrium configurations, whereas those to the right of the peak, correspond to unstable equilibrium configurations. From the graph, we see that no equilibrium configurations exist for values of Γ greater than 0.0134. This result implies that solid-solid contact must occur whenever $\Gamma > 0.0134$. Further, once solid-solid contact occurs, the contact region grows without bound (i.e., the free energy, which now includes a solid-solid contribution, decreases monotonically with increasing contact radius).

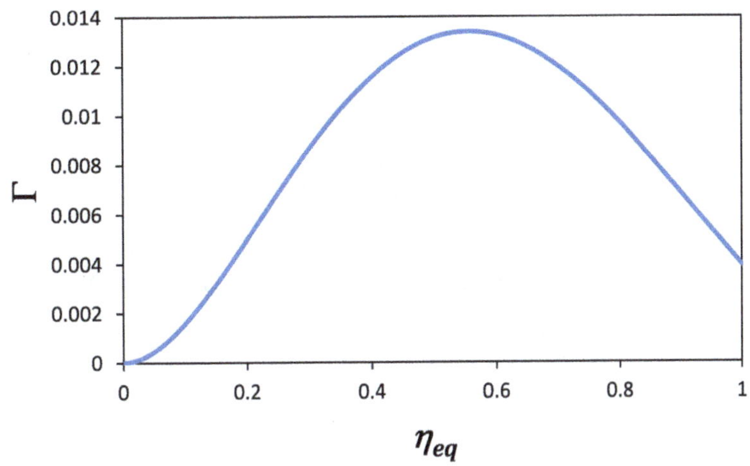

Figure 7. Equilibrium configurations for two half-spaces bridged by a liquid film.

Using Figure 7, one can determine the adhesive force. Letting the subscript "eq" identify values corresponding to a stable equilibrium configuration, it can be shown using Eqs. (35)–(37), (43), and (48), that

$$b_{eq} = \sqrt{\frac{V_o}{\pi H\left(1 - \frac{8}{3\pi}\eta_{eq}\right)}} \tag{49}$$

$$h_{eq}(b_{eq}) = H\left(1 - \frac{2}{\pi}\eta_{eq}\right) \tag{50}$$

$$\Delta p_{eq} = \frac{\gamma(\cos\theta_1 + \cos\theta_2)}{h_{eq}(b_{eq})} \tag{51}$$

Then, the adhesive force is given by

$$F_{ad} = \pi b_{eq}^2 \Delta p_{eq} = \frac{\gamma(\cos\theta_1 + \cos\theta_2)V_o}{H^2\left(1 - \frac{8}{3\pi}\eta_{eq}\right)\left(1 - \frac{2}{\pi}\eta_{eq}\right)} \tag{52}$$

2.5. Liquid film between smooth, elastic spheres

When a liquid bridges two elastic spheres [30], as illustrated in Figure 8, the situation is similar to the case of two elastic half-spaces (discussed above), but with an added feature due the surface curvature. The displacement profile is still given by Eq. (35), but the film thickness profile is now given by

$$h(r) = H + \frac{r^2}{2R} - u(r) \tag{53}$$

where R is the composite radius of curvature, defined by $1/R = 1/R_1 + 1/R_2$. Thus, upon integrating, the liquid volume takes the form

$$V_o = \pi H b^2 - \frac{16}{3E'}\Delta p b^3 + \frac{\pi b^4}{4R} \tag{54}$$

Note that the expressions for the elastic strain energy and surface energy are the same as those for the two half-spaces, so that the total free energy is still given by Eq. (41). In addition to non-dimensional parameters η and Γ (Eqs. 43 and 45), we introduce a dimensionless volume according to

$$\Psi \equiv \frac{V_o}{RH^2} \tag{55}$$

and use a different form for the dimensionless free energy

$$U_T^* \equiv \frac{U_T - \pi R_o^2\left(\gamma_{SV_1} + \gamma_{SV_2}\right)}{\pi E' H^{5/2} R^{1/2}} \tag{56}$$

This results in

$$U_T^* = 4\sqrt{\frac{\Gamma}{\pi}}\left[1 - \frac{8}{3\pi}\eta - \sqrt{\left(1 - \frac{8}{3\pi}\eta\right)^2 + \frac{\Psi}{\pi}}\right] + \frac{2}{3\pi}\eta^2\sqrt{-2\left(1 - \frac{8}{3\pi}\eta\right) + \sqrt{\left(1 - \frac{8}{3\pi}\eta\right)^2 + \frac{\Psi}{\pi}}} \tag{57}$$

Setting $dU_T^*/d\eta = 0$ leads to the following necessary condition on η for a stable equilibrium to be achieved

$$\eta^2\left[\frac{2}{3\pi}\eta + \sqrt{\left(1 - \frac{8}{3\pi}\eta\right)^2 + \frac{\Psi}{\pi}}\right]^2\left[1 - \frac{8}{3\pi}\eta + \sqrt{\left(1 - \frac{8}{3\pi}\eta\right)^2 + \frac{\Psi}{\pi}}\right] = \frac{32\Gamma}{\pi} \tag{58}$$

It can readily be shown that for $\Psi=0$, Eq. (58) reduces to Eq. (48), which is applicable to two half-spaces. This result is expected because for a finite liquid volume and a finite, non-zero value of surface spacing, an infinite value of sphere radius (which corresponds to a flat-flat interface) causes Ψ to vanish as per Eq. (55).

The solution space for Eq. (58) is plotted in Figure 9 for several values of dimensionless volume Ψ. As observed, the smallest chosen value of Ψ yields a curve that is quite close to the half-space solution (Figure 7). It turns out that stable equilibrium configurations exist without solid-solid contact for a range of Γ values that depends on the value of Ψ. This relationship is summarized in Figure 10, which reveals regions with and without solid-solid contact within $\Psi-\Gamma$ space. The boundary curve is given by $\Gamma_c=0.03007\Psi^{1.3955}+0.01336$. If, for a given Ψ, $\Gamma<\Gamma_c$ then there is no solid-solid contact, whereas for $\Gamma\geq\Gamma_c$ the spheres must experience contact.

For equilibrium configurations that do not involve solid-solid contact, the pressure drop is given by Eq. (51), but with the gap at the free-surface given by

$$h_{eq}(b_{eq}) = H\left[\frac{2}{3\pi}\eta_{eq} + \sqrt{\left(1-\frac{8}{3\pi}\eta_{eq}\right)^2 + \frac{\Psi}{\pi}}\right] \tag{59}$$

and the wetted radius given by (via solution of Eq. 54)

$$b_{eq} = \left\{\frac{2V_o}{\pi H\left[1-\frac{8}{3\pi}\eta_{eq} + \sqrt{\left(1-\frac{8}{3\pi}\eta_{eq}\right)^2 + \frac{V_o}{\pi R H^2}}\right]}\right\}^{\frac{1}{2}} \tag{60}$$

.

Thus, the adhesive force then becomes

$$F_{ad} = \pi b_{eq}^2 \Delta p_{eq} = \frac{2\gamma(\cos\theta_1+\cos\theta_2)V_o}{H^2\left[1-\frac{8}{3\pi}\eta_{eq} + \sqrt{\left(1-\frac{8}{3\pi}\eta_{eq}\right)^2 + \frac{V_o}{\pi R H^2}}\right]\left[\frac{2}{3\pi}\eta_{eq} + \sqrt{\left(1-\frac{8}{3\pi}\eta_{eq}\right)^2 + \frac{\Psi}{\pi}}\right]} \tag{61}$$

The above force represents the external, separating force (over and above the weight of the sphere) required to maintain the spheres at the given configuration (i.e., with undeformed separation, H).

In cases where $\Gamma > \Gamma_c$, the solids come into contact over some contact radius a, with the contact region surrounded by an annulus of liquid. The presence of contact modifies the form of the free energy, which becomes [31]

$$U_T = \pi\left(R_o^2 - b^2\right)\left(\gamma_{SV_1} + \gamma_{SV_2}\right) - \pi\left(b^2 - a^2\right)\left(\gamma_{SL_1} + \gamma_{SL_2}\right) + \pi a^2 \gamma_{S_{12}} \tag{62}$$

where $\gamma_{S_{12}}$ is the surface tension associated with the solid-solid interface.

The dimensionless formulation involves two additional ratios [31]:

$$m \equiv \frac{b}{a} \tag{63}$$

$$\Phi \equiv \frac{\Delta\gamma}{\gamma\left(\cos\theta_1 + \cos\theta_2\right)} - 1 \tag{64}$$

where $\Delta\gamma$ is the well-known work of adhesion and is given by [8]

$$\Delta\gamma = \gamma_{SV_1} + \gamma_{SV_2} - \gamma_{S_{12}} \tag{65}$$

The equilibrium solution, for given values of Γ and Ψ, is now expressed in terms of both η and m. In the case of solid-solid contact, no analytical expressions exist for the elastic strain energy, owing to the unknown contact solid-solid pressure distribution. Therefore, the equilibrium configurations and associated adhesive forces must be acquired through a numerical process [31].

It can be shown [31] that the advent of solid-solid contact introduces hysteresis, just as in the case of the JKR contact model [32], which applies to dry contact. Thus, the set of configurations that the interface would pass through when breaking the contact, such as during a controlled separation process, would be different from those experienced upon its formation. For example, the value of H at which the solid-solid contact is lost during a separation process is different from the value of H that corresponds to the formation of solid-solid contact during an approach process. Put another way, there is a jump-on instability at a certain H upon approach, where the interface goes suddenly from no contact to contact, as well as a jump-off instability upon separation (at a larger H), where the interface proceeds suddenly from having a contact radius a to having no solid-solid contact. One convenient experimental measure of the strength of an adhesive contact is the pull-off force, which can take on different values depending upon how the pull-off process is conducted. When the separation H (which is defined by the minimum gap between the undeformed sphere contours) is specified and increased quasi-statically, the interface will reach a configuration that is unstable and then abruptly lose contact. The magnitude of external, separating force required to reach this point of instability during separation is defined as the pull-off force during a controlled separation process.

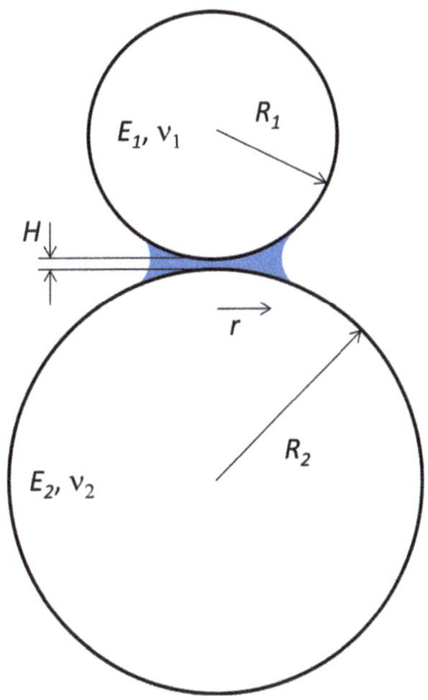

Figure 8. A liquid film bridging two smooth, elastic half-spaces with no solid-solid contact.

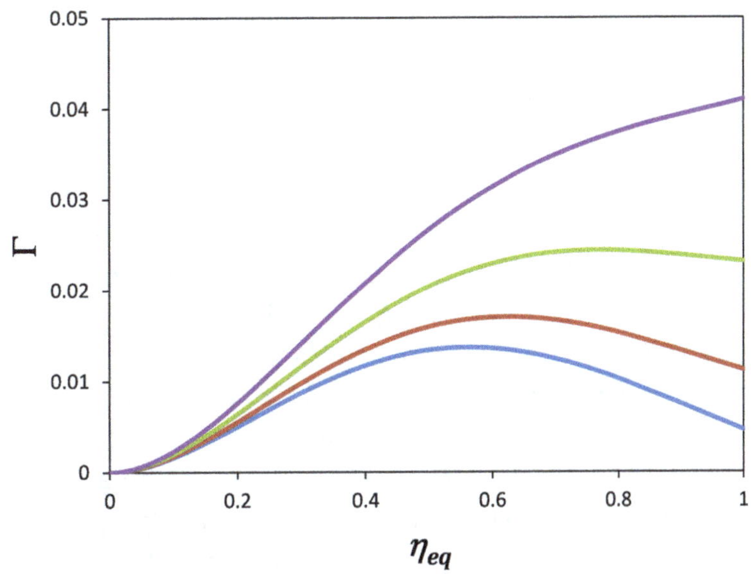

Figure 9. Equilibrium configurations for two spheres bridged by a liquid film without solid-solid contact.

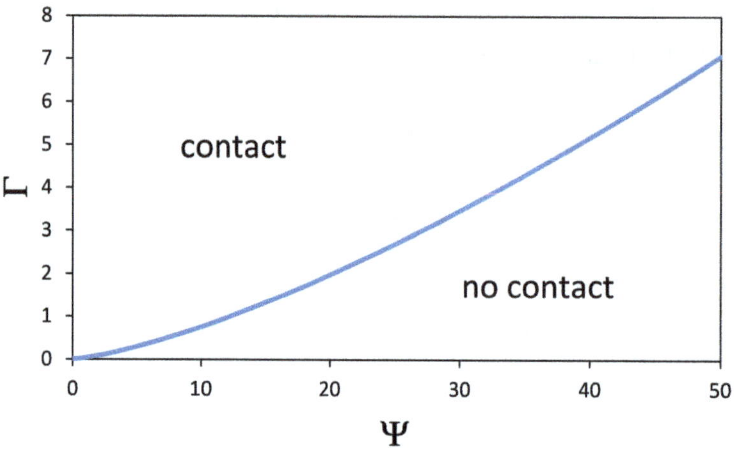

Figure 10. Regions in Γ-Ψ space showing regions with and without solid-solid contact.

2.6. Liquid film between contacting rough, elastic surfaces

Adhesive forces arising due to the presence of a liquid film between rough, elastic (or elastic-plastic) surfaces have been the subject of several recent works [17, 19–21, 33–37]. Figure 11 depicts a situation where two rough, elastic surfaces are in contact in the presence of an intervening liquid film. Taking into consideration a three-dimensional geometry, the assumption here is that the liquid film is continuous, so that there are no regions of liquid completely encased within a zone of solid-solid contact. Now in the case where the liquid wets the surfaces (i.e., the contact angles are less than 90°), the free surface of the liquid is concave and the film pressure is sub-ambient. Assuming that the lateral dimensions are much greater than the liquid film thickness, the pressure drop across the free surface is given by

$$\Delta p = p_a - p = \frac{\gamma(\cos\theta_1 + \cos\theta_2)}{h_{fs}} \tag{66}$$

where h_{fs} is the film thickness at the location of the free surface. For a continuous liquid film in static equilibrium, the pressure throughout the film must be the same, so Eq. (66) suggests that the periphery of the liquid film is at a constant height. The overall equilibrium shape of the film will depend upon the details of the gap distribution within the surfaces, which itself will be modulated by surface deformation due to the tensile stresses exerted by the liquid film. In general, the equilibrium configuration of the liquid film will not be axisymmetric. However, for surfaces that are nearly flat aside from a small-scale roughness, we can expect that the equilibrium film shape will be nearly axisymmetric. Assuming an axisymmetric liquid film, one can describe the establishment of equilibrium as follows: First, suppose a quantity of liquid is placed upon a surface that will serve as the lower surface of the contacting pair. Then, the other surface is placed in contact with the lower surface (and the liquid film) under some external load P. As solid-solid contact is first formed at the mutual asperity peaks, the liquid film will be quickly squeezed out to a radius that is determined by the given liquid volume and the average gap between the surfaces within the wetted region. However, as capillary forces take effect, the elastic surfaces will further deform, thereby reducing the mean gap between the surfaces and causing an increase in the wetted radius. An increase in the wetted radius in conjunction with a decreasing interfacial gap will cause a greater tensile force F_t and greater surface deformation. Hence, there is the possibility that the rate of increase of the capillary force (with expanding wetted radius) will exceed the rate of increase of the compressive force coming from the contacting asperities. In such a situation, the interface is expected to collapse, whereby the solid surfaces come into complete or nearly complete contact.

One numerical model of such an interface appears in [35]. Here it is assumed that the liquid film is axisymmetric and that deformation of the asperities is modeled according to the multi-scale contact model of [38]. Thus, the surface topography is characterized by its spectral content and algebraic formulas are applied to compute the effects of external and capillary forces on the average spacing within the interface. Another important assumption is that the mean spacing $\bar{h}$ within the wetted region is a good approximation to the spacing at the free surface h_{fs} so that pressure drop within the film is approximated by

$$\Delta p = \frac{\gamma(\cos\theta_1 + \cos\theta_2)}{\bar{h}} \tag{67}$$

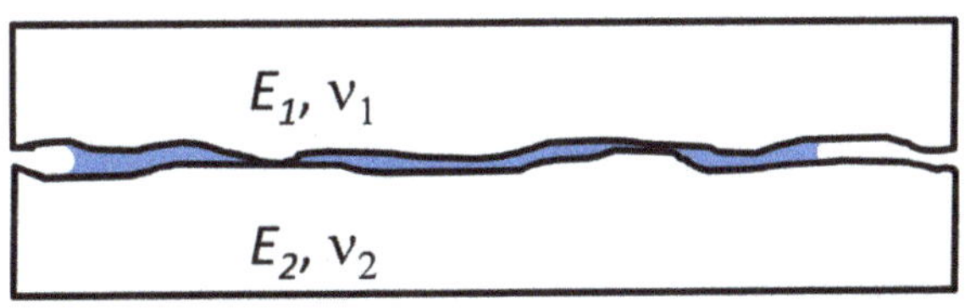

Figure 11. A liquid film bridging two rough, elastic surfaces.

Thus, the tensile force (F_t) becomes

$$F_t = \pi b^2 \frac{\gamma(\cos\theta_1 + \cos\theta_2)}{\bar{h}} \tag{68}$$

where b is the radius of the wetted region. If one defines the adhesive force as the tensile contribution to the net force exerted on either solid body, then F_t is the just the adhesive force (F_{ad}). An alternative definition for the adhesive force would be the value of the tensile external load required to maintain static equilibrium, or required to achieve a certain separation and separation rate. The latter definition views the adhesive force as the interfacial tensile force less the interfacial compressive force.

Sample results of the analysis are displayed in Figure 12, for the following input parameters: $V_o = 0.1$ mm^3, $\gamma = 72.7$ mN/m, $\sigma = 0.4$ μm, and $A_n = 4$ cm^2, where σ is the r.m.s. surface roughness of a 3D isotropic surface with a Gaussian height distribution, and A_n is the nominal contact area of the interface (i.e., the projected area of the interface in Figure 11). Figure 12 shows the influence of external load on several contact parameters, including tensile force and contact area (Fig. 12a) as well as average gap and wetted radius (Fig. 12b). The tensile force is seen to

increase of tensile force with load approaches infinity. The attainment of a near vertical slope in the curve suggests that the interface is unstable: no equilibrium configurations could be found for values of external load beyond the critical value. Analogous results are found for the average gap, tensile radius and solid-solid contact area. Such behavior suggests interface collapse, whereby beyond the critical point, the surfaces come into complete and near complete contact [33–36]. By introducing certain dimensionless parameters, the results can be generalized. Let an adhesion parameter Γ be defined according to

$$\Gamma \equiv \frac{\gamma\left(\cos\theta_1 + \cos\theta_2\right)V_o}{\pi\sqrt{2}A_n^{1/2}E'\sigma^3} \tag{69}$$

and let the dimensionless versions of external load, tensile force, and liquid volume be defined respectively as

$$P^* \equiv \frac{P}{\gamma V_o\left(\cos\theta_1 + \cos\theta_2\right)/\sigma^2} \tag{70}$$

$$F_t^* \equiv \frac{F_t}{\gamma V_o\left(\cos\theta_1 + \cos\theta_2\right)/\sigma^2} \tag{71}$$

$$V_o^* \equiv \frac{V_t}{A_n\sigma} \tag{72}$$

The results for dimensionless tensile force versus the adhesion parameter are depicted in Figure 13 at several values of dimensionless volume. This figure reveals that, for each dimensionless volume considered, there is a critical value of the adhesion parameter whereby the force curve becomes vertical, suggesting the onset of surface collapse.

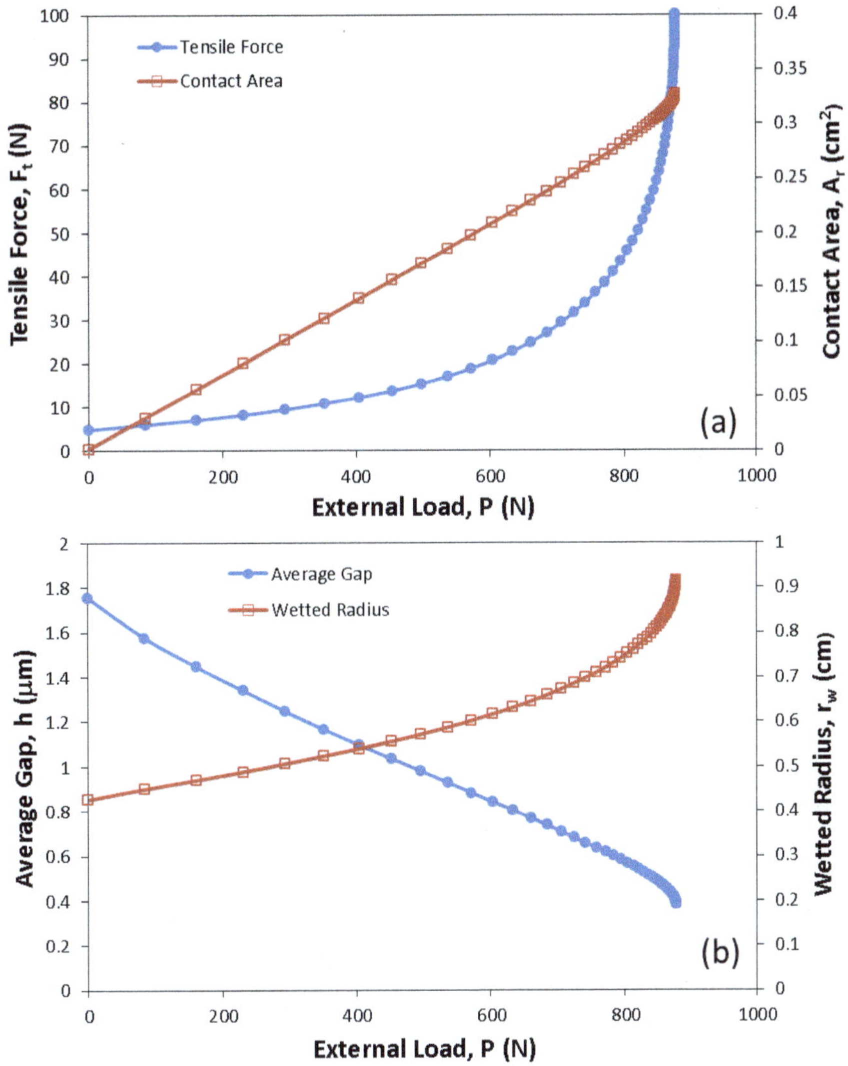

Figure 12. The effect of external load: (a) tensile force and contact area; (b) average gap and wetted radius.

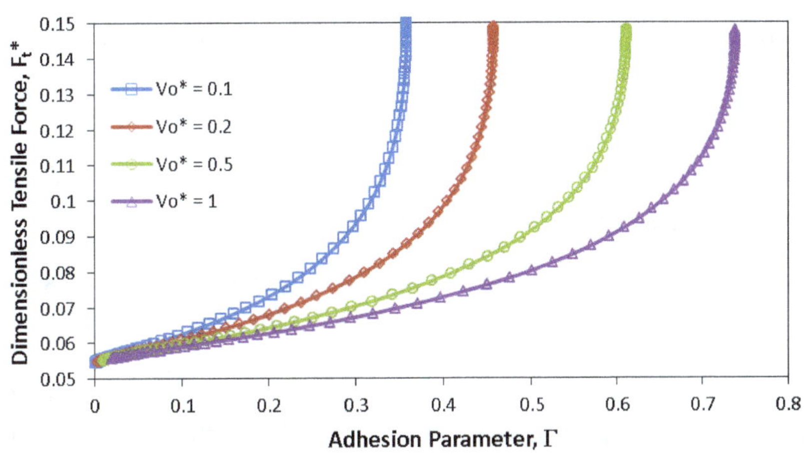

Figure 13. Dimensionless tensile force as a function of adhesion parameter for several dimensionless liquid volumes.

Acknowledgements

The author would like to thank the National Science Foundation (US) for support of this work and Amir Rostami, a graduate research assistant, for performing some calculations used herein.

Author details

Jeffrey L. Streator

Address all correspondence to: jeffrey.streator@me.gatech.edu

G.W. Woodruff School of Mechanical Engineering, Georgia Institute of Technology, Atlanta, GA, USA

References

[1] Bhushan B. Adhesion and stiction: Mechanisms, measurement techniques, and methods for reduction. Journal of Vacuum Science & Technology B. 2003;21(6):2262-2296.

[2] Van Spengen WM, Puers R, De Wolf I. A physical model to predict stiction in MEMS. Journal of Micromechanics and Microengineering. 2002;12(5):702.

[3] Van Spengen WM, Puers R, De Wolf I. On the physics of stiction and its impact on the reliability of microstructures. Journal of Adhesion Science and Technology. 2003;17(4):563-582.

[4] Komvopoulos K. Adhesion and friction forces in microelectromechanical systems: Mechanisms, measurement, surface modification techniques, and adhesion theory. Journal of Adhesion Science and Technology. 2003;17(4):477-517.

[5] Liu H, Bhushan B. Adhesion and friction studies of microelectromechanical systems/ nanoelectromechanical systems materials using a novel microtriboapparatus. Journal of Vacuum Science & Technology A. 2003;21(4):1528-1538.

[6] Lee S-C, Polycarpou AA. Adhesion forces for sub-10 nm flying-height magnetic storage head disk interfaces. Journal of Tribology. 2004;126(2):334-341.

[7] Yang SH, Nosonovsky M, Zhang H, Chung K-H. Nanoscale water capillary bridges under deeply negative pressure. Chemical Physics Letters. 2008;451(1):88-92.

[8] Adamson AW, Gast AP. Physical chemistry of surfaces: JWiley and Sons. New York; 1967.

[9] Cai S, Bhushan B. Meniscus and viscous force during separation of hydrophilic and hydrophobic smooth/rough surfaces with symmetric and asymmetric contact angles. Philosophical Transactions of the Royal Society A. 2008;366:1627-1547. DOI: 10.1098/rsta.2007.2176.

[10] Hamrock BJ. Fundamentals of fluid film lubrication: McGraw-Hill; 1994.

[11] Munson BR, Okiishi TH, Huebsch WW, Rothmayer AP. Fundamentals of fluid mechanics, 7th ed. New York: Wiley; 2015.

[12] Sedgewick SA, Travena DH. Limiting negative pressure of water under dynamic stressing. Journal of Physics D. 1976; 9(14):1983-1990.

[13] Sun DC, Brewe DE. A high speed photography study of cavitation in a dynamically load journal bearing. Journal of Tribology. 1991;113(2):287-292.

[14] Braun MJ, Hannon WM. Cavitation formation and modeling for fluid film bearings: A review. Journal of Engineering Tribology. 2010;224(9):839-863.

[15] Roemer DB, Johansen P, Pedersen HC, Anderson TO. Fluid stiction modeling for quickly separating plates considering the liquid tensile strength. Journal of Fluids Engineering. 2015;137:061205:1-8

[16] Israelachvilli JN. Intermolecular and surface forces, 2nd ed. Academic Press, London; 1992.

[17] de Boer MP. Capillary adhesion between elastically hard rough surfaces. Experimental Mechanics. 2007;47:171-183. DOI: 10.1007/s11340-006-0631-z.

[18] [18]de Boer, MP, de Boer, PCT. Thermodynaics of capillary adhesion between rough surfaces. Journal of Colloid and Interface Science. 2007;311:171-185. DOI: 10.1007/s11340-006-0631-z.

[19] DelRio FW, Dunn ML, de Boer MP. Capillary adhesion model for contacting micromachined surfaces. Scripta Materialia. 2008;59:916-920. DOI: 10.1016/j-scriptamat.2008.02.037.

[20] Persson BNJ. Capillary adhesion between elastic solids with randomly rough surfaces. Journal of Physics of Condensed Matter. 2008;20:315007. DOI: 10.10088/0953-8984/20/31/314007.

[21] Peng YF, Guo YB, Hong YQ. An adhesion model for elastic-contacting fractal surfaces in presence of meniscus. Journal of Tribology. 2009;131:024504.

[22] Butt H-J, Barnes WJP, del Campo A, Kappl M, Schonfeld F. Capillary forces between soft elastic spheres. Soft Matter. 2010;6:5930-5936.

[23] Men Y, Zhang Z, Wenchuan W. Capillary liquid bridges in atomic force microscopy: Formation, rupture and hysteresis. Journal of Chemical Physics. 2009;131:184702.

[24] Rabinovich YI, Esayanur MS, Moudgil BM. Capillary forces between two spheres with a fixed volume liquid bridge: Theory and experiment. Langmuir. 2005;21:10992-10997.

[25] Streator JL. Analytical instability model for the separation of a sphere from a flat in the presence of a liquid. Proceedings of the ASME/STLE International Joint Tribology Conference, IJTC 2004. 2004; Part B: 1197-1204.

[26] Chan DYC, Horn RG. The drainage of thin liquid films between solid surfaces. Journal of Chemical Physics.1985;83(10);4311-5324.

[27] Matthewson M. Adhesion of spheres by thin liquid films. Philosophical Magazine A. 1988;57(2):207-216.

[28] Zheng J, Streator J. A liquid bridge between two elastic half-spaces: A theoretical study of interface instability. Tribology Letters. 2004;16(1-2):1-9.

[29] Johnson KL, Johnson KL. Contact mechanics: Cambridge University Press; 1987.

[30] Zheng J, Streator JL. A micro-scale liquid bridge between two elastic spheres: Deformation and stability. Tribology Letters. 2003;15(4):453-464.

[31] Zheng J, Streator J. A generalized formulation for the contact between elastic spheres: Applicability to both wet and dry conditions. Journal of Tribology. 2007;129:274.

[32] Johnson KL, Kendall, K, Roberts, AD. Surface energy and the contact of elastic solids. Proceedings of the Royal Society (London), Series A. 1971;324(1558):301-313.

[33] Streator JL. A model of liquid-mediated adhesion with a 2D rough surface. Tribology Interntional. 2009;42:1439-1447.

[34] Streator JL, Jackson RL. A model for the liquid-mediated collapse of 2-D rough surfaces. Wear. 2009;267(9):1436-1445.

[35] Rostami A, Streator JL. Study of Liquid-mediated adhesion betweeen 3D rough surfaces: A spectral approach. Tribology International. 2015;58(2):1-13.

[36] Rostami A, Streator JL. A deterministic approach to studying liquid-mediated adhesion between rough surfaces. Tribology Letters. 2015;84:36-47. DOI: 10.1007/s11249-015-0497-2.

[37] Wang L, Regnier S. A more general capillary adhesion model including shape index: Single-asperity and multi-asperity cases. Tribology Transacations. 2015;58:106-112. DOI: 10.1080/10402004.2014.951751.

[38] Jackson RL, Streator JL. A multi-scale model for contact between rough surfaces. Wear. 2006;261(11):1337-1347.

8

Lipid Self-Spreading on Solid Substrates

Irep Gözen, Paul Dommersnes and Aldo Jesorka

Abstract

This chapter is dedicated to wetting and fracturing processes involving molecular phospholipid films and high-energy solid surfaces. In these systems, wetting of planar surfaces occurs in an aqueous environment by means of self-spreading of phospholipid membranes from artificially generated lipid sources, which range from manually deposited single sources (multilamellar liposomes) to liposome suspensions of different particle sizes, which are directly pipetted onto the substrate. The most prominent of the molecular lipid films is the phospholipid bilayer, which constitutes the fundamental structure of the biological cell membrane. Lipid membranes have peculiar characteristics, are highly dynamic, feature two-dimensional fluidity, and can accommodate functional molecules. Understanding the interactions of lipid films with solid interfaces is of high importance in areas like cell biology, biomedical engineering, and drug delivery.

Keywords: Phospholipid bilayer, phospholipid double bilayer, phospholipid monolayer, supported bilayer, lipid self-spreading

1. Introduction

This chapter highlights recent advances in wetting of solid surfaces by the self-spreading of phospholipid biomembranes upon deposition of lipid reservoirs [1, 2]. It should provide researchers with the necessary material to understand and evaluate spontaneous propagation of lipid monolayers [2, 3] and double bilayers [2, 4] on solid supports, which occurs when lipid reservoirs are brought in contact with low- and high-energy surfaces, respectively, in an aqueous environment. The first section provides a brief introduction of surfaces and interfaces, the second section is dedicated to the mechanism of interaction of lipid films with the supporting surface during wetting, and the third section introduces the formation of ruptures in double bilayers caused by that interaction.

Biological membranes organize cellular complexity, and thus establish and promote structure in the living world [5]. They compartmentalize the cell, form transport networks, organize proteins, serve as a smart barrier for molecules and ions, and establish the chemical identity of the cell. The fundamental structure of the cellular membrane is the phospholipid bilayer, consisting of a large number of individual phospholipid molecules, which organize themselves spontaneously in a self-assembly process. The membrane has peculiar characteristics, is highly dynamic, and features two-dimensional fluidity [6]. It can accommodate proteins and other functional molecules which fulfil important functions such as recognition, signal transduction, and transport of chemical entities through the membrane.

Biomembrane models, designed to capture some of the features of the cell membrane in a simplified setting, have become a popular research subject [7, 8]. They are naturally less complex than their biological counterpart, but can be relatively easily assembled, for example, on suitable flat solid surfaces (supported membranes), or as spherical membrane compartments, often referred to as vesicles or liposomes [9]. In the past few decades, a large number of model systems of increasing sophistication have been introduced, often with the purpose to identify and study the role and function of lipids and other membrane components in the cell [7, 10]. In particular, the two-dimensional fluidity of the membrane and their ability to harbour proteins have been in the centre of attention.

Although self-organization of lipid molecules to lipid membranes occurs spontaneously, which is frequently exploited to assemble membranes from lipid mixtures in solution in experimental settings, controlled assembly and preparation of stable, well-defined phospholipid films on supporting surfaces, such as glass, metals and metal oxides, are still challenging engineering tasks.

The deposition of lipid reservoirs onto various solid surfaces leads to formation of self-spreading surface-supported lipid films [1, 11]. Lipid reservoirs range from manually deposited single sources (multilamellar liposomes) to liposome suspensions of different particle sizes [10, 12], which can be directly transferred onto the substrate, either manually with glass microneedles [13] or by means of sophisticated microfluidic instrumentation [10, 14]. The following chapter is dedicated to the formation of lipid mono- and bilayer membranes by means of self-spreading from a lipid source. The result of lipid spreading is typically a solid-supported self-assembled membrane, i.e., a continuous supramolecular structure with two-dimensional fluidity. Each of the three forms of supported membranes mentioned above represents a real biological structure. Monolayers surround lipid droplets in cells, which are small stocks of lipid molecules regulating lipid metabolism of the cell [15]. Single bilayers are – in terms of composition – simplified versions of the plasma membrane of cells, which resides on top of the cytoskeleton. Spreading double bilayers are reminiscent of spreading cell membranes, which in addition to a lipid membrane contain actin filaments [16]. Accordingly, each of the supported membrane types can be utilized in experimental studies to address different questions. The supported bilayer is the most commonly used model system, as it directly resembles the membrane of the biological cell. Supported bilayer structures can be easily prepared in an aqueous environment on available surfaces that are compatible with microscopy experiments, such as glass, mica, or sapphire plates. Nowadays, with microfabrication and micromanipulation equipment being commonplace, a greater variety of surfaces is available, for example, silicon or aluminium oxide coated glass, which opened pathways for the generation of self-spreading double bilayers from lipid reservoirs. In particular, amorphous fluoropolymers and the epoxy photoresist SU-8, which can be utilized to coat and pattern a variety of surfaces with nanoscopic polymer films [17, 18], have enabled experiments where single reservoirs are manually deposited, allowing for the controlled generation of self-spreading lipid monolayers. Table 1 gives an overview over the most relevant publications [19 - 54] in the area of lipid film research from 1985 to present. It covers fabrication, biophysical characterization, and utilization in, for example, membrane protein studies.

	Article title	Authors	Source title	Year	Reference	Key points
1	*Wetting: Statics and dynamics*	De Gennes, P.G.	Rev. Modern Phys.	1985	[19]	One of the first articles in wetting of liquids on solid substrates, cited 3700 times
2	*Translational and rotational drag coefficients for a disk moving in a liquid membrane associated with a rigid substrate.*	Evans E. And Sackmann S.	Journal of Fluid Mechanics	1988	[20]	First theoretical framework of spreading/wetting membranes and relation to friction (drag) coefficient; supports the current model explained throughout the chapter.
3	*Phenomenology and kinetics of lipid bilayer spreading on hydrophilic surfaces.*	Rädler J. *et al.*	Langmuir	1995	[2]	Pioneer study, kinetics of lipid wetting via self-spreading lipid reservoirs: bilayer, double bilayer membranes
4	*Supported membranes: scientific and practical applications.*	Sackmann, E.	Science	1996	[8]	First examples of polymer cushioned protein incorporated bilayers, double bilayers, supported lipid bilayer-based biosensors
5	*Wetting films of lipids in the development of sensitive interfaces. An electrochemical approach.*	Kochev V. And Karabaliev M.	Advances in Colloid and Interface Science	2004	[21]	Review of basic principles underlying the techniques of formation, as well as the conditions of the films stability.
6	*Growth of giant membrane lobes mechanically driven by wetting fronts of phospholipid membranes at water-solid interfaces.*	Suzuki K. and Masuhara H.	Langmuir	2005	[22]	Multilayer self-spreading lipid bilayers

	Article title	Authors	Source title	Year	Reference	Key points
7	*Wetting fibers with liposomes.*	Borghi N. *et al.*	J. of Colloid and Interface Sci.	2005	[23]	Coating fibers with lipids through lipid wetting
8	*Following the formation of supported lipid bilayers on mica: A study combining Afm, Qcm-D, and ellipsometry.*	Richter R.P. *et al.*	Biophysical Journal	2005	[24]	Study of bilayer formation dynamics by a combination approach of different analytical surface techniques
9	*On the kinetics of adsorption and two-dimensional self-assembly of annexin A5 on supported lipid bilayers.*	Richter R.P. *et al.*	Biophysical Journal	2005	[25]	Supported membrane protein studies
10	*Formation of solid-supported lipid bilayers: An integrated view.*	Richter R.P. *et al.*	Langmuir	2006	[26]	Study of bilayer formation dynamics
11	Supported lipid bilayer self-spreading on a nanostructured silicon surface.	Furukawa K. *et al.*	Langmuir	2006	[27]	Self-spreading in combination with nanostructured surfaces
12	*Direct immobilization of cholesteryl-TEG-modified oligonucleotides onto hydrophobic SU-8 surfaces.*	Erkan Y. *et al.*	Langmuir	2007	[28]	Lipid monolayer self-spreading/wetting
13	*Controlled formation and mixing of two-dimensional fluids.*	Czolkos I. *et al.*	Nano Letters	2007	[3]	Lipid monolayer self-spreading/wetting
14	*Supported lipid bilayer/ carbon nanotube hybrids.*	Zhou X. *et al.*	Nature Comm.	2007	[29]	Lipid wetting over single-walled carbon nanotube transistors
15	*Membrane lipids: where they are and how they behave.*	Van Meer G. *it al.*	Nature Rev. Mol. Cell Biol.	2008	[30]	Review on plasma membrane lipids: structure, phase behavior, function
16	*Plasma membrane area Increases with spread area by exocytosis of a GPI-anchored protein compartment.*	Gauthier N.C. *et al.*	Mol. Biology of the Cell	2009	[31]	Cell membrane spreading similar to double lipid bilayer spreading
17	*Protrusive growth and periodic contractile motion in surface-adhered vesicles induced by Ca^{2+}-gradients.*	Lobovkina T. *et al.*	Soft Matter	2010	[11]	Double lipid bilayer self-spreading/localized Marangoni flow

	Article title	Authors	Source title	Year	Reference	Key points
18	*Fractal avalanche Ruptures in biological membranes*	Gözen I. et al.	Nature Materials	2010	[4]	Double lipid bilayer self-spreading, wetting through rupturing
19	*Quartz crystal microbalance with dissipation monitoring of supported lipid bilayers on various substrates*	Cho N.J. *et al.*	Nature Protocols	2010	[32]	QCM-D as a measure of wetting of solid surfaces by lipid bilayers
20	*Changes in wetting and energetic properties of glass caused by deposition of different lipid layers.*	Gołabek M., Hołysz L.	Applied Surface Science	2010	[33]	Investigation of wetting and energetic properties of different lipid layers deposited on the glass surface, carried out by contact angles measurements and determination of the apparent surface free energy .
21	*A Self-assembly route for double bilayer lipid membrane formation*	Han X. *et al.*	ChemPhysChem	2010	[34]	Bilayer on bilayer formation using fusogenic reagents
22	*Evolution of supported planar lipid bilayers on step-controlled sapphire surfaces*	Isono T., Ikeda T., and Ogino T.	Langmuir	2010	[35]	Supported bilayer wetting on varying topology
23	*Templating membrane assembly, structure, and dynamics using engineered interfaces*	Oliver A.E. and Parikh A.N.	Biochim. Biophys. Acta - Biomembranes	2010	[36]	Monolayer/bilayer/interfaces
24	*Calcium-ion-controlled nanoparticle-induced tubulation in supported flat phospholipid vesicles*	Gözen I. *et al.*	Soft Matter	2011	[37]	Lipid wetting is tuned with Ca^{2+} ions, leading to tubulation
25	*Control of dynamics and molecular distribution in a self-spreading lipid bilayer using surface-modified metal nanoarchitectures.*	Nabika H. *et al.*	Phys. Chem. Chem. Phys.	2011	[38]	Self-spreading bilayers on nanostructures
26	*Using patterned supported lipid membranes to investigate the role of receptor organization in intercellular signaling*	Nair P.M. *et al.*	Nature Protocols	2011	[39]	Protocol supported bilayer, SUV rupture

	Article title	Authors	Source title	Year	Reference	Key points
27	*Mechanism of lipid nanodrop spreading in a case of asymmetric wetting.*	Mohamad, S. *et al.*	Physical Review Letters	2012	[40]	Involves surface-enhanced ellipsometric contrast microscopy to observe the spreading of egg phosphatidylcholine nanodroplets on a hydrophilic substrate
28	*Instrumental methods to characterize molecular phospholipid films on solid supports*	Gözen I. and Jesorka A.	Analytical Chemistry	2012	[41]	Review on characterization methods and tools for surface-supported membranes
29	*Evidence for membrane flow through pores in stacked phospholipid membranes*	Gozen I. *et al.*	Soft Matter	2012	[42]	Wetting of proximal bilayers via rupturing of distal bilayers
30	*Mechanical feedback between membrane tension and dynamics*	Gauthier N.C. *et al.*	Trends in Cell Biology	2012	[43]	Cell spreading
31	*Glycans pattern the phase behaviour of lipid membrane.*	Subramaniam A.B. *et al.*	Nature Materials	2013	[44]	How self-spreading model membranes can be used to understand membrane-mediated transport processes. Findings indicate an intimate coupling between cellular lipidomes and glycomes.
32	*Multiplexed biomimetic lipid membranes on graphene by dip-pen nanolithography*	Hirtz M. *et al.*	Nature Comm.	2013	[45]	Lipid bi-/multilayer wetting using dip-pen lithography
33	*Repair of large area pores in supported double bilayers*	Gözen I. *et al.*	Soft Matter	2013	[46]	Different cause of bilayer-on-bilayer wetting
34	*Mechanics of spreading cells probed by atomic force microscopy*	Pietuch A., and Janshoff A.	Open Biology	2013	[47]	Cell spreading
35	*Lab on a Biomembrane: Rapid prototyping and manipulation of 2D fluidic lipid bilayer circuits*	Ainla A. *et al.*	Scientific Reports	2013	[10]	Printed/spreading bilayers deposited by a microfluidic pipette
36	*Spatial organization of cytokinesis*	Nguyen P.A. *et al.*	Science	2014	[48]	Supported bilayers used to study the effect of (artificial)

	Article title	Authors	Source title	Year	Reference	Key points
	signaling reconstituted in a cell-free system					centrosomes, an application example
37	*Effective Brownian ratchet separation by a combination of molecular filtering and a self-spreading lipid bilayer system.*	Motegi T. *et al.*	Langmuir	2014	[49]	Nanofabricated obstacle-based molecular filtering of self-spreading lipid bilayer membrane
38	*Arrayed lipid bilayer chambers allow single-molecule analysis of membrane transporter activity*	Watanabe R. *et al.*	Nature Comm.	2014	[50]	Lipid wetting from a lipid reservoir, this time not onion-like but dissolved lipids in solvent used for single-molecule studies
39	Solvent-assisted lipid bilayer formation on silicon dioxide and gold	Tabaei S. *et al.*	Langmuir	2014	[51]	Impact of solvent in membrane wetting type
40	*Nanopatterning of mobile lipid monolayers on electron-beam-sculpted teflon AF surfaces*	Shaali M. e*t al.*	ACS Nano	2015	[52]	Precise patterning of self-spreading monolayers using E-beam lithography
41	*Discovery of the migrasome, an organelle mediating release of cytoplasmic contents during cell migration*	Ma L. *et al.*	Cell Research	2015	[53]	Plasma membrane dewetting on solid support during cell migration
42	*A versatile nano display platform from bacterial spore coat proteins*	Wu, I.-L. *et al.*	Nature Comm.	2015	[[54]]	Solid support as micron-size silica beads (spherical) instead of flat surfaces

Table 1. Selected publications in the field of lipid self-spreading/wetting

2. Surfaces and interfaces

Supported lipid membranes form on solid surfaces, where many of the membrane properties are depending on the properties of the underlying surface, including material and composition, surface charge, roughness, and surface tension. The wetting phenomena observed in double bilayers only occur on solid high-energy surfaces, such as silicon oxide, or aluminium oxide films. In contrast, the wetting phenomena leading to lipid monolayers occur exclusively on low-energy hydrophobic polymer surfaces. The surface tension is a central concept, that is, in principle, a measure of how much energy is associated with a surface per unit area. If in a thought experiment a cube with the side length *a*, consisting of a solid material, is split into two parts of identical size, work of cohesion W must be exerted, equal to the difference in free energy G_2–G_1, which characterize the system before and after the split. If $A = a^2$ is the area of the separation, then $F = W/A$ is the surface free energy of the newly created surfaces. The surface tension σ of the material can be defined, which is a measure of how much work is required to create a surface of area A [55]:

$$\sigma = F + A\frac{\partial F}{\partial A} \tag{1}$$

High-energy surfaces ($\sigma > 500$ Nm/m) are composed of atoms which are attached to each other by covalent, ionic, or metallic bonds. Examples include metals, diamond, silica, glasses, and ceramics. Molecules of low-energy surfaces, such as various polymers, are attached to each other by the weaker van der Waals interactions or hydrogen bonds. It is evident that the intermolecular forces which determine the surface energy of a material also determine latent heat, melting, and boiling point. Materials with high boiling points ($T > 2000$°C) usually also have high-surface energies.

In water (and likewise in other monomolecular fluids), the molecules in the bulk are surrounded by other water molecules, where they are energetically indistinguishable. Only at solid and gas interfaces (e.g., the walls of a container, or the water vapour/water interface in an open container), there are molecules which experience different forces. The resulting energetic differences with respect to the bulk molecules are reflected by the interfacial tension of the solid–liquid interface σ_{SL} , and the surface tension of the liquid σ_{LV} at the liquid–vapour interface. Accordingly, the surface tension of a solid in equilibrium with a liquids vapour is σ_{SV}. A typical value of σ_{LV} for water is 73 mN/m [56, 57]. Temperature increase and added solutes lead to a noticeable decrease in surface tension. At a contact line between liquid, solid, and fluid at equilibrium (e.g., a fluid droplet in saturated air on a planar surface), the three interfacial tensions are force-balanced. The Young equation describes this force balance, and introduces a contact angle ϑ between the fluid/vapour and fluid/solid interface as a convenient measure of surface hydrophobicity/hydrophilicity:

$$\sigma_{SV} - \sigma_{SL} = \sigma_{LV}\cos\vartheta \tag{2}$$

Hydrophobic (or high-energy) surfaces feature small contact angles of $\vartheta < 90$ °, since the system is minimizing its free energy by spreading the droplet across the surface. The dynamics of wetting and spreading of fluids on solid surfaces have been extensively reviewed in other publications [58] and will not be further discussed here. However, the Young equation allows only direct observation of σ_{LV} and ϑ. Direct comparison of solid surfaces by means of their surface tensions is not possible, since these values cannot be obtained in a straightforward manner. By introducing the concept of a critical surface tension σ_c, at which complete wetting occurs (cos ϑ = 1), solids can be well compared with respect to their wetting behaviour. Materials with strong cohesion, such as oxides and metals, can be categorized under high critical surface tension materials (σ_c >> 100 mN/m), whereas polymers and other organic films typically feature low critical surface tensions (σ_c < 50 mN/m). The nature of lipid films formed on such surfaces depends very much on the surface energy.

In addition, the roughness of a surface is of considerable importance to its wetting properties, which was discovered by Wenzel in 1936 [59]. In nature, topographical surfaces with regular or irregular features on the micro- or nanoscale have a water-repelling function. The most prominent examples are the leaves of the Sacred Lotus, which has a nanopatterned surface to protect it from water, and the skin of sharks, which give the animal the ability to move faster in water. These superhydrophobic surface features, which are characterized by very large contact angels of $\vartheta > 150°$, cannot be described by the Young equation, but by the Cassie-Baxter surface model [59]. According to that model, which is applicable to extreme surface texture, water droplets cannot wet the surface, since they are forced to remain on top of the structures. The prominent topographical features prevent wettings, because the surface would have to bend down into the gaps, which would result in high local curvatures and high Laplace pressures, but also in an increased surface-to-volume ratio. Both surface energy and roughness have an influence on the wetting properties of lipid material when it comes in contact with the surface. For example, fluorocarbon films, generated by plasma decomposition and deposition of perfluorinated hydrocarbons, are more hydrophobic than epoxy polymer films, but their greater roughness makes them comparatively less suitable for large area spreading of lipid monolayers [60].

3. Lipid self-spreading

A lipid reservoir brought into contact with a solid substrate leads to wetting of the surface by lipids in the form of a molecularly thin phospholipid film. Lipid reservoirs can be considered as stocks of phospholipid molecules and can exist in various forms. Rädler et al. have used "solid lipid sources" as reservoirs, which they have described as irregularly wrapped (entangled) layers of lipid bilayers [2]. The reservoirs have been referred to as "lipid lumps" in a few other studies and illustrated as liquid drops in which the individual lipid molecules are randomly oriented [22]. Multilamellar vesicles (MLVs) containing several hundreds of lipid bilayers packed in a compact sphere can also be employed as lipid reservoirs [4, 11]. Because of their dense, layered structure, such vesicles are also referred to as "onion shell vesicles" [61]. The internal molecular structure of such lipid reservoirs can be complex and the mechanisms of initiation of wetting from such reservoirs have not yet been fully understood. We will further comment on the potential impact of the lipid reservoir structure on wetting dynamics in the later paragraphs of this chapter.

Due to wetting, individual lipids originating from a lipid source self-assemble on a solid surface as a planar lipid film, extending the surface area over a distance of several tens to hundreds of microns. The wetting motion of lipids on solid substrates is therefore commonly referred to as "lipid spreading". Since the spreading is generally not initiated due to an external stimulus, but begins rather spontaneously, ensuring energetically the most favourable conditions, the spreading is further defined as "lipid self-spreading".

In context of this chapter, lipid wetting of a solid substrate occurs in biologically relevant conditions, e.g., in water-based physiological buffers. Let us consider the spreading of a lipid monolayer. When a lipid source is deposited on a hydrophobic substrate in the presence of an aqueous solution, surface tension at two interfaces may play a role in wetting. One is the tension at interface of lipids with the solid substrate (σ_L), the second is the tension at the interface of aqueous buffer (σ_A) with the solid substrate. In other words, the surface can be wetted by either phospholipids emerging from the lipid source or by the water molecules in the buffer. Lipid spreading will initiate spontaneously if the surface free energy of the system would be reduced during spreading. This is possible if σ_L is lower than σ_A. Accordingly, lipids would migrate from the lipid source to wet the substrate.

Structure of the wetting lipid film depends on the nature of the solid surface, e.g., if it is of high or low energy, a hydrophilic or hydrophobic substrate. Single bilayers tend to form on moderately hydrophilic substrates such as glass. Double bilayer wetting can be observed on silicon oxides and multiple metal oxide surfaces which are generally considered to be high-energy substrates. On hydrophobic surfaces like fluoropolymers or the epoxy SU-8, lipids spread as monolayers, i.e., as a single leaflet of a lipid bilayer. This can be expected since the hydrocarbon chains of lipid molecules would tend to face towards the hydrophobic surface.

The mechanism of interaction with the surface during wetting varies, depending on the nature of the substrate. The monolayer lipids form due to the hydrophobic interactions with the surface. Spreading monolayers screen along the buffer–surface interface, establishing a direct contact with the hydrophobic substrate (Figure 1a). During single bilayer spreading on glass or mica, there remains a thin (few nm) lubricating layer beneath the membrane (Figure 1b). The bilayers spread by "sliding" on top of the water layer. The interaction between the bilayer and the surface is governed by hydration and the van der Waals forces. Double bilayer spreading on highly oxidized silica or other metals is mostly under the influence of electrostatic interactions. Physiological buffers have high ionic strength and contain multivalent ions, for example, Ca^{2+} or Mg^{2+}. These positively charged entities in the ambient solution act as "fusogenic agents" and establish bridging connections between the negatively charged phospholipid headgroups and the negatively charged surface (Figure 1c). This facilitates the spreading of lipid films on oxidized surfaces. Concentrations of only a few mM/L [62] of divalent cations and 10^{-5} M of trivalent cations have a dramatic effect on surface potentials. For instance, they can neutralize a negatively charged surface and even start to accumulate positive charges [62]. A few studies have reported on multi-bilayer spreading featuring up to seven bilayers on Si/SiO_2 surfaces [22, 63], but very little is known about the spreading dynamics.

Self-spreading of multilamellar vesicles on solid substrates leads to circular lipid patches (Figure 1d). The main reason for the circular geometry lies in the spreading dynamics which is characterized by Darcy flow. Briefly, the spreading front can be described by the same equation that is used to describe a Saffman-Taylor instability (a specific form of Darcy flow), but with the opposite sign. Therefore, any perturbation from a circular shape during spreading will be rapidly damped. We will explain thoroughly why membrane flow is a form of Darcy flow in the next section.

The overall tension (σ_S) during lipid wetting is expected to be:

$$\sigma_S = \sigma_A - \sigma_L \tag{3}$$

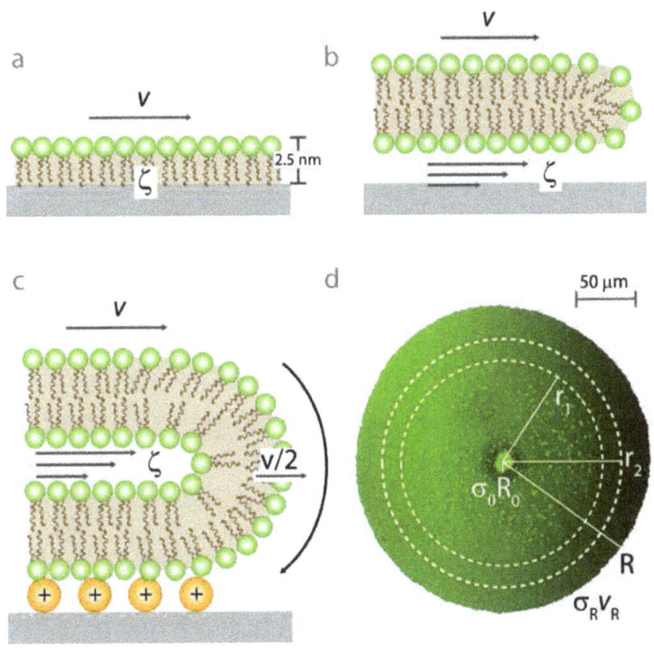

Figure 1. Lipid self-spreading. Illustrations showing in a side view the cross-sectional edge profiles of a spreading (a) monolayer, (b) single bilayer, and (c) double bilayer. The spreading monolayer is directly in contact with the substrate. The single bilayer exhibits sliding motion on a thin lubricating water layer. During double bilayer spreading, positively charged ions bridge the proximal (lower) bilayer to the surface, where the distal (upper) bilayer slides on a layer of water trapped in between the two layers. ζ indicates the region where the friction is effective during spreading. v is the velocity of the spreading membrane. (d) Laser scanning confocal micrograph of a self-spreading double bilayer (top view). The bright spot at the centre of the circular patch is the multilamellar lipid reservoir, from which lipid material is drawn.

Lipid self-spreading is therefore considered to be a form of *Marangoni flow*, which is the mass transfer between two fluids along an interface in a surface tension gradient. During spreading, there is mass transfer of lipids towards the buffer solution along the solid substrate.

The lipid reservoir has also an internal tension (σ_0) which will resist the spreading. The spreading power (S) can thus be formulated as the difference between overall interfacial tension and the tension of the reservoir:

$$S = \sigma_S - \sigma_0 = \sigma_A - \sigma_L - \sigma_0 \tag{4}$$

Note that the terms in Eq. (3) and (4) may not fully represent the single and the double bilayers. As described above, during monolayer spreading the lipids replace the water molecules on the surface; therefore, terms σ_L and σ_A are appropriately describing the driving forces in lipid monolayer spreading. For the other two types of spreading, more complicated surface interactions may play a role. For instance, because of the existing lubricating water layer between the surface and the membrane during single and double bilayer spreading, the water molecules are not essentially replaced by lipids. Therefore, σ_S can simply be defined as the adhesion energy between the membrane and the surface, instead of the surface tension differences of σ_A and σ_L .

Different wetting modes of phospholipids can be distinguished by a kinetic spreading coefficient, β ; which is directly related to velocity (v) of the spreading membrane:

$$v(t) = \sqrt{\frac{\beta}{t}} \tag{5}$$

The spreading power (S), which quantifies the driving force for the spreading process, is composed of the spreading and the friction coefficients (ζ):

$$\beta = \frac{S}{2\zeta} \tag{6}$$

Next, we will formulate the spreading coefficients for different modes of lipid self-spreading on solid supports. These modes include monolayers, single bilayers, and double bilayers. We will first adapt a one-dimensional model which had been originally proposed by Rädler et al. for single lipid bilayer spreading [2].

We make two main assumptions to establish our model, which would be valid for all three types of lipid spreading:

i. membrane flow can be described by a two-dimensional Stokes equation:

$$\zeta\bar{v} = \nabla\sigma - \mu\nabla^2\bar{v} \tag{7}$$

where μ is the membrane viscosity (10^{-3} dyn s cm^{-1}) [64] and v is the membrane velocity. The first term describes the frictional forces and the second term the viscous shear forces. The viscous forces are only influential in membrane dynamics if the length of the membrane is below 30 nm. We will discuss in the next section in detail how the critical length (L_c), at which the impact of viscous forces is comparable to the frictional forces, has been determined. Since the lengths of the supported membranes are way above 30 nm, we now eliminate the second term in Eq. (7) and take into account only the first term:

$$\zeta\bar{v} = \nabla\sigma \Rightarrow \zeta v = \frac{d\sigma}{dx} \tag{8}$$

This means that the tension gradient ($\nabla \sigma$) which drives the spreading process is balanced by the frictional stress. For a monolayer or a single bilayer, ζ applies to the region between the surface and the lipid film. For a double bilayer membrane, ζ is effective between the proximal (lower) and the distal (upper) bilayers. This is due to the tank-thread-like motion of the double bilayer. In this type of motion, the proximal layer is immobilized as soon as it is fused to the surface via the multivalent ions and the distal bilayer slides over the proximal bilayer and lays onto the substrate. This is why the self-spreading double bilayer has been called by Rädler et al. "a rolling bilayer" [2].

ii. To a first degree of approximation, lipid membranes are incompressible [20].

$$\nabla \vec{v} = 0 \Rightarrow \frac{dv}{dx} = 0 \tag{9}$$

Combining Eq. (8) and (9) leads to:

$$\frac{d^2\sigma}{dx^2} = 0 \Rightarrow \sigma = ax + b \tag{10}$$

$\sigma(R)=\sigma_s$, where σ_s is the tension due to the adhesion to the surface. $\sigma(0)=\sigma_0$ is the tension of the lipid reservoir.

$$\sigma = \frac{\sigma_s - \sigma_0}{R} x + \sigma_0 \tag{11}$$

Combining Eq. (10) and (11) leads to:

$$\zeta v = \frac{\sigma_s - \sigma_0}{R} \text{ or } \zeta v R = \sigma_s - \sigma_0 \tag{12}$$

This means that the membrane flow caused by the tension difference over a radius R is dissipated by the friction. The velocity at the spreading edge of a double bilayer (v_R) will be half the velocity of the membrane (v) since the spreading edge exhibits rolling motion, where only half of the membrane material is laid upon the substrate. The membrane velocity of a monolayer or a single bilayer will be equal to the velocity at the spreading edge since such rolling motion is not performed. Because of this distinction among the spreading dynamics of different types of membranes, we will explain the model from this point on under two groups: (1) monolayer and single bilayers, (2) double bilayers. The equations for the calculation of β for group 1 will be denoted as "a" in the left column, and for group 2 as "b" for the right column.

The membrane velocity with respect to the velocity at the spreading edge at a distance R from the center is:

$$\begin{array}{ll}\text{One-dimensional monolayer/single bilayer} & \\ v = \dfrac{dR}{dt} & \text{(a)} \\ \text{One-dimensional double bilayer} & \\ v = 2\dfrac{dR}{dt} & \text{(b)}\end{array} \tag{13}$$

Eq. (12a) and (12b) can be inserted into Eq. (12), which leads to:

$$\begin{array}{ll}\text{One-dimensional monolayer/single bilayer} & \\ \zeta \dfrac{dR}{dt} R = \sigma_s - \sigma_0 & \text{(a)} \\ \text{One-dimensional double bilayer} & \\ \zeta 2\dfrac{dR}{dt} R = \sigma_s - \sigma_0 & \text{(b)}\end{array} \tag{14}$$

During spreading, the radius of the circular lipid patch would grow from R_0 (the reservoir radius) at time t_0 to $R(t)$ at time t. Integrating the left side of Eq. (13a) and (13b), from $R(0)$ to $R(t)$ and the right side of the equations from time t=0 to t gives:

$$\begin{array}{ll}\text{One-dimensional monolayer/single bilayer} & \\ \zeta \dfrac{1}{2} R^2(t) = (\sigma_s - \sigma_0)t & \text{(a)} \\ \text{One-dimensional double bilayer} & \\ \zeta R^2(t) = (\sigma_s - \sigma_0)t & \text{(b)}\end{array} \tag{15}$$

The radius of the spread at $t = 0$ is R (0) = 0. Rearranging Eq. (14a) and (14b) gives:

$$\begin{array}{ll}\text{One-dimensional monolayer/single bilayer} & \\ R(t) = \sqrt{\dfrac{2(\sigma_s - \sigma_0)t}{\zeta}} & \text{(a)} \\ \text{One-dimensional double bilayer} & \\ R(t) = \sqrt{\dfrac{(\sigma_s - \sigma_0)t}{\zeta}} & \text{(b)}\end{array} \tag{16}$$

The derivative of $R(t)$ in Eq. (15a) and (15b) with respect to t is the velocity of the membrane v :

$$\begin{aligned}&\text{One-dimensional monolayer/single bilayer}\\ &v=\frac{dR}{dt}=\sqrt{\frac{(\sigma_s-\sigma_0)}{2\zeta t}} \qquad \text{(a)}\\ &\text{One-dimensional double bilayer}\\ &v=\frac{dR}{dt}=\sqrt{\frac{(\sigma_s-\sigma_0)}{4\zeta t}} \qquad \text{(b)}\end{aligned} \tag{17}$$

As mentioned above, the spreading power $S=\sigma_s-\sigma_0$, is the product of the spreading coefficient and the friction coefficient:

$$\beta=\frac{S}{2\zeta} \quad (6)$$

Inserting Eq. (6) into Eq. (16a) and (16b) will provide the spreading coefficient β for monolayer–single bilayer and double bilayer spreading, respectively.

$$\begin{aligned}&\text{One-dimensional monolayer/single bilayer}\\ &v=\sqrt{\frac{\beta}{t}} \qquad \text{(a)}\\ &\text{One-dimensional double bilayer}\\ &\frac{v}{2}=v_R=\sqrt{\frac{\beta}{t}} \qquad \text{(b)}\end{aligned} \tag{18}$$

After determining the spreading coefficient with regards to membrane velocity by adapting a one-dimensional model, we will now determine the relationships for a two-dimensional model. The two-dimensional model would provide more insights for lipid self-spreading in correlation with experiments, since the two-dimensional wetting of a lipid membrane can be monitored experimentally, for example, via confocal microscopy (Figure 1d). In such experiments, a self-spreading membrane doped with a membrane-attached fluorophore can be observed from top view as a circular patch with quite a distinct circumference (Figure 1d). On this circular patch, we imagine an arbitrary ring with an inner radius r_1 and the outer radius r_2 (Figure 1). During the spreading of an incompressible membrane, the number of lipid molecules within this ring should be constant. The law of mass conservation would require the flux – rate of lipid flow per cross-sectional area – at distance r_1 (J_1) and r_2 (J_2) to be equal.

$$J_1=J_2$$

The lipid molecules coming from the reservoir will pass through r_1 with velocity v_1, and subsequently r_2 with velocity v_2 :

$$2\pi r_1 \rho v_{r_1} = 2\pi r_2 \rho v_{r_2}$$

where ρ is the particle density. Note that we had earlier assumed the membrane to be incompressible, thus ρ is constant. Therefore, rv_r is constant. This means that the velocity v_r of the lipid molecules at any arbitrary circle with radius r can be related to the velocity v_R of the lipids at the spreading edge as following:

$$rv_r = Rv_R \tag{19}$$

We will initiate the formulation of our two-dimensional model by rearranging Eq. (19): $v_r = \frac{R}{r} v_R$. As explained above, during the determination of the one-dimensional model, the dynamics of spreading monolayers-single bilayers and the double bilayers will be formulated separately due to the variations in the spreading edge velocity. We will use notation "a" for the equations defining the monolayers/single bilayers and "b" for the equations defining the double bilayers.

$$\begin{aligned} &\text{Two-dimensional monolayer and single bilayer} \\ &v_r = \frac{R}{r}\frac{dR}{dt} \qquad \text{(a)} \\ &\text{Two-dimensional double bilayer} \\ &v_r = 2\frac{R}{r}\frac{dR}{dt} \qquad \text{(b)} \end{aligned} \tag{20}$$

The velocity of the spreading edge of the double bilayer membrane is again half the velocity of the membrane (v_r). The radial tension gradient drives the spreading process and it is balanced by the frictional stress (Eq. 7):

$$\zeta \bar{v} = \nabla \sigma \Rightarrow \zeta v_r = \frac{d\sigma}{dr}$$

Inserting Eq. (19a) and (19b) into Eq. (8) gives:

$$\begin{aligned} &\text{Two-dimensional monolayer and single bilayer} \\ &\zeta v_r = \frac{R}{r}\frac{dR}{dt} = \frac{d\sigma}{dr} \qquad \text{(a)} \\ &\text{Two-dimensional double bilayer} \\ &\zeta v_r = \frac{R}{r}\frac{dR}{dt} = 2\frac{d\sigma}{dr} \qquad \text{(b)} \end{aligned} \tag{21}$$

Integrating Equations (20a) and (20b) from R_0 (the reservoir radius at time t_0) to R (the radius of the growing circular lipid film at time t) gives:

Two-dimensional monolayer and single bilayer

$$\zeta R\frac{dR}{dt}ln(\frac{R}{R_0})=\sigma_s-\sigma_0 \qquad (a)$$

Two-dimensional double bilayer

$$2\zeta R\frac{dR}{dt}ln(\frac{R}{R_0})=\sigma_s-\sigma_0 \qquad (b) \qquad (22)$$

The spreading power ($S=\sigma_s-\sigma_0$) is the product of spreading coefficient and friction coefficient. Eq. (21a) and (21b) are therefore rearranged as following to obtain the two-dimensional spreading coefficients:

Two-dimensional monolayer and single bilayer

$$R\frac{dR}{dt}ln(\frac{R}{R_0})=\frac{\sigma_s-\sigma_0}{\zeta}=2\beta \qquad (a)$$

Two-dimensional double bilayer

$$R\frac{dR}{dt}ln(\frac{R}{R_0})=\frac{\sigma_s-\sigma_0}{2\zeta}=2\beta \qquad (b) \qquad (23)$$

By monitoring the self-spreading of a fluorescently labelled monolayer on a SU-8 surface over time under a microscope, β has been calculated to be $1-3\ \mu m^2/s$. [3] Note that SU-8 is a hydrophobic epoxy polymer that is commonly used as substrate, due to its favourable processing properties [3, 60]. In content of the same study, Eq. (22 a/b) have been solved numerically and shown to yield a good fit with the experimental data. The tension of the reservoir is considered to be on the order of the lysis tension of a single bilayer membrane ($6\ mN/m$). "Lysis" can be thought of as the process of breaking down a single bilayer. Since a multilamellar reservoir consists of several bilayers, monolayer spreading can only be initiated by the lysis of a single bilayer within the reservoir. The spreading power S is the difference in the adhesion energy at the membrane-SU-8 interface (σ_S) and the reservoir (σ_0): $\sigma_S-\sigma_0$. For simplicity, S has been assumed to be in the order of the reservoir tension. Taking $\beta\approx 3\ \mu m^2/s$ and S as $6\ mN/m$, ζ for monolayers spreading on a SU-8 has been estimated to be $\sim 10^9\ Pas/m$. This value is comparable to the friction coefficients estimated between the monolayer leaflets within a bilayer ($\sim 10^8-10^9\ Pas/m$) [3, 65].

Rädler et al. have reported the experimental values of β for a sliding bilayer to be up to $43\ \mu m^2/s$. [2] The high β values indicate a much faster spreading motion compared to the monolayers. This is predictable since the presence of the lubricating water layer (25 $\mathring{A}$) enhances the sliding by reducing friction.

The friction coefficient, which characterizes the lower monolayer leaflet of the distal bilayer and the upper monolayer leaflet of the proximal bilayer during bilayer rolling (double bilayer spreading), has been experimentally determined to be $5x10^7\ Pa.s/m$. [4] However, a spreading coefficient for double bilayers has not yet been revealed. This may be due to the relatively complicated spreading dynamics of double bilayer membranes. The few assumptions we have made above to develop a theoretical model for the calculation of spreading coefficients may not necessarily be fully satisfied for double bilayers. For example, we assumed that the tension of the reservoir was constant at all times during spreading. However, there is evidence confirming that the inter-bilayer defects naturally exist or spontaneously form in such reservoirs [42, 66]. In this case, the reservoir tension would be expected to vary, as it cannot be maintained. Similarly, Rädler et al. have reported ceasing of double bilayer self-spreading due to the exhaustion of the lipid source [2]. Another variable parameter during double bilayer spreading would be the thickness of the water layer, on which the distal bilayer slides. During the spreading of single bilayers, the inter-bilayer distance is regulated by the hydration forces and van der Waals interactions, as mentioned above. For a distal bilayer sliding on a proximal bilayer, no such forces apply. Additionally, surface roughness, irregularities or defects may cause the thickness of the water layer to alter, which would in turn influence the magnitude of friction.

4. Lipid membrane rupturing

In the previous sub-chapter, we described the details of the spreading motion of lipids on solid supports in various forms, without reporting on the eventual outcome of such spreading. Spreading continues until the reservoir is depleted, or exhausted due to defects. This will cause termination of lipid supply from the source to the spreading membrane, so that the spreading motion will slow down and eventually almost stop. In the meantime, the adhesion energy remains constant. This means that regardless of the insufficient lipid supply, the spreading edge of the membrane will favour lipid wetting and tend towards adhering. Note that a lipid membrane cannot stretch more than 5% of its surface area [67]. The fate of the spreading from this point on will differ depending on the type of the surface and corresponding spreading mode.

When the reservoir is completely consumed and the spreading velocity reaches zero, the circumference of a spreading monolayer starts to "evaporate" [68]. Evaporation occurs when the hydrophobic tails of individual lipids lay open and completely adhere to the substrate. If the membrane is tagged with fluorophores, the evaporating rim of the membrane can be observed as a fuzzy edge rather than a distinct one. The driving force for the evaporation of monolayers is the increase in entropy which in turn minimizes the Gibbs free energy.

Similarly, single bilayer spreading simply comes to an end when the reservoir is depleted. It has recently been shown that an additional stock of lipids can be provided to a supported bilayer lacking a reservoir, by using a microfluidic pipette device for continuous supply [10]. Briefly, at the tip of the microfluidic pipette, a virtual flow cell provides a steady hydrodynamically confined flow, featuring a low Reynolds (Re) and a high Péclet (Pe) number. This flow cell can therefore deliver liquid cargo locally to a surface under highly controlled laminar flow conditions. By using the microfluidic pipette, it is possible to supply and fuse small unilamellar lipid reservoirs (vesicles ~100 *nm*) into an existing lipid bilayer patch. This can lead to interesting wetting behaviour. If the new supply of lipids are zwitterionic with a net charge of zero (e.g., POPC), no further effect in wetting is observed and the size of the bilayer lipid patch remains constant over time. If the lipid stock provided to the single bilayer patch consists of cationic transfection lipids (e.g., DOTAP), the single bilayer starts to further wet the substrate. This is possible due to the intercalation of DOTAP lipids into the membrane, which are exposed via the microfluidic pipette. By means of the microfluidic device, spreading of membranes can be precisely controlled, and patches of desired size and composition can be easily generated. This provides a means of prototyping supported membranes in a rapid and reproducible fashion.

For a double bilayer membrane, the free energy can further be minimized, if the distal bilayer ruptures and adheres on the substrate. Rupturing of the distal membrane is possible if the tension of the membrane increases due to the continuous adhesion of the membrane edge to the substrate, and exceeds the lysis tension. In 2010, two forms of ruptures occurring in the distal bilayer of spreading double bilayer membranes have been reported. The "floral" ruptures, named after pore morphologies resembling flowers, can be observed mostly at the centre of a circular lipid spread propagating towards its periphery (Figure 2a). Such pores continuously grow until the double bilayer membrane entirely transforms into a single bilayer membrane on the solid support. The second rupture type appears in "fractal" patterns, most frequently at the circumference of the lipid patch, and develops inwards in the form of avalanches (Figure 2b). Within the fractal ruptures there remain "islands", the entrapped regions of the distal membrane which are strongly pinned to the proximal bilayer. Except for the islands, the lipid material of the upper bilayer migrates towards the edges (cf. Figure 1c) and is deposited on the surface in the same manner as during floral rupturing. Since the fractal ruptures propagate in form of avalanches, the wetted area on the surface increases step-wise over time. Both types of rupture formation are spontaneous, and occur when the lipid reservoir is exhausted. Discrimination among the two rupture types has been attributed to the amount of pinning between the two bilayers. The pinning can be established by means of Ca^{2+} or other multivalent cations present in the ambient buffer. A high number of pinning sites is assumed to favour fractal morphology. Precise control of the number and location of pinning sites during spreading/rupture experiments has not yet been achieved.

We want to dedicate this sub-chapter to membrane ruptures, as the displacement of the membrane on the surface during rupturing is a form of wetting of the solid substrate, and also a simultaneous form of de-wetting of the proximal bilayer membrane (Figure 2c). Next, we will describe a mathematical analogy between the dynamics of floral ruptures and the dynamics of flow in conventional "porous media". A porous medium can be depicted as a fluidic compartment packed regularly with particles, for example, micron-size beads. Such fluidic media can consist of, for example, glycerol, oil, or water. If now a secondary fluid of lower viscosity is pushed through the porous medium to displace the existing fluid, a morphologically instable interface between the two immiscible fluids is formed. Common injection fluids in flow experiments in porous media are water and air. Water can be injected only into liquids with higher viscosity, e.g., glycerol, but is often used as the main medium itself, if the injected fluid is air. The instability at the boundary of two immiscible fluids formed during the flow in a porous medium exhibits complex, finger-like patterns, therefore, is referred to as "viscous fingering" [69]. Membrane ruptures highly resemble these fingering instabilities. A membrane flow causing edge instabilities is comprehensible since the lipid membranes are considered as two-dimensional fluids. One interesting aspect regarding the similarities between the viscous fingering and the membrane pore edge instabilities is the difference in length scales: membrane ruptures are of micrometer size where fingering patterns is in the order of centimetres.

Viscous fingering instabilities can be observed in a "Hele-Shaw cell", which is an experimental set up explicitly designed to simulate the flow in a three-dimensional porous media in a two-dimensional environment. The cell consists of two flat glass plates, positioned in parallel and separated with an infinitesimally small distance h. If h is very small, the flow inside the cell becomes incompressible ($\nabla \vec{v}=0$). Eq. (7), which describes the force balance for spreading solid-supported membranes, is in fact mathematically identical to the Brinkman equation used to describe flow in porous media. Correspondingly, a characteristic length scale L_c, which can be determined from Eq. (7), is equivalent to the Brinkman length scale.

Eq. (7), which describes the membrane flow ($\zeta\vec{v}=\nabla\sigma$), is the same as the Hele-Shaw equation, a form of the Darcy equation specific to the Hele-Shaw flow. This flow can be expressed as:

$$\vec{v} = -\frac{h^2}{12\eta}\nabla P \tag{24}$$

where $\vec{v}$ is the velocity field. This vector mathematically describes the motion of the fluid where the length of the vector field is the flow speed, h is the distance between the two plates, η is the viscosity of the fluid, and ∇P is the pressure gradient. η may depict the viscosity of the existing in the Hele-Shaw cell (η_1) or the subsequently introduced fluid (η_2).

The inviscid edge at the moving boundary of the fluids is balanced by the pressure of the invading liquid:

$$P = -\gamma_L c \tag{25}$$

where P is the pressure, γ_L is the surface tension of the boundary, and c is the edge curvature. If the viscosity of the injected fluid, such as air, is significantly lower than the viscosity of the accommodating liquid, e.g., water or oil. In practice, the fluid can be considered to be inviscid (zero viscosity). When a pore opens in the membrane, the water in the ambient buffer penetrates into the ruptured areas of the membrane. However, mainly the friction influences the flow dynamics of a supported membrane and the effect of water viscosity is considered to be insignificant. Hence, the opening of a pore in the membrane is equivalent to air injection to a Hele-Shaw cell. At the boundary of a rupture, the membrane tension σ is balanced by the line tension γ :

$$\sigma = \gamma c \tag{26}$$

where c is the edge curvature of the rupture. The line tension emerges when lipids curve at any membrane edge to avoid the exposure of the hydrophobic fraction of the lipids to the aqueous environment.

The instability of a membrane pore edge, where the pore void represents an inviscid fluid, is mathematically analogous to a Saffman-Taylor instability [69]. A periodic membrane edge modulation can be expressed as $u=\varepsilon\ (t)\sin(qx)$, where for a wavelength of L , ε is the peak amplitude of the modulation. If Eq. (7) and (26) are solved with this boundary condition, where $\nabla \vec{v}=0$, the resulting wavelength gives the dynamics picture equivalent to the Saffman-Taylor instability observed in a Hele-Shaw cell. The growth rate of an amplitude $\varepsilon(q)$ with wave number $q = \frac{2\pi}{L}$ therefore is:

$$\frac{\partial \varepsilon}{\partial t} = q(V - \frac{\gamma q^2}{\zeta})\varepsilon \tag{27}$$

Here, γ is the edge tension of the pore which corresponds to γ_L . V is the mean velocity of the interface at the pore edge.

In a basic Hele-Shaw cell the porosity is regular. An inhomogeneous porosity can be introduced into the cell by placing grains, e.g., glass beads at random locations. This leads to irregular permeability where the capillary forces become significantly effective and eventually cause fractal displacements termed invasion percolation clusters. The percolation clusters form when invading fluid chooses the "path of least resistance", entrapping islands of the displaced fluid. This can be achieved, for example, by injecting air into water. The displacements appear in characteristic bursts with a broad size distribution, known as Haines jumps, which are similar to the avalanches observed during fracturing of the lipid membranes. The islands within the clusters are comparable to the islands surrounded by the fractal ruptures. Another similarity is the fractal dimension (D) of the ruptures: 1.71. This is lower than the theoretically estimated dimension of a percolation cluster D = 1.83, but equal to typical experimental values (1.70–1.71) [70].

A Hele-Shaw cell contains particles or beads which provide the effect of porosity. In between the proximal and the distal membranes, there are no particles, but a corresponding effect is established by "pinning". The pinning can be due to the Ca^{2+} ions which bridge bits of the two bilayers together [71]. The pinned regions become visible during formation of floral ruptures in forms of thin threads at the pore edges [4]. The pinning points, where the fluidity is reduced, act like solid particles and play the role of grains in a Hele-Shaw cell. One other reason for pinning can be the surface structure. Nanometer-sized grains of silicon dioxide (SiO_2) are known to create incisions in solid-supported lipid membranes. The granules of surface therefore can punch through the proximal bilayer and act like particles placed in between the two bilayers. The flow of lipids during rupturing is therefore considered to be through a porous medium.

A clearly defined analogy between the fractal ruptures and the invasion percolation instabilities, as we have previously shown for the floral ruptures and the Saffman-Taylor instabilities, has not yet been established. However, it is possible to estimate a characteristic length scale for membrane pores, within which they are not expected to exhibit instabilities.

Free standing membranes produce circular pores with straight edges [72]. An instable pore edge in such a membrane can be pictured as a wave or a modulation. The excess energy of an instability compared to a straight edge ($u = 0$) over one wavelength is $E = \gamma \int_0^L dx \frac{1}{2}\left(\frac{du}{dx}\right)^2 = \pi^2 \gamma \frac{\varepsilon^2}{L}$.

The instable membrane pore edge will over time relax to a straight edge due to dissipation which emerges from the two-dimensional Stokes flow of surfactants in the membrane. This means we have to take into account the membrane viscosity (μ) regarding dissipation of the excess energy at a pore edge. The dissipation scales as $T\dot{S} \sim \mu L^2 \left(\frac{V}{L}\right)^2 \sim \mu V^2 \sim \mu \dot{\varepsilon}^2$. This has been obtained by integrating over one wavelength L, the second term of the dissipation function of spreading membranes (Eq. (32)), on which we will elaborate in the paragraphs further down.

Balancing the excess edge energy with dissipation ($T\dot{S} + \dot{E} = 0$) leads (on a scaling level) to $\dot{\varepsilon} \sim \frac{1}{\tau}\varepsilon$. Therefore, the relaxation time (τ) is:

$$\tau \sim \frac{\mu}{\gamma} L \tag{28}$$

In a supported membrane, τ is much longer and increases rapidly with the wavelength of the modulation: the dissipation due to sliding friction, over one wavelength L, scales as $T\dot{S}$ = $\int dA \frac{1}{2} \zeta v^2 \sim \zeta L^2 \dot{\varepsilon}^2$, where ζ is the friction coefficient between the proximal and the distal bilayers.

Balancing the dissipation and the excess energy of the instability gives:

$$\tau \sim \frac{\zeta}{\gamma} L^3 \tag{29}$$

A rupture propagating at velocity v exposes a membrane edge length L in time $\tau = \frac{L}{v}$. By inserting τ into Eq. (29) gives a characteristic length (L_c):

$$L_c = \sqrt{\frac{\gamma}{\zeta v}} \tag{30}$$

For modulations larger than L_c, the ruptures will propagate before the edge shape modulation can possibly relax. v has been determined from experimental observations of fractal rupture [4] to be 20−30 $\frac{\mu m}{s}$. Taking $\gamma \approx$ 10 pN and $\zeta \sim 5x10^7 \frac{Pa.s}{m}$, [4] L_c is calculated as 0.1 μm. Above this length, pore edge dynamics is too slow to secure a stable pore edge. The size of fractal ruptures is in the order of several tens of micrometers and is therefore in agreement with the prediction.

Biological cells can also spread their membrane material on solid supports, often in order to be able to migrate. In some instances, a 200-nm thick lamellipodial sheet protrudes from the cell body onto the substrate. [73] The sheet includes a double layer of plasma membrane in addition to actin filaments sandwiched in between the layers. The detailed mechanism of cellular wetting is a subject still under debate. Whether the proceeding edge is rolling or sliding driven by actin polymerization is not yet known. The lamellipodia-based cellular wetting, however, have been found to follow a similar power law as discussed in Section 1.3. [73] Spreading of cells can also be promoted by introducing trivalent ions to the substrate, for example, Eu^{3+} ions onto SiO_2. [4] In such conditions, Chinese Hamster Ovary (CHO) cells continuously adhere onto the substrate. Interestingly, the adhesion leads to the formation of fractal ruptures [4] as well as the islands, in distal plasma membrane of spreading CHO cells. Upon rupturing, the wetted area on the substrate suddenly increases in a step-wise manner, similar to the fractal ruptures of the self-spreading double bilayers. The plasma membranes are connected to the underlying scaffolding layer (cytoskeleton) via linking molecules. The membrane flows around the linkers; the plasma membrane flow on cytoskeleton, therefore, can be considered as a two-dimensional porous media flow, i.e., Darcy flow. The tension causing the ruptures is still moderate and in the range of plasma membrane adhesion to the cytoskeleton, or membrane–membrane adhesion, suggesting that such ruptures can in fact occur in vivo.

Another form of lipid wetting on solid substrates involving ruptures is accommodated by inter-membrane "defects" or "fusion pores" (Figure 2d). The fusion pores are nanometer-sized circular conduits connecting two membranes in the shape of an hour-glass. The dimension of pores (~ nm) makes direct observations with the current microscopy technologies tremendously challenging. An alternative way is to study the flow phenomena, indirectly, by the fluorescent intensity-based wetting area analysis of membranes. [42] In the following paragraphs, we will talk about double bilayer membranes, 1% of which consist of fluorescently tagged lipids. From the top view, such fluorescent distal membranes emit twice as intensely as the proximal membranes; therefore, the two bilayers can be easily distinguished.

The double bilayer membranes mentioned above exhibiting floral or fractal ruptures consists of two bilayers which are intact, performing a rolling motion. In some occasions, the proximal and distal bilayers split along the circumference. After splitting, the proximal bilayer continues to wet the surface, which can be observed by increasing the wetted area on the substrate. The area of the distal bilayer membrane remains unchanged over the time period of a few minutes, followed by sudden decreases caused by instant avalanche ruptures. The decrease in the area of the distal bilayer simultaneously causes an increase in the wetted area by the proximal bilayer. This supports the notion of a physical connection among the two bilayers, through which the lipids are transferred. The outer border of the distal membrane does not expand along the peripheries upon rupturing, as occurs for the floral and fractal ruptures, indicating that the distal membrane is not dragged by the proximal membrane along the circumference. The stretching and eventual rupturing of the distal membrane can therefore be caused by the downwards lipid flow towards the proximal bilayer, most likely through narrow vertical channels. [42]

Defects among lipid membranes may have already formed during swelling of MLVs or form dynamically during spreading as a response to physical or chemical cues. We had briefly mentioned in Section 1.3 the defects existing in onion vesicles or MLVs. Since the lipid layers packed in the reservoir later spreads on the surface, these defects can be transferred to the supported membranes. Additionally, changes in membrane tension may cause instantaneous (~ nanoseconds) formation of defects in membranes. Fusion is a thermally agitated process and alterations in line tension can promote the formation of fusion pores. Formation of defects can also be induced by the presence of multivalent ions, e.g., Ca^{2+}, in the ambient buffer. 2 mM of Ca^{2+} has been reported to advance the formation of defects among myelin sheaths of neurons, 80% of which consists of lipid material. The lipid transfer among the bilayers can as well be enhanced by hemi-fusion pores through which the dynamics change from bilayer sliding to monolayer sliding. As we had mentioned in Section 1.3, the sliding friction coefficient for monolayers is much higher than bilayers. Furthermore, the transformation of hemi-fusion to fusion pores is very energy-intensive and complex. For the simplicity of calculations, we will only assume defects to be in the shape of fusion pores in this section. We will discuss the monolayer sliding among the bilayers further in this section in another context.

The dynamics of lipid transfer via a fusion pore can be characterized semi-quantitatively via a "dissipation" function. Our model consists of a circular proximal bilayer, a circular distal bilayer and as an initial assumption: a single fusion pore with a diameter of 10 nm connecting these two membranes. Lipids flow from distal to proximal bilayer through the 10 nm defect, driven by the continuous adhesion of proximal bilayer to the substrate (Figure 2d). In this flow, there would be two separate forms of energy loss (dissipation). The first one is friction. The friction applies to the region (1) in between the proximal bilayer and the surface, (2) in between two bilayers since distal bilayer is de-wetting the proximal bilayer. The second form of dissipation is due to the viscous flow around the fusion pore. The lipid flow is expected to be different in remote areas of the membrane compared to the proximity of the pore. This is because the pore is small and the surfactants flowing through the pore collide with each other more intensely than they would in distant areas. Next, we will quantify and compare these two types of dissipations. If the magnitudes are compatible in relevant time scales, we will conclude that a single pore of 10 nm is sufficient to accommodate such a flow in between the membranes. If viscous flow (Stokes flow) cause dissipation that is much stronger than the frictional, this will indicate that several pores are required.

This brings us back to Eq. (7), where the difference of frictional and viscous forces determines the tension gradient across the membrane. A characteristic length scale where the frictional forces are of the same order as the viscous forces can be obtained based on Eq. (7) as: $\zeta V \sim \mu \frac{V}{L^2}$. V is the characteristic velocity to determine the characteristic length scale. The cross-over length scale $L_c = \sqrt{\frac{\mu}{\zeta}}$ can help us to compare the viscous forces to the frictional forces. One can think of L_c as the "Reynolds number" of fluid mechanics, which is the ratio of inertial forces to viscous forces within a fluid. Low Reynolds numbers indicates laminar, high Reynolds numbers the turbulent flow. Similarly, above L_c the frictional forces are dominating membrane flow dynamics, whereas below L_c, the viscous forces are effective. In the following we will estimate L_c in order to be able to confirm that it is rational to take into account the viscosity, considering a membrane flow around a pore of 10 nm.

The friction coefficient is:

$$\zeta = \frac{\eta_w}{d_w} \tag{31}$$

where η_w is viscosity and d_w is the thickness of the water layer on which the bilayer is sliding. There may be two bilayers sliding on a water layer under the spreading conditions described above: the distal bilayer slides on the proximal and the proximal slides on the surface. If we take η_w as 8.9×10^{-3} dyn s cm^{-1} and d_w as 1 nm, we can determine L_c to be ~ 30 nm. The thickness of the water layer between the bilayers can be slightly higher than the one between the surface and the proximal bilayer. With increasing d_w, the order of magnitude for L_c would not change. Above $L_c = 30$ nm, the sliding friction dominates over viscosity. Since we assume the pore size to be ~ 10 nm, it can be concluded that the pore vicinity will high likely be under the influence of the viscous forces.

For two-dimensional incompressible flow of the membrane, Eq. (7) leads to the following dissipation function [20]:

$$T\dot{S} = \int dA[\ \zeta \vec{v}^2 + 2\mu(\frac{\partial v_i}{\partial x_k} + \frac{\partial v_k}{\partial x_i})^2] \tag{32}$$

v_i, v_k are the components of velocity and x_k, x_i are the components of position, where i and k change from 1 to 2 in a two-dimensional membrane. We assume that the membrane flow is radially symmetric, where $v = \frac{r_p}{r} v_p$ around a pore with a membrane velocity of v_p in the pore channel. Integrating Eq. (32) from the pore radius r_p to the radius of the distal membrane island patch R (Figure 2d) leads to:

$$T\dot{S} = 2\mu \int dr 2\pi r \left[\left(\frac{\partial v}{\partial r} \right)^2 + \left(\frac{v}{r} \right)^2 \right] = 8\pi\mu v_p^2 r_p^2 \int_{r_p}^{R} \frac{dr}{r^3} \approx 4\pi\mu v_p^2 \qquad (33)$$

The integral Eq. (33) converges rapidly when R increases. This means that the viscous dissipation is intense in the pore area and is not as significantly effective in remote areas of the membrane. Taking into account the dissipation on both the proximal and the distal bilayer side of the pore, we obtain $T\dot{S} = 8\pi\mu v_p^2$, which is the total dissipation due to the viscous flow around a circular pore. The membrane viscosity μ is in the order of water viscosity ($\mu = \eta$ $(water\ viscosity) x1\ \mu m$).

The dissipation caused by the sliding friction is composed of two parts: The sliding between the two bilayers (ζ) is expressed by the first term and the sliding of the proximal bilayer on the surface (ζ_s) by the second term of Eq. (34):

$$T\dot{S} = 2\pi\zeta (2v_p)^2 r_p^2 \ln\left(\frac{R}{r_p} \right) + 2\pi\zeta_s v_p^2 r_p^2 \ln\left(\frac{R_s}{r_p} \right) \qquad (34)$$

R_s is the radius of the proximal bilayer (Figure 2d). Unlike viscous flow, the dissipation due to sliding friction is not local and depends logarithmically on the size of the membrane patch. ζ (5×10^5 Pa.s.m^{-1}) and ζ_s (5×10^7 Pa.s.m^{-1}) are not material constants and their values depend on the amount of water trapped between the bilayers, as well as on the degree of adhesion. Assuming $\zeta = 5x105\ \frac{Pa.s}{m}$, $\zeta_s = 5x107\ \frac{Pa.s}{m}$, $R_s = 100\ \mu m$, $R = 10\ \mu m$ and $r_p = 10\ nm$ the viscous dissipation around a 10 nm defect turns out to be spatially localized and high, but it is not expected to exceed the dissipation due to the sliding friction of the membrane. As a result, only a single pore can be sufficient to accommodate the rapid lipid transfer.

One can think of opening of large area pores in distal membranes as the process of de-wetting of the proximal bilayer. It is possible to reverse de-wetting by treating the pores with chemical 'repair' agents. When pores open in the membrane, the multivalent ions in the ambient buffer such as Ca^{2+} can penetrate through the pore edges and eventually pin fractions of the upper and lower bilayers together. At this instant, chelators such as 1,2-bis (o-aminophenoxy)ethane-N,N,N',N'-tetraacetic acid (BAPTA) can be introduced to the ambient buffer to target and deplete the Ca^{2+} ions. The chelators can bind to Ca^{2+} with high affinity and remove them from the pore edges and from the surface. This in turn frees the pore edges and reduces the overall membrane tension. Eventually, the membrane ruptures heal due to the pore edge tension (γ) and the distal membrane completely re-wets the proximal membrane. The lipid material

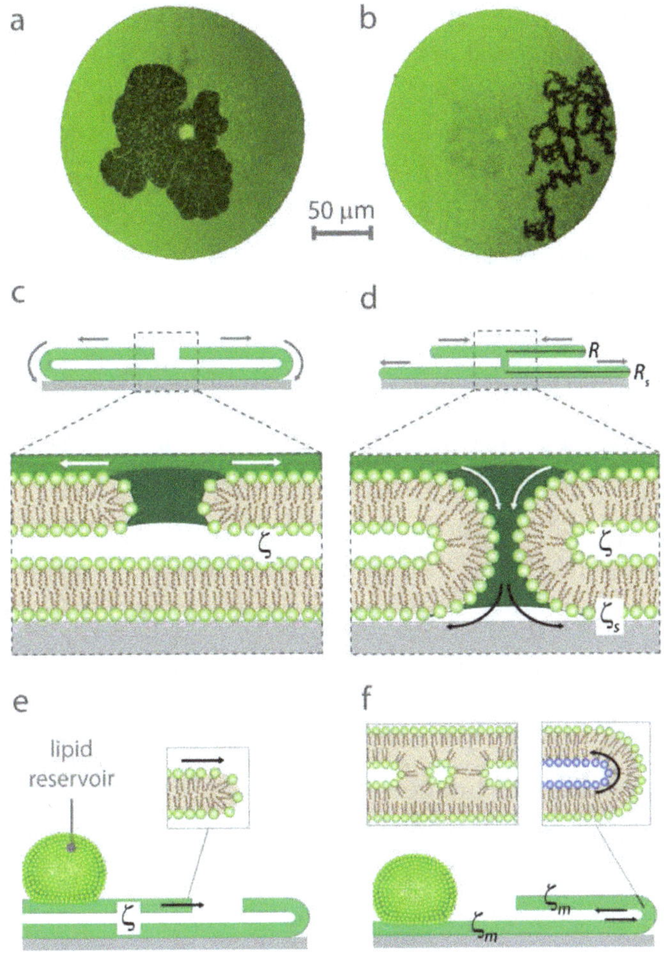

Figure 2. Biomembrane ruptures as a cause of wetting. Confocal fluorescence micrographs of (a) floral (b) fractal membrane ruptures (top view). The double bilayer membrane areas can be visualized as twice as intense as the single bilayer areas. Darker regions are the proximal membrane which is visible through the ruptures in the distal membrane. (c) Cross-sectional schematic view of a rupturing distal membrane, representing (a) or (b). (d) Cross-sectional schematic view of a fusion pore connecting proximal and distal bilayers. ζ and ζ_s are the friction coefficients which apply to the region between the bilayers and between the proximal bilayer and the surface, respectively. Pore/rupture repair via (e) bilayer sliding (f) monolayer sliding. In (e), there is a direct connection of the distal membrane to the lipid reservoir. In (f), the pore is around the reservoir which does not have any direct connection to the distal bilayer. The lipid transfer to the pore region can be via inter-monolayer sliding through the inverted micelle like defects (inset to the left) or simply by reverse sliding of the inner monolayer leaflet. The latter is represented by the inset to the right, in which the sliding monolayer is depicted in blue.

sealing the pores originates from the reservoir and re-locates around the pore area by means of two different mechanisms. In the following, we will explain these mechanisms in detail.

The large area pores can form at different locations in the distal membrane. A fraction of the ruptures appear towards the edges of the distal bilayer as we had described above (Figure 2 b,e). There, one side of the pore edge maintains a physical connection to the MLV through the distal bilayer. The sealing of pores becomes possible through "bilayer-on-bilayer sliding" from the reservoir towards the pore region. Alternatively, rupturing may occur around the MLV (Figure 2a,f). In this case, there remains no direct contact of the distal bilayer to the lipid reservoir. The re-location of lipids towards the pore area can only be through the proximal bilayer. The free positive surface charges in the areas which are not wetted by the lipid membrane are mostly removed by BAPTA. This means depletion of fusogenic agents and termination of spreading. On the other hand, the access into the confined region between the proximal bilayer and the surface is impeded and chelation of the ions in the region above surface and underneath the membrane is expected to take significantly longer time. The total wetted area of the membrane on the solid substrate therefore can remain constant over several hours, confirmed by the experimental analysis [42]. While the edges of the spreading patch is pinned, reverse bilayer rolling becomes unlikely and one can presume the monolayer sliding to be the dominant flow mechanism for repair of such pores (Figure 2f).

To calculate the dissipation ($T\dot{S}$) during bilayer-on-bilayer sliding, we can neglect the second term in Eq. (32), and integrate the friction term from r_p to R_c. It is complicated to calculate dissipation regarding a pore at an arbitrary position in a circular spread since in that case the flow field would not be radial. Thus, we assume that the pore is centred, and consider R_c as the cut-off characteristic length scale. R_c is comparable to the size of the spread R_S. The integral will give $\pi\zeta r_p^2 \dot{r}_p^2 \ln(\frac{R_c}{r_p})$, where ζ is the sliding friction coefficient between the distal and the proximal bilayers, v is the velocity of the sliding distal bilayer.

The dissipation ($T\dot{S}$) caused by the bilayer-on-bilayer sliding friction will be balanced by the edge tension energy $\dot{E}$ of the pore: $(2\pi\gamma\dot{r}_p)$ [74]:

$$\zeta r_p^2 \dot{r}_p \, ln\left(\frac{R_c}{r_p}\right) = -2\gamma \tag{35}$$

where $\dot{r}_p = \frac{dr_p}{dt}$. Integrating the left side of Eq. (35) from R_p (pore radius at $t = 0$, i.e., initial pore radius) to $r_p = 0$ (when the pore is closed), and the right side of the equation from 0 to τ (the time required to relax the pore) leads to:

$$\tau = \frac{\zeta R_p^{\ 3}}{\gamma} \cdot \frac{1}{6}\left[\ln\left(\frac{R_s}{R_p}\right) + \frac{1}{3}\right] \sim \frac{\zeta R_p^{\ 3}}{\gamma} \tag{36}$$

The radius of the proximal membrane R_s in Eq. (36) is only a logarithmic factor. This means that the size of the membrane spread on the surface does not determine the pore relaxation time and τ is rather highly depending on the initial pore size R_p. Note that Eq. (36) is a different form of Eq. (29) and can therefore be used to describe pores in cell membranes.

Now we will calculate the dissipation for the second scnerio, where the membrane flow towards the pore area is through the monolayer sliding. The monolayer sliding occurs between the two leaflets within a bilayer and will be opposed by friction. The friction coefficient (ζ_m) can be calculated by rearranging the first term of Eq. (32) as below:

$$T\dot{S} = \int_I dA \frac{1}{2}\zeta_m v_I^2 + \int_{II} dA \frac{1}{2}\zeta_m v_{II}^2 \tag{37}$$

Here, index I refers to the surface area of the proximal bilayer; index II of the distal bilayer. v is the velocity of the sliding monolayer. Integrating the first term of Eq. (37) from R_L (effective radius of the lipid source, $R_L \sim R_0$) to R_S (radius of the spread) and the second term from r_p (pore radius) to R_S gives:

$$T\dot{S} = 4\pi\zeta_m r_p^2 \dot{r}_p^2 \ln\left(\frac{R_s^2}{R_L r_p}\right) \tag{38}$$

where $\dot{r}_p = \frac{dr_p}{dt}$. Balancing Eq. (38) with the line tension energy of the pore ($T\dot{S} = \dot{E}$) gives

$$2\frac{\zeta_m r_p^2}{\gamma} \dot{r}_p \ln\left(\frac{R_s^2}{R_L r_p}\right) = 1 \tag{39}$$

Based on experimental values inserted into Eq. (39), ζ_m has been calculated to be 10^5 to $10^6 \frac{Ns}{m}$. This friction coefficient is at least two orders of magnitude lower than typical values reported in the literature: 10^8 to $10^9 \frac{Ns}{m}$ [3, 65]. This means that a possible inter-leaflet monolayer sliding in the experimental conditions described above is occurring at much faster speed. Even though unlikely circumstances where a monolayer can slide under such rates have been previously simulated [75], alternatively, the "hemi-fusion pores" or stalks can accommodate the relocation of surfactants via monolayer sliding. One can think of hemi-fusion pores as intermediate forms of fusion pores where the two merged bilayers are continuous but have not yet evolved into an aqueous connection. A portion of pore closure mechanisms based on monolayer sliding can be assisted by hemi-fusion pores or defects resembling inverted-micelles (Figure 2f). Monolayer sliding in some areas may also be complemented by bilayer sliding which can explain the relatively low sliding frictions observed.

5. Summary

Research on model membranes has been conducted for decades, and the understanding of the dynamics of lipid films has reached advanced levels. However, enabled by the rapid advances in micro- and nano-technologies and analytical capabilities, new phenomena are frequently discovered, such as the occurrence of the fractal membrane ruptures in double bilayer membranes, which created a new, exciting link between solid materials and the biological soft matter world. The discovery of this rupture phenomenon was closely related to the spontaneous wetting of high-energy surfaces, which was experimentally established in a microenvironment under the microscope. This and the other wetting phenomena described in the previous sub-chapters are feature-rich and have possible implications not only for future technological advancements, such as membrane protein studies, cell migration, but also for very advanced applications such as chemistry confined to two dimension. The double bilayer, which was at the heart of these investigations, can be easily classified as a new membrane model, which adds to the mono-, bi-, and cushioned bilayers. One can perhaps also view it, on one hand, as a self-cushioning bilayer, but on the other hand, it is essentially a flat giant unilamellar vesicle, with an approximately 10 nm thin water layer encapsulated between the two bilayer sheets. The thus encapsulated volume is on the order of a few hundred femtoliters. It bears a richness in possibilities for application in nanofluidics and artificial cell models, and potentially allows through its spreading an rupturing dynamics greater insights into, for example, the membrane-related mechanisms of cell migration and chemotaxis. We have provided in this chapter an overview over the wetting and rupturing properties and features of phospholipid monolayers and double bilayers on solid support, which should provide the foundation for the design of new experiments, and in many cases the prediction of their outcome. The dynamics of pores in membranes and associated materials transport phenomena, which are also accompanied with wetting phenomena, are also discussed. There are some particular points where further research is required. For example, attempts to establish a relationship for the spreading coefficient to quantitatively describe the spreading dynamics of double bilayers have so far been unsuccessful, which leaves the spreading approach to lipid film formation in this case still not entirely predictable.

Author details

Irep Gözen[1], Paul Dommersnes[2] and Aldo Jesorka[3*]

*Address all correspondence to: aldo@chalmers.se

1 Department of Engineering, Harvard University, Harvard School of Engineering and Applied Sciences, Pierce Hall, Cambridge, MA, USA

2 Department of Physics, Norwegian University of Science and Technology, Trondheim, Norway

3 Biophysical Technology Laboratory, Department of Chemistry and Chemical Engineering, Chalmers University of Technology, Göteborg, Sweden

References

[1] Czolkos I, Jesorka A, Orwar O. Molecular phospholipid films on solid supports. Soft Matter 2011;7 (11):4562.

[2] Radler J, Strey H, Sackmann E. Phenomenology and kinetics of lipid bilayer spreading on hydrophilic surfaces. Langmuir 1995;11 (12):4539.

[3] Czolkos I, Erkan Y, Dommersnes P, et al. Controlled formation and mixing of two-dimensional fluids. Nano Letters 2007;7 (7):1980.

[4] Gozen I, Dommersnes P, Czolkos I, et al. Fractal avalanche ruptures in biological membranes. Nature Materials 2010;9 (12):908.

[5] van Meer G, Voelker DR, Feigenson GW. Membrane lipids: where they are and how they behave. Nature Reviews Molecular Cell Biology 2008;9 (2):112.

[6] Bloom M, Evans E, Mouritsen OG. Physical properties of the fluid lipid-bilayer component of cell membranes - a perspective. Quarterly Reviews of Biophysics 1991;24 (3):293.

[7] Pomorski TG, Nylander T, Cardenas M. Model cell membranes: Discerning lipid and protein contributions in shaping the cell. Advances in Colloid and Interface Science 2014;205:207.

[8] Sackmann E. Supported membranes: Scientific and practical applications. Science 1996;271 (5245):43.

[9] Walde P, Ichikawa S. Enzymes inside lipid vesicles: preparation, reactivity and applications. Biomolecular Engineering 2001;18 (4):143.

[10] Ainla A, Goezen I, Hakonen B, et al. Lab on a Biomembrane: Rapid prototyping and manipulation of 2D fluidic lipid bilayer circuits. Scientific Reports 2013;3:2743.

[11] Lobovkina T, Gozen I, Erkan Y, et al. Protrusive growth and periodic contractile motion in surface-adhered vesicles induced by Ca2+-gradients. Soft Matter 2010;6 (2): 268.

[12] Reimhult E, Kasemo B, Höök F. Rupture pathway of phosphatidylcholine liposomes on silicon dioxide. International Journal of Molecular Sciences 2009;10 (4):1683.

[13] Jesorka A, Stepanyants N, Zhang H, et al. Generation of phospholipid vesicle-nanotube networks and transport of molecules therein. Nature Protocols 2011;6 (6):791.

[14] Ainla A, Jeffries GDM, Brune R, et al. A multifunctional pipette. Lab on a Chip 2012;12 (7):1255-61.

[15] Thiam AR, Farese Jr RV, Walther TC. The biophysics and cell biology of lipid droplets. Nature Reviews Molecular Cell Biology 2013;14 (13):775.

[16] Li J, Han D, Zhao Y-P. Kinetic behaviour of the cells touching substrate: the interfacial stiffness guides cell spreading. Sci Rep 2014;4:3910.

[17] Czolkos I, Hakonen B, Orwar O, et al. High-resolution micropatterned Teflon AF substrates for biocompatible nanofluidic devices. Langmuir 2012;28 (6):3200.

[18] Erkan Y, Halma K, Czolkos I, et al. Controlled release of chol-TEG-DNA from nano- and micropatterned SU-8 surfaces by a spreading lipid film. Nano Letters 2008;8 (1): 227.

[19] de Gennes PG. Wetting: statics and dynamics. Reviews of Modern Physics 1985;57 (3):827.

[20] Evans E, Sackmann E. Translational and rotational drag coefficients for a disk moving in a liquid membrane associated with a rigid substrate. Journal of Fluid Mechanics 1988;194:553.

[21] Kochev V, Karabaliev M. Wetting films of lipids in the development of sensitive interfaces. An electrochemical approach. Advances in Colloid and Interface Science 2004;107 (1):9.

[22] Suzuki K, Masuhara H. Growth of giant membrane lobes mechanically driven by wetting fronts of phospholipid membranes at water-solid interfaces. Langmuir 2005;21 (2):537.

[23] Borghi N, Alias K, de Gennes PG, et al. Wetting fibers with liposomes. Journal of Colloid and Interface Science 2005;285 (1):61.

[24] Richter RP, Brisson AR. Following the formation of supported lipid bilayers on Mica: A study combining AFM, QCM-D, and ellipsometry. Biophysical Journal 2005;88 (5): 3422.

[25] Richter RP, Lai Kee Him J, Tessier B, et al. On the kinetics of adsorption and two-dimensional self-assembly of annexin A5 on supported lipid bilayers. Biophysical Journal 2005;89 (5):3372.

[26] Richter RP, Bérat R, Brisson AR. Formation of solid-supported lipid bilayers: An integrated view. Langmuir 2006;22 (9):3497.

[27] Furukawa K, Sumitomo K, Nakashima H, et al. Supported lipid bilayer self-spreading on a nanostructured silicon surface. Langmuir 2007;23 (2):367.

[28] Erkan Y, Czolkos I, Jesorka A, et al. Direct immobilization of cholesteryl-TEG-modified oligonucleotides onto hydrophobic SU-8 surfaces. Langmuir 2007;23 (11):5259.

[29] Zhou X, Moran-Mirabal JM, Craighead HG, et al. Supported lipid bilayer/carbon nanotube hybrids. Nat Nano 2007;2 (3):185.

[30] van Meer G, Voelker DR, Feigenson GW. Membrane lipids: where they are and how they behave. Nature Reviews Molecular Cell Biology 2008;9 (2):112.

[31] Gauthier NC. Plasma membrane area increases with spread area by exocytosis of GPI anchored protein compartment. Biophysical Journal;96 (3):151a.

[32] Cho N-J, Frank CW, Kasemo B, et al. Quartz crystal microbalance with dissipation monitoring of supported lipid bilayers on various substrates. Nature Protocols 2010;5 (6):1096.

[33] Gołąbek M, Hołysz L. Changes in wetting and energetic properties of glass caused by deposition of different lipid layers. Applied Surface Science 2010;256 (18):5463.

[34] Han X, Achalkumar AS, Cheetham MR, et al. A self-assembly route for double bilayer lipid membrane formation. ChemPhysChem 2010;11 (3):569.

[35] Isono T, Ikeda T, Ogino T. Evolution of supported planar lipid bilayers on step-controlled sapphire surfaces. Langmuir 2010;26 (13):9607.

[36] Oliver AE, Parikh AN. Templating membrane assembly, structure, and dynamics using engineered interfaces. Biochimica et Biophysica Acta (BBA) - Biomembranes 2010;1798 (4):839.

[37] Gozen I, Billerit C, Dommersnes P, et al. Calcium-ion-controlled nanoparticle-induced tubulation in supported flat phospholipid vesicles. Soft Matter 2011;7 (21): 9706.

[38] Nabika H, Oowada M, Murakoshi K. Control of dynamics and molecular distribution in a self-spreading lipid bilayer using surface-modified metal nanoarchitectures. Physical Chemistry Chemical Physics 2011;13 (14):5561.

[39] Nair PM, Salaita K, Petit RS, et al. Using patterned supported lipid membranes to investigate the role of receptor organization in intercellular signaling. Nature Protocols 2011;6 (4):523.

[40] Mohamad S, Noël O, Buraud J-L, et al. Mechanism of lipid nanodrop spreading in a case of asymmetric wetting. Physical Review Letters 2012;109 (25):248108.

[41] Gözen I, Jesorka A. Instrumental methods to characterize molecular phospholipid films on solid supports. Analytical Chemistry 2012;84 (2):822.

[42] Gozen I, Dommersnes P, Orwar O, et al. Evidence for membrane flow through pores in stacked phospholipid membranes. Soft Matter 2012;8 (24):6220.

[43] Gauthier NC, Masters TA, Sheetz MP. Mechanical feedback between membrane tension and dynamics. Trends in Cell Biology 2012;22 (11):527.

[44] Subramaniam AB, Guidotti G, Manoharan VN, et al. Glycans pattern the phase behaviour of lipid membranes. Nature Materials 2013;12 (2):128.

[45] Hirtz M, Oikonomou A, Georgiou T, et al. Multiplexed biomimetic lipid membranes on graphene by dip-pen nanolithography. Nature Communications 2013;4:2591.

[46] Gozen I, Ortmen B, Poldsalu I, et al. Repair of large area pores in supported double bilayers. Soft Matter 2013;9 (11):2787.

[47] Pietuch A, Janshoff A. *Mechanics of spreading cells probed by atomic force microscopy,* 2013:3 (7):130084.

[48] Nguyen PA, Groen AC, Loose M, et al. Spatial organization of cytokinesis signaling reconstituted in a cell-free system. Science 2014;346 (6206):244.

[49] Motegi T, Nabika H, Fu Y, et al. Effective brownian ratchet separation by a combination of molecular filtering and a self-spreading lipid bilayer system. Langmuir 2014;30 (26):7496.

[50] Watanabe R, Soga N, Fujita D, et al. Arrayed lipid bilayer chambers allow single-molecule analysis of membrane transporter activity. Nature Communications 2014;5:4519.

[51] Tabaei SR, Choi J-H, Haw Zan G, et al. Solvent-assisted lipid bilayer formation on silicon dioxide and gold. Langmuir 2014;30 (35):10363.

[52] Shaali M, Lara-Avila S, Dommersnes P, et al. Nanopatterning of mobile lipid monolayers on electron-beam-sculpted Teflon AF surfaces. ACS Nano 2015;9 (2):1271.

[53] Ma L, Li Y, Peng J, et al. Discovery of the migrasome, an organelle mediating release of cytoplasmic contents during cell migration. Cell Research 2015;25 (1):24.

[54] Wu IL, Narayan K, Castaing J-P, et al. A versatile nano display platform from bacterial spore coat proteins. Nature Communications 2015;6:7548.

[55] Shuttleworth R. The surface tension of solids. Proceedings of the Physical Society of London Section A 1950;63 (365):444.

[56] Zhai L, Berg MC, Cebeci FC, et al. Patterned superhydrophobic surfaces: Toward a synthetic mimic of the Namib Desert beetle. Nano Letters 2006;6 (6):1213.

[57] Sedev R, Fabretto M, Ralston J. Wettability and surface energetics of rough fluoropolymer surfaces. Journal of Adhesion 2004;80 (6):497.

[58] Degennes PG. Wetting: statics and dynamics. Reviews of Modern Physics 1985;57 (3): 827.

[59] Dorrer C, Ruehe J. Some thoughts on superhydrophobic wetting. Soft Matter 2009;5 (1):51.

[60] Czolkos I. Micro- and nano-scale devices for controlling two-dimensional chemistry. PhD Thesis, Chalmers University of Technology, 2009.

[61] Chenevier P, Grandjean C, Loing E, et al. Grafting of synthetic mannose receptor-ligands onto onion vectors for human dendritic cells targeting. Chemical Communications 2002 (21):2446.

[62] Israelachvili JN. *Intermolecular and surface forces*. 3 ed: Elsevier, 2011.

[63] Yokota K, Toyoki A, Yamazaki K, et al. Behavior of raft-like domain in stacked structures of ternary lipid bilayers prepared by self-spreading method. Japanese Journal of Applied Physics 2014;53 (5):05FA11.

[64] Evans EA, Hochmuth RM. Membrane viscoelasticity. Biophysical Journal 1976;16 (1): 1.

[65] Shkulipa SA, den Otter WK, Briels WJ. Thermal undulations of lipid bilayers relax by intermonolayer friction at submicrometer length scales. Physical Review Letters 2006;96 (18):178302.

[66] Leng J, Nallet F, Roux D. Swelling kinetics of a compressed lamellar phase. European Physical Journal E 2001;4 (1):77.

[67] Rawicz W, Olbrich KC, McIntosh T, et al. Effect of chain length and unsaturation on elasticity of lipid bilayers. Biophysical Journal 2000;79 (1):328.

[68] Czolkos I, Guan J, Orwar O, et al. Flow control of thermotropic lipid monolayers. Soft Matter 2011;7 (16):6926.

[69] Homsy GM. Viscous fingering in porous-media. Annual Review of Fluid Mechanics 1987;19:271.

[70] Crandall D, Ahmadi G, Ferer M, et al. Distribution and occurrence of localized-bursts in two-phase flow through porous media. Physica a-Statistical Mechanics and Its Applications 2009;388 (5):574.

[71] Akashi K, Miyata H, Itoh H, et al. Formation of giant liposomes promoted by divalent cations: Critical role of electrostatic repulsion. Biophysical Journal 1998;74 (6): 2973.

[72] Zhelev DV, Needham D. Tension-stabilized pores in giant vesicles - Determination of pore size and pore line tension. Biochimica Et Biophysica Acta 1993;1147 (1):89.

[73] Dobereiner HG, Dubin-Thaler B, Giannone G, et al. Dynamic phase transitions in cell spreading. Physical Review Letters 2004;93 (11):108105.

[74] Gozen I, Dommersnes P. Pore dynamics in lipid membranes. European Physical Journal-Special Topics 2014;223 (10):1813.

[75] den Otter WK, Shkulipa SA. Intermonolayer friction and surface shear viscosity of lipid bilayer membranes. Biophysical Journal 2007;93 (2):423.

Permissions

All chapters in this book were first published in SE, by InTech Open; hereby published with permission under the Creative Commons Attribution License or equivalent. Every chapter published in this book has been scrutinized by our experts. Their significance has been extensively debated. The topics covered herein carry significant findings which will fuel the growth of the discipline. They may even be implemented as practical applications or may be referred to as a beginning point for another development.

The contributors of this book come from diverse backgrounds, making this book a truly international effort. This book will bring forth new frontiers with its revolutionizing research information and detailed analysis of the nascent developments around the world.

We would like to thank all the contributing authors for lending their expertise to make the book truly unique. They have played a crucial role in the development of this book. Without their invaluable contributions this book wouldn't have been possible. They have made vital efforts to compile up to date information on the varied aspects of this subject to make this book a valuable addition to the collection of many professionals and students.

This book was conceptualized with the vision of imparting up-to-date information and advanced data in this field. To ensure the same, a matchless editorial board was set up. Every individual on the board went through rigorous rounds of assessment to prove their worth. After which they invested a large part of their time researching and compiling the most relevant data for our readers.

The editorial board has been involved in producing this book since its inception. They have spent rigorous hours researching and exploring the diverse topics which have resulted in the successful publishing of this book. They have passed on their knowledge of decades through this book. To expedite this challenging task, the publisher supported the team at every step. A small team of assistant editors was also appointed to further simplify the editing procedure and attain best results for the readers.

Apart from the editorial board, the designing team has also invested a significant amount of their time in understanding the subject and creating the most relevant covers. They scrutinized every image to scout for the most suitable representation of the subject and create an appropriate cover for the book.

The publishing team has been an ardent support to the editorial, designing and production team. Their endless efforts to recruit the best for this project, has resulted in the accomplishment of this book. They are a veteran in the field of academics and their pool of knowledge is as vast as their experience in printing. Their expertise and guidance has proved useful at every step. Their uncompromising quality standards have made this book an exceptional effort. Their encouragement from time to time has been an inspiration for everyone.

The publisher and the editorial board hope that this book will prove to be a valuable piece of knowledge for researchers, students, practitioners and scholars across the globe.

List of Contributors

Meicheng Li, Yingfeng Li and Joseph Michel Mbengue
State Key Laboratory of Alternate Electrical Power System with Renewable Energy Sources, School of Renewable Energy, North China, Electric Power University, Beijing, China

Navaz Homayun, Zand Ali and Atkinson Theresa
Kettering University, Flint, USA

Gat Amir
Technion - Israel Institute of Technology, Haifa, Israel

Calin Jianu
Department of Food Science, Faculty of Food Processing Technology, Banat's University of Agricultural Sciences and Veterinary Medicine, Timişoara, Romania

Kwangseok Seo, Minyoung Kim and Do Hyun Kim
Department of Chemical and Biomolecular Engineering, Korea Advanced Institute of Science and Technology, Daejeon, South Korea

Daeyoung Kim
Department of Electrical Engineering, Korea Army Academy, Yeong-cheon, Kyungsangbuk- do, South Korea

Jeong-Bong Lee
Department of Electrical Engineering, The University of Texas at Dallas, Richardson, TX, USA

Gregor Filipič, Harinarayanan Puliyalil and Uroš Cvelbar
Jožef Stefan International Postgraduate School, Ljubljana, Slovenia
Jožef Stefan Institute, Ljubljana, Slovenia

Jeffrey L. Streator
G.W. Woodruff School of Mechanical Engineering, Georgia Institute of Technology, Atlanta, GA, USA

Irep Gözen
Department of Engineering, Harvard University, Harvard School of Engineering and Applied Sciences, Pierce Hall, Cambridge, MA, USA

Paul Dommersnes
Department of Physics, Norwegian University of Science and Technology, Trondheim, Norway

Aldo Jesorka
Biophysical Technology Laboratory, Department of Chemistry and Chemical Engineering, Chalmers University of Technology, Göteborg, Sweden

Index

www.ingramcontent.com/pod-product-compliance
Lightning Source LLC
LaVergne TN
LVHW081032210926
848485LV00004B/1007